New Directions in Statistical Physics

Springer
Berlin
Heidelberg
New York
Hong Kong
London
Milan
Paris
Tokyo

Physics and Astronomy ONLINE LIBRARY

springeronline.com

Luc T. Wille (Ed.)

New Directions in Statistical Physics

Econophysics,
Bioinformatics,
and Pattern Recognition

With 134 Figures
Including 8 Color Figures

 Springer

Professor Luc T. Wille
Department of Physics
Florida Atlantic University
777 Glades Road
Boca Raton, FL 33431
USA

Library of Congress Cataloging-in-Publication Data: New directions in statistical physics : econophysics, bioinformatics, and pattern recognition / Luc T. Wille (ed.). p. cm. Includes bibliographical references and index. ISBN 3-540-43182-9 (alk. paper) 1. Statistical physics. I. Wille, Luc T. QC174.8.N49 2004 530.15'95–dc22 2003044274

ISBN 3-540-43182-9 Springer-Verlag Berlin Heidelberg New York

This work is subject to copyright. All rights are reserved, whether the whole or part of the material is concerned, specifically the rights of translation, reprinting, reuse of illustrations, recitation, broadcasting, reproduction on microfilm or in any other way, and storage in data banks. Duplication of this publication or parts thereof is permitted only under the provisions of the German Copyright Law of September 9, 1965, in its current version, and permission for use must always be obtained from Springer-Verlag. Violations are liable for prosecution under the German Copyright Law.

Springer-Verlag is a part of Springer Science+Business Media

springeronline.com

© Springer-Verlag Berlin Heidelberg 2004
Printed in Germany

The use of general descriptive names, registered names, trademarks, etc. in this publication does not imply, even in the absence of a specific statement, that such names are exempt from the relevant protective laws and regulations and therefore free for general use.

Typesetting: Data prepared by the editor using a Springer TEX macro package
Final layout: LeTeX, Leipzig
Cover design: *design & production* GmbH, Heidelberg

Printed on acid-free paper   SPIN 10852051   57/3141/ba   5 4 3 2 1 0

Preface

Statistical physics has its roots in thermodynamics, a field largely developed for practical reasons such as a desire to improve the operation and efficiency of steam engines. From such 'hot air', through the work of pioneers like Carnot, Gibbs, and Boltzmann, a theory emerged that was anything but hot air, and that could rightly take its place alongside the other cornerstones of modern physics.

The theoretical foundations of the subject are subtle and the necessary calculations often require mathematical ingenuity, two factors that are irresistibly attractive to many physicists. Small wonder then that around the beginning of the 20th century nearly all the greats – Planck, Einstein, and Schrödinger, to name just a few – contributed to its development.

Around the middle of the previous century, there was major progress through the work of Onsager on exactly solvable models and through the contributions of Yang and Lee which laid the foundations for the modern theory of critical phenomena. This led in turn to a series of breakthroughs providing practitioners with such profound notions as scaling and universality. With the concomitant development of the powerful mathematics of fractal geometry and non-linear dynamics, as well as the widespread availability of large-scale computing, the scene was set for a veritable golden age which continues to this day.

Not only have the last few years seen numerous novel ideas and applications to physics and its closely related fields, there has also been an increasing realization that the tools and techniques of statistical physics have exceptionally wide applicability. Thus physicists have enthusiastically tackled problems in economics, sociology, biology, medicine, meteorology, and computer science, to name just a few. Initially this activity was mainly viewed with amusement by the experts in those fields, especially as the 'simple models' approach, so common to the thinking of physicists, is often not part of the culture of other academic areas. As statistical physicists learned to speak the local lingo, truly appreciate the important problems, and as they notched up their first successes, a fertile interaction between the various fields emerged.

It is the purpose of this volume to illustrate some of these accomplishments and, it is hoped, to point the way to future interactions. Thus, an eclectic mix of papers is gathered here, with plenty of success stories, but equally many pointers to remaining questions and avenues for further investigation. The authors come from a variety of fields, but there is clearly a great deal of overlap and a commonality of problems and solutions, which they have taken great pains to indicate.

Graduate students should find some really exciting stories here, with ample pointers to possible research areas for further study and exploration. Moreover, the level of mathematics is generally not beyond their ability and there are numerous references to further background information. Professional physicists may glean novel ideas or possible applications of their pet theories, while other researchers may find that statistical physics has something to offer to their fields. Even non-scientists may after perusal of this volume develop an appreciation for the excitement and sweep of scientific research and how it has a lasting impact on society.

It is a great pleasure for the editor to sincerely thank all the contributors for their efforts. They are in particular to be congratulated for going the extra mile to make their papers as accessible as possible to a wide audience. Sincere thanks are also due to Drs. Angela Lahee and Claus Ascheron at Springer Verlag, for their cheerful cooperation and gentle encouragement.

Boca Raton, August 2003 *Luc T. Wille*

Contents

Part I Fundamental Aspects

Predicting the Direction of a Time Series
Dimitrios D. Thomakos 3

1   Introduction 3
2   Embedding in Direction Space 4
3   Predicting the Direction 7
4   Empirical Examples 11
5   Concluding Remarks 14
References 14

On the Variability of Timing in a Spatially Continuous System with Heterogeneous Connectivity
Viktor K. Jirsa 17

1   Introduction 17
2   Spatiotemporal Dynamics and Integral Equations 18
3   Influence of Connectivity: A Two-Point Connection 21
4   Variability of the Timing of Distant Sites 23
    4.1   Homogeneous Connectivity Only 25
    4.2   Homogeneous Connectivity and Projection from A to B 26
    4.3   Homogeneous Connectivity and Bilateral Pathway
          Between A and B 26
    4.4   Heterogeneous Pathways Only 27
5   Conclusions 28
References 29

First Passage Time Problem:
A Fokker-Planck Approach
Mingzhou Ding and Govindan Rangarajan 31

1   Introduction 31
2   FPT Distribution for Brownian Motion 32
3   FTP Distribution for Continuous Time Random Walks 38
4   Summary 45
References 45

First- and Last-Passage Algorithms in Diffusion Monte Carlo
James A. Given, Chi-Ok Hwang, and Michael Mascagni 47

1   Introduction 47
2   The Angle-Averaging Method 52
3   The Simulation-Tabulation (ST) Method 53
4   The Feynman-Kac Method 56
5   Last Passage Methods for Diffusion Monte Carlo 58
6   Conclusions and Suggestions for Further Study 64
References 65

Part II Econophysics

An Updated Review of the LLS Stock Market Model: Complex Market Ecology, Power Laws in Wealth Distribution and Market Returns
Sorin Solomon and Moshe Levy 69

1   Introduction to the Levy-Levy-Solomon (LLS) Model 69
2   Crashes, Booms and Cycles 71
3   Predation, Competition and Symbiosis Between Trader Species... 73
    3.1   Market Ecologies with Two Trader Species 73
    3.2   Three Investor Species 79
4   LLS with Many Species: Realistic Dynamics of Market Returns .. 80
    4.1   Return Autocorrelations: Momentum and Mean-Reversion ... 80
    4.2   Excess Volatility 81
    4.3   Heavy Trading Volume 81
    4.4   Volume is Positively Correlated with Absolute Returns .. 81
5   The Emergence of Pareto's Law in LLS 82
6   Market Efficiency, Pareto Law and Thermal Equilibrium 84
7   Leptokurtic Market Returns in LLS 86
8   Summary 89
References 89

Patterns, Trends and Predictions in Stock Market Indices and Foreign Currency Exchange Rates
Marcel Ausloos and Kristinka Ivanova 93

1   An Introduction with Some Historical Notes as "Symptoms" 93
    1.1   Tulipomania 95
    1.2   Monopolymania 96
    1.3   WallStreetmania 97
2   Econophysics of Stock Market Indices 98
    2.1   Methodology and Data Analysis 101
    2.2   Aftershock Patterns 104
3   Foreign Currency Exchange Rates 107

	3.1 DFA Analysis	107
	3.2 Data and Analysis	108
	3.3 Probing the Local Correlations	110
4	Conclusions	112
References		112

Toward an Understanding of Financial Markets Using Multi-agent Games
Neil F. Johnson, David Lamper, Paul Jefferies, and Michael L. Hart ... 115

1	Introduction	115
2	The Basic MG	115
3	Grand Canonical Minority Game	119
4	Next Timestep Prediction	121
5	Corridors for Future Price Movements	123
6	Real-World Risk	124
7	Conclusion	126
References		127

Towards Understanding the Predictability of Stock Markets from the Perspective of Computational Complexity
James Aspnes, David F. Fischer, Michael J. Fischer,
Ming-Yang Kao, and Alok Kumar 129

1	Introduction	129
2	A Basic Market Model	130
	2.1 Defining the DSMC Model	131
	2.2 Computer Simulation on the DSMC Model	133
3	A General Market Model	134
4	Predicting the Market	135
	4.1 Markets as Systems of Linear Constraints	136
	4.2 An Easy Case for Market Prediction: Many Traders but Few Strategies	138
	4.3 A Hard Case for Market Prediction: Many Strategies	141
5	Future Research Directions	149
References		150

Patterns in Economic Phenomena
H. E. Stanley, P. Gopikrishnan, V. Plerou, and M. A. Salinger 153

1	Introduction to Patterns in Economics	153
2	Classic Approaches to Finance Patterns	156
3	Patterns in Finance Fluctuations	157
4	Patterns Resembling "Diffusion in a Tsunami Wave"	161
5	Patterns Resembling Critical Point Phenomena	162
6	Cross-Correlations in Price Fluctuations of Different Stocks	164
7	Patterns in Firm Growth	164

X   Contents

8   Universality of the Firm Growth Problem 165
9   "Take-Home Message" 166
References 167

Part III   Bioinformatics

New Algorithms and the Physics of Protein Folding
Ulrich H.E. Hansmann... 173

1   Introduction 173
2   The Generalized-Ensemble Approach 175
     2.1   Multicanonical Sampling............................. 175
     2.2 $1/k$-Sampling 177
     2.3   Simulated Tempering................................. 178
     2.4   Other Generalized Ensembles......................... 178
     2.5   Parallel Tempering.................................. 179
3   The Thermodynamics of Folding 180
     3.1   Helix-Coil Transitions in Homopolymers 180
     3.2   Energy Landscape Analysis of Peptides 185
4   Structure Prediction of Proteins 188
5   Conclusion 190
References 190

Sequence Alignment in Bioinformatics
Yi-Kuo Yu... 193

1   Introduction to Sequence Alignment 193
     1.1   The Holy Grail 194
     1.2   Alignment Algorithms 195
     1.3   Score Statistics 200
     1.4   Substitution (Scoring) Matrices..................... 201
2   Some Recent Developments 204
     2.1   Optimal Alignments.................................. 204
     2.2   Hybrid Alignment 205
     2.3   Open Problems....................................... 208
References 211

Resolution of Some Paradoxes
in B-Cell Binding and Activation: A Computer Study
Gyan Bhanot... 213

1   Introduction 213
2   Brief Description of Human Immune System 214
3   The Dintzis Experimental Results and the Immunon Theory ... 217
4   Modeling the B-Cell Receptor Binding to Antigen:
     Our Computer Experiment 217

5	Results	219
References		224

Proliferation and Competition in Discrete Biological Systems
Yoram Louzoun and Sorin Solomon 225

1	Introduction	225
2	Dynamics of Discrete Proliferating Agents	227
3	How Well Do Different Methods Deal with Discreteness?	229
4	Single S Analysis	230
5	RG Analysis	232
6	Mechanisms Limiting Population Growth	233
	6.1 Local Competition	234
	6.2 Global Competition	235
	6.3 Emergence of Complexity	238
7	Discussion	239
	7.1 Dimensionality	240
	7.2 Inter-Scale Information Flow	240
References		241

Privacy and Data Exchanges
Bernardo A. Huberman 243

1	Introduction	243
2	A Lightning Review of Cryptographic Techniques	245
3	Secret Matching of Data Sets	246
4	Private Surveys in the Public Arena	247
5	Conclusion	250
References		250

Part IV   Pattern Recognition

Statistical Physics and the Clustering Problem
Sebastiano Stramaglia, Carmela Marangi,
Luigi Nitti, and Mario Pellicoro 253

1	Introduction	253
2	Hierarchical Clustering for Phylogeny Reconstruction	255
	2.1 Coupled Map Clustering (CMC) Algorithm	255
	2.2 Distance Measures	257
	2.3 Experiment	259
	2.4 Discussion	260
3	The Auto-encoder Frame	261
	3.1 Cost Functions	261
	3.2 Deterministic Annealing	264
	3.3 Experiments	264

The Challenges of Clustering High Dimensional Data
Michael Steinbach, Levent Ertöz, and Vipin Kumar 273

3.4	Resampling Technique for Unsupervised Estimation of the Number of Classes	266
3.5	Discussion	268
4	Conclusions	271
References		271

1	Introduction	273
2	Basic Concepts and Techniques of Cluster Analysis	274
2.1	What Cluster Analysis Is	274
2.2	What Cluster Analysis Is Not	275
2.3	The Data Matrix	275
2.4	The Proximity Matrix	276
2.5	The Proximity Graph	276
2.6	Some Working Definitions of a Cluster	276
2.7	Measures (Indices) of Similarity and Dissimilarity	279
2.8	Hierarchical and Partitional Clustering	281
2.9	Specific Partitional Clustering Techniques: K-Means	282
2.10	Specific Hierarchical Clustering Techniques: MIN, MAX, Group Average	283
3	The "Curse of Dimensionality"	284
4	Recent Work in Clustering High Dimensional Data	288
4.1	Clustering via Hypergraph Partitioning	288
4.2	Grid Based Clustering Approaches	289
4.3	Noise Modeling in Wavelet Space	296
4.4	A "Concept-Based" Approach to Clustering High Dimensional Data	297
5	Conclusions	307
References		307

Part V Other Applications

Some Statistical Physics Approaches for Trends and Predictions in Meteorology
Kristinka Ivanova, Marcel Ausloos, Thomas Ackerman,
Hampton Shirer, and Eugene Clothiaux 313

1	Introduction	313
1.1	Techniques of Time Series Analysis	315
2	Experimental Techniques and Data Acquisition	315
3	Nonstationarity and Spectral Density	317
4	Roughness and Detrended Fluctuation Analysis	319
5	Time Dependence of the Correlations	322

6	Multi-affinity and Intermittency	324
7	Conclusions	326
Appendix		327
References		328

An Initial Look at Acceleration-Modulated Thermal Convection
Jeffrey L. Rogers, Michael F. Schatz, Werner Pesch, and Oliver Brausch... 331

1	Introduction		331
2	Laboratory		334
	2.1	Experimental Apparatus	334
	2.2	Numerical Methods	336
3	Onset, Time-Dependence, and Typical Patterns		337
	3.1	Onset Measurements	337
	3.2	Confirmation of Time-Dependence	338
	3.3	Harmonic Patterns at Onset	339
	3.4	Harmonic Patterns away from Onset	340
	3.5	Subharmonic Patterns at Onset	343
	3.6	Subharmonic Patterns away from Onset	343
4	Direct Harmonic-Subharmonic Transition		344
	4.1	Transition from Pure Harmonics to Coexistence	345
	4.2	Transition from Pure Subharmonics to Coexistence	347
5	Superlattices		349
	5.1	Observations near Bicriticality	349
	5.2	Observations away from Bicriticality	351
	5.3	Resonant Tetrads	352
	5.4	Other Frequencies	354
6	Discussion		356
References			356

Index 359

List of Contributors

Thomas Ackerman
Pacific Northwest National
Laboratory
Richland, WA 99352
USA
Thomas.Ackerman@pnl.gov

Leonardo Angelini
Dipartimento di Fisica
Universitá di Bari
Via Amendola, 173
I-70126 Bari
Italy
leonardo.angelini@ba.infn.it

James Aspnes
Department of Computer Science
Yale University
New Haven, CT 06520-8285
USA
aspnes@cs.yale.edu

Marcel Ausloos
SUPRAS & GRASP, B5
University of Liège
B-4000 Liège
Belgium
marcel.ausloos@ulg.ac.be

Gyan Bhanot
IBM Research
Yorktown Hts., NY 10598
USA
gyan@us.ibm.com

Oliver Brausch
Physikalisches Institut
der Universität Bayreuth
D-95440 Bayreuth
Germany
ob112@web.de

Eugene Clothiaux
Department of Meteorology
603 Walker Building
University Park, PA 16802
USA
cloth@essc.psu.edu

Mingzhou Ding
Center for Complex Systems and
Brain Sciences
Florida Atlantic University
Boca Raton, FL 33431
USA
ding@walt.ccs.fau.edu

Levent Ertöz
Department of Computer Science
and Engineering
University of Minnesota
Minneapolis, MN 55455
USA
ertoz@csumn.edu

David F. Fischer
Class of 1999
Yale College
New Haven, CT 06520-8285
USA
fischer@aya.yale.edu

Michael J. Fischer
Department of Computer Science
Yale University
New Haven, CT 06520-8285
USA
fischer-michael@cs.yale.edu

James A. Given
Angle Inc.
7406 Alban Station Court
Suite A112
Springfield, VA 22150
USA
given@angleinc.com

P. Gopikrishnan
Goldman Sachs and Co.
10 Hanover Square
New York, NY 10005
USA

Ulrich H.E. Hansmann
Department of Physics
Michigan Technological University
Houghton, MI 49931-1295
USA
hansmann@mtu.edu

Michael L. Hart
Oxford Center for Computational
Finance (OCCF), and
Physics Department
Oxford University
Oxford, OX1 3PU
U.K.
michael.hart@physics.ox.ac.uk

Bernardo A. Huberman
HP Laboratories
Palo Alto, CA 94304
USA
huberman@exch.hpl.hp.com

Chi-Ok Hwang
Computational Electronics Center
Inha University
Incheon 402-751
Korea
chwang@hsel.inha.ac.kr

Kristinka Ivanova
Pennsylvania State University
University Park, PA 16802
USA
kristy@essc.psu.edu

Paul Jefferies
Oxford Center for Computational
Finance (OCCF), and
Physics Department
Oxford University
Oxford, OX1 3PU
U.K.
linc0227@herald.ox.ac.uk

Viktor K. Jirsa
Center for Complex Systems and
Brain Sciences
Florida Atlantic University
Boca Raton, FL 33431
USA
jirsa@ccs.fau.edu

Neil F. Johnson
Oxford Center for Computational
Finance (OCCF), and
Physics Department
Oxford University
Oxford, OX1 3PU
U.K.
n.johnson@physics.ox.ac.uk

Ming-Yang Kao
Department of Computer Science
Northwestern University
Evanston, IL 60201
USA
kao@cs.northwestern.edu

Alok Kumar
School of Management
Yale University
New Haven, CT 06520
USA
alok.kumar@yale.edu

Vipin Kumar
Department of Computer Science
and Engineering
University of Minnesota
Minneapolis, MN 55455
USA
kumar@cs.umn.edu

David Lamper
Oxford Center for Computational
Finance (OCCF), and
OCIAM
Oxford University
Oxford, OX1 3LB
U.K.
lamper@maths.ox.ac.uk

Moshe Levy
School of Business Administration
Mount Scopus 91905
Hebrew University of Jerusalem
Jerusalem
Israel
mslm@mscc.huji.ac.il

Yoram Louzoun
Department of Mathematics
Bar Ilan University
Ramat Gan 52900
Israel
louzouy@math.biu.ac.il

Carmela Marangi
Dipartimento di Fisica
Universitá di Bari
Via Amendola, 173
I-70126 Bari
Italy
Carmela.Marangi@ba.infn.it

Michael Mascagni
Department of Computer Science
Florida State University
203 Love Building
Tallahassee, FL 32306-4530
USA
mascagni@cs.fsu.edu

Luigi Nitti
Istituto di Fisica Medica
Universitá di Bari
Via Amendola, 173
I-70126 Bari
Italy
nitti@ba.infn.it

Mario Pellicoro
Dipartimento di Fisica
Universitá di Bari
Via Amendola, 173
I-70126 Bari
Italy
mario.pellicoro@ba.infn.it

Werner Pesch
Physikalisches Institut
der Universität Bayreuth
D-95440 Bayreuth
Germany
werner.pesch@uni-bayreuth.de

V. Plerou
Center for Polymer Studies, and
Department of Physics
Boston University
Boston, MA 02215
USA
plerou@cgl.bu.edu

Govindan Rangarajan
Department of Mathematics
Indian Institute of Science
Bangalore 560 012
India
rangaraj@math.iisc.ernet.in

Jeffrey L. Rogers
HRL Laboratories
3011 Malibu Canyon Road
Malibu, CA 90265
USA
jeff@hrl.com

M. A. Salinger
Dept. of Finance and Economics
School of Management
Boston University
Boston, MA 02215
USA
salinger@acs.bu.edu

Michael F. Schatz
Center for Nonlinear Science and
School of Physics
Georgia Institute of Technology
Atlanta, GA 30332
USA
michael.schatz
@physics.gatech.edu

Hampton Shirer
Department of Meteorology
518 Walker Building
University Park, PA 16802
USA
hns@psu.edu

Sorin Solomon
Racah Institute of Physics
The Hebrew University of Jerusalem
Givat Ram
Jerusalem 91904
Israel
sorin@vms.huji.ac.il

H. E. Stanley
Center for Polymer Studies, and
Department of Physics
Boston University
Boston, MA 02215
USA
hes@bu.edu

Michael Steinbach
Department of Computer Science
and Engineering
University of Minnesota
Minneapolis, MN 55455
USA
steinbac@cs.umn.edu

Sebastiano Stramaglia
Dipartimento di Fisica
Universitá di Bari
Via Amendola, 173
I-70126 Bari
Italy
sebastiano.stramaglia@ba.infn.it

Dimitrios D. Thomakos
Department of Economics
Florida International University
Miami, FL 33199
USA
dimitrios.thomakos@fiu.edu

Luc T. Wille
Department of Physics
Florida Atlantic University
Boca Raton, FL 33431
USA
willel@fau.edu

Yi-Kuo Yu
National Center for Biotechnology
Information
National Library of Medicine
National Institutes of Health,
Bethesda, MD 20894
USA
yyu@ncbi.nlm.nih.gov

Part I

Fundamental Aspects

Predicting the Direction of a Time Series

Dimitrios D. Thomakos

This chapter proposes and analyzes a new method for predicting the direction of a timeseries, that is, the relative position of future observations with respect to past coordinates, a problem of obvious interest to financial forecasters. The method involves two steps: an embedding step from real-valued observations to discrete values and a prediction step based on statistical inference. Both of these are explained in detail and rigorously justified. Finally, the method is applied to two illustrative time series: the daily closing prices of the S&P500 market index (for the period 1995-2001) and the quarterly growth rates for the US gross domestic product from 1959 till 2000. The results obtained for these two cases are extremely encouraging.

1 Introduction

Prediction is of prime importance in time series analysis. In this paper we try to broaden the scope of standard time series predictive methods by moving away from point and interval predictions. Instead, we propose a method in which what is being predicted is the *direction* of a time series. That is the relative *position* of future observations with respect to some predetermined interval of past coordinates. Our main motivation for such a method mainly comes from the practices of financial market analysts. Quite frequently they are only interested in the future direction of the markets rather than in point forecasts of market indices or individual equities. In fact, those analysts that perform what is known as "technical analysis" explicitly try to predict "market trends" and "turning points", which do not necessarily involve particular point or interval values. In this context we have terms like an "up trend", a "down trend" or a "sideways trend" that describe the past (and hopefully future) path of a particular time series. However, the methods underlying technical analysis do not involve statistical inference, they are descriptive in nature and they are not devoid of controversy as to their real value and their success rate. In any event, what we mean to say is that there is interest among practitioners not only in point or interval forecasts but also in forecasts about the more general concept of direction of a time series.

The method we propose for predicting the direction of a time series involves two steps. The first step is to take the original, real-valued observations and to map them into discrete-valued ones. Our mapping involves a reference interval (or point) of past coordinates of the original observations and three (or two) discrete values which can be thought of as underlying "direction states"; we call this step "embedding in direction space". The second step is to model the new, discrete-valued observations using inference results about nonstationary, categorical time series and then construct probability-based predictors for the future direction of the original time series; we call this step "predicting the direction". While this method does involve statistical inference and the use of prior information (conditioning) it has the standard drawback: the potential probability models used in the second step are only approximations of the "true" underlying probabilities attached to each "direction state". However, the proposed method does offer an inference-based alternative to time series forecasting when interest focuses only on direction rather than on point or interval forecasts.

Related to our approach are the "state-switching" or "regime-switching" or "Markov-switching" models in [7].[1] These models are traditional time series models that involve changes in the underlying data generating process (DGP) due to particular "states" or "regimes" that form a Markov chain. These DGP changes are manifested in changes in some of the models' parameters. The inference theory is based on the states following a Markov chain[2] and the DGP following some type of mixture distribution.

The rest of the paper is organized as follows. The next two sections describe the "embedding in direction space" and "predicting the direction" steps respectively. Section 4 has two empirical examples that illustrate the proposed method while section 5 offers some concluding remarks.

2 Embedding in Direction Space

Consider the measurable space $(\Omega, \mathcal{A})$ together with a family of parametric probability measures $\{\mathcal{P}_\theta : \theta \in \Theta \subseteq \mathbb{R}^k\}$. The triplet $(\Omega, \mathcal{A}, \mathcal{P}_\theta)$ is said to be a statistical model. In the context of this statistical model, we are interested in analyzing a real-valued time series:

$$\{y_t(\omega) : \omega \in \Omega , t \in \mathbb{N}_+\} \tag{1}$$

Suppose that we have available a sample realization of n observations, say $\boldsymbol{y} = \{y_t\}_{t=1}^n$. Define $\mathcal{A}_0 = (\emptyset, \Omega)$, $\mathcal{A}_t = \sigma(y_s, s \leq t) \subset \mathcal{A}$ and consider the nested sequence of (sub) σ-algebras $\{\mathcal{A}_t\}_{t=1}^n$. We denote the restriction of $\mathcal{P}_\theta$ on $\{\mathcal{A}_t\}_{t=1}^n$ by $\{\mathcal{P}_{t\theta}\}_{t=1}^n$. The sample realization induces the probability measure $\mathsf{P}_{n\theta}$ on the (n-dimensional) Borel space $(\mathbb{R}^n, \mathcal{B}^n)$ and defines the (joint) cumulative density function (cdf) of the observations as:

$$F_n(\boldsymbol{y}; \theta) = \mathsf{P}_{n\theta}(B) = \mathcal{P}_{n\theta}(A) \tag{2}$$

---
[1] [8] and [9] are two of the earlier references on this type of models.
[2] However, [3] have considered regime-switching models with time varying probabilities.

for any two sets A and B such that:

$$\{B \in \mathcal{B}^n : B = (-\infty, y_1] \times (-\infty, y_2] \times \cdots (-\infty, y_n] \,,\, A = \boldsymbol{y}^{-1}(B) \in \mathcal{A}_n\}$$

We assume that $F_n(\boldsymbol{y}; \theta)$ is "well behaved", in the sense that is it can be integrated and differentiated an appropriate number of times. The corresponding (joint) density of the observations (that is, the likelihood function) is then given by:

$$\mathcal{L}_n(\boldsymbol{y}; \theta) = \frac{\partial F_n(\boldsymbol{y}; \theta)}{\partial \boldsymbol{y}} \equiv \prod_{t=1}^n f_t(y_t | \mathcal{A}_{t-1}; \theta) \tag{3}$$

where $f_1(y_1 | \mathcal{A}_0; \theta) = f_1(y_1; \theta)$ is the unconditional density of the first observation and where $f_t(y_t | \mathcal{A}_{t-1}; \theta)$ is the conditional density of the tth observation.[3]

Note that in the above set-up we are not concerned with the temporal dependence properties of the time series under study. We are only interested in obtaining the underlying joint density of the observations so as to later establish the existence of certain conditional, transition probabilities.

We would like to make predictions about the direction of future observations of the time series y_t. That is, we are interested in predicting the position of future observations relative to a past coordinate or an interval of past coordinates. While there is an advantage of using an interval of past coordinates, rather than a single coordinate, our definition allows for direction to be defined with respect to a single coordinate as well. There could be many ways of successively partitioning the range of values of the sample realization $\boldsymbol{y}$. Below we consider one such way that, we believe, is both useful and intuitive. We have:

Definition 1. Let $m \geq 1$ be an integer and consider the subset of m observations $\mathcal{Y}_{t,m} = \{y_{t-m+1}, \ldots, y_t\}$. Let $y_{t,m}^{\min} = \min \mathcal{Y}_{t,m}$ and $y_{t,m}^{\max} = \max \mathcal{Y}_{t,m}$ denote the extreme value statistics of the subset $\mathcal{Y}_{t,m}$. Then, define the interval of reference coordinates as $r_{t,m} = \left[y_{t,m}^{\min}, y_{t,m}^{\max}\right]$, for $t = m, \ldots n$. For $m = 1$ we have that the interval $r_{t,m}$ becomes a single coordinate, that is $r_{t,1} = y_t$.

The above reference intervals effectively capture the range of oscillation for overlapping, rolling windows of m observations each, that cover the whole length of the realization. For a sample of size n we have $n - m + 1$ such

---

[3] More precisely, the likelihood function is defined as the Radon-Nikodym derivative of $\mathcal{P}_{n\theta}(A)$ with respect to a σ-finite, dominant probability measure $\mathcal{P}$ (that does not depend on the parameter vector θ). That is:

$$\mathcal{P}_{n\theta}(A) = \int_A \mathcal{L}_n(\boldsymbol{y}; \theta) d\mathcal{P} \quad \forall A \in \mathcal{A}_n$$

windows. The value of m needs to be chosen before the analysis, though we will suggest a straightforward, data-dependent method for selecting it in the next section. How are these reference intervals useful? They allow for a simple definition of direction; in fact, assuming that $m > 1$, there can be only three modes of direction relative to any reference interval $r_{t,m}$. For any integer $h \geq 1$ (the forecasting horizon) we say that that whenever $y_{t+h} \in r_{t,m}$ the time series moves *sideways*, whenever $y_{t+h} > y_{t,m}^{\max}$ the time series exhibits an *upward local trend* and whenever $y_{t+h} < y_{t,m}^{\min}$ the time series exhibits a *downward local trend*. Direction can effectively be treated as an underlying "state" of the time series. Thus, we can embed y_t in an appropriate (finite) state space, which we may term *local trend space* or *direction space* and transform the real-valued time series y_t into a discrete, categorical time series, say $x_{t+h,m}$. Modeling the state series $x_{t+h,m}$ can lead to probability forecasts about the future direction of the original time series y_t for any chosen forecasting horizon h. We define the direction state series $x_{t+h,m}$ in:

Definition 2. Let $h \geq 1$ be an integer. Given a sequence of reference coordinates $\{r_{t,m}\}_{t=m}^{n-h}$, the *local trend* or *direction* state series for the original series y_{t+h}, taking values on the set $\mathbb{S} = \{-1, 0, 1\}$, is given by:

$$x_{t+h,m} = \text{sign}\left(y_{t+h} - y_{t,m}^{\max}\right)\left[1 - I(y_{t+h} \in r_{t,m})\right] \quad (4)$$

where $I(\cdot)$ is the indicator function.

i. The unconditional probabilities (UP) of being in a particular direction state $j \in \mathbb{S}$ at time $t + h$, denoted by $\pi_{j,t+h}(\theta) = \mathsf{P}(x_{t+h,m} = j; \theta)$, can be obtained using the individual, unconditional cdfs and they are:

$$\begin{aligned} \pi_{1,t+h}(\theta) &= 1 - F_{t+h}(y_{t,m}^{\max}; \theta) \\ \pi_{0,t+h}(\theta) &= F_{t+h}(y_{t,m}^{\max}; \theta) - F_{t+h}(y_{t,m}^{\min}; \theta) \\ \pi_{-1,t+h}(\theta) &= F_{t+h}(y_{t,m}^{\min}; \theta) \end{aligned} \quad (5)$$

ii. The conditional transition probabilities (CTP) to direction state $j \in \mathbb{S}$ at time $t + h$, given $\mathcal{G}_{t,m} = \sigma(\{y_s, x_{s,m}\}, s \leq t)$ denoted by $p_{j,t+h}(\theta) = \mathsf{P}(x_{t+h,m} = j | \mathcal{G}_{t,m}; \theta)$ can also be obtained using the individual, conditional cdfs and they are:

$$\begin{aligned} p_{1,t+h}(\theta) &= 1 - F_{t+h}(y_{t,m}^{\max} | \mathcal{G}_{t,m}; \theta) \\ p_{0,t+h}(\theta) &= F_{t+h}(y_{t,m}^{\max} | \mathcal{G}_{t,m}; \theta) - F_{t+h}(y_{t,m}^{\min} | \mathcal{G}_{t,m}; \theta) \\ p_{-1,t+h}(\theta) &= F_{t+h}(y_{t,m}^{\min} | \mathcal{G}_{t,m}; \theta) \end{aligned} \quad (6)$$

iii. The individual cdfs $F_{t+h}(\cdot; \theta)$ and the conditional cdfs $F_{t+h}(\cdot | \mathcal{G}_{t,m}; \theta)$ are well defined in lieu of the well defined joint cdf $F_n(\boldsymbol{y}; \theta)$ and can be obtained by appropriate integrations.

The unconditional probabilities given above are time-varying: they depend on the extreme value statistics of the tth window of observations and we do not expect that these statistics will be the same in all windows. Thus, $x_{t+h,m}$ cannot, in general, be modeled as a stationary categorical time series or a homogeneous Markov chain. In the following section, we will adapt results from the categorical time series literature in order to make inferences about these CTPs.

Note that, unlike standard time series prediction, the proposed method allows us to assess the probability of an event (direction) given a reference interval whose width does change as the forecasting horizon increases.[4] To see this, note that the reference intervals depend only on m and not on h. Once m is selected one can obtain the sequence of $r_{t,m}$ intervals; then, different forecasting horizons h lead to different direction series $x_{t+h,m}$, which of course lead to different probability forecasts for direction.

3 Predicting the Direction

The discussion of the previous section implies that the embedded direction series $x_{t+h,m}$ can be modeled as a categorical time series, with CTPs given by (6). The importance of (6) is that it shows that, for any well defined parametric time series model, the CTPs of the embedded series $x_{t+h,m}$ exist and are functions of some underlying cdf and of a vector of unknown parameters θ. In order to make predictions about the future direction of y_t we thus need to estimate θ, given knowledge of the conditional densities $F_{t+h}(\cdot|\mathcal{G}_{t,m};\theta)$. However, the exact form of underlying cdf $F_n(\boldsymbol{y};\theta)$ (from which the conditional densities can be obtained) is rarely known. Moreover, even if the density of the data was known there are other, more efficient, methods for prediction. Thus, we now proceed under the assumption that the underlying density of the data is not known and that the CTPs of the embedded series can be well approximated by an appropriate function $\psi(\cdot;\gamma)$, $\gamma \in \Gamma \subseteq \mathbb{R}^q$, where it is possible that the new parameter vector $\gamma = g(\theta)$, for some function $g(\cdot) : \mathbb{R}^k \to \mathbb{R}^q$. We note that while the elements of the original parameter vector θ can have meaningful interpretations (e.g. mean, variance or some other moment), the same cannot be necessarily said about the elements of the new parameter vector γ.

There exist results in inference theory for categorical time series that are modeled like the direction series that we consider in this study. We rely on the findings of [4], [10] and [5]. The broadest treatment is that in [5], as the authors make no Markovian assumptions (in contrast to [4] and [10]) about the state series, which considerably extends the modeling framework. The approach in all three references is an extension of the class of Generalized Linear Models (GLIM) to categorical time series.[5] Our subsequent discussion

---

[4] Assuming a fixed forecasting origin. As new observations become available the width of the reference intervals will, of course, change.
[5] See [11] for a comprehensive treatment of GLIM models or chapter 19 in [6] for a summary of limited dependent variable models.

draws extensively on [10] and, especially, [5]. Further details and technical assumptions can be found in these references.

Consider the sample realization of the direction series $\{x_{t+h,m}\}_{t=m}^{n-h}$, for some fixed integer m and horizon h. The CTPs are now parametrized through the function $\psi(.;\gamma)$ as follows. Consider the (2×1) vector $\boldsymbol{w}_{t+h} = (w_{j,t+h})$ defined as:

$$w_{j,t+h} = \begin{cases} 1 \text{ if state } j \text{ occurs} \\ 0 \quad \text{otherwise} \end{cases} \tag{7}$$

Thus, the vector $\boldsymbol{w}_{t+h}$ takes values depending on which state j is observed. Which state $j^* \neq j$, out of the three considered here, will be left out is of no consequence: for example, suppose that we decide to leave out the "downward local trend" state, i.e. $j^* = -1$. Then, take $w_{1,t+h} = 1$ when $x_{t+h,n} = 1$, $w_{0,t+h} = 1$ when $x_{t+h,m} = 0$ and compute $w_{-1,t+h} = 1 - \sum_{j=0,1} w_{j,t+h}$. Similarly, define a (2×1) vector of transition probabilities $\boldsymbol{p}_{t+h}(\gamma)$ with j^{th} element given by $p_{j,t+h}(\gamma)$. Again, the transition probability of the left out state j^* can be computed as $1 - \sum_{j=0,1} p_{j,t+h}(\gamma)$.

It is clear from (6) that the CTPs depend both on the past history of the original observations (through the extreme value statistics of the reference intervals) and on the past history of the embedded state series. Thus, it is not unreasonable to assume that the function $\psi(\cdot;\gamma)$ depends on a $(q \times 2)$ matrix of stochastic, time dependent covariates $\boldsymbol{Z}_t$ and write $\psi(\boldsymbol{Z}_t;\gamma)$. In our context these covariates could be any combination of current and lagged values of the original as well as the embedded series.[6] Since $\boldsymbol{Z}_t$ will be used to model the CTPs at time $t+h$ it is considered predetermined at time t. A natural choice for the lag length used on the covariate matrix in $\boldsymbol{Z}_t$ is m, since the embedded direction series at $t+h$ depends on the past m observations of the original series y_{t+h}.[7] In the case where no interactions between the original and embedded series are used in $\boldsymbol{Z}_t$, the dimension of the parameter vector γ can vary between $1 \leq q \leq 2m$. Following the paradigm of the GLIM models, it is further assumed that the CTPs are affected, through $\psi(\boldsymbol{Z}_t;\gamma)$, by a linear combination of the covariates and the parameters, that is we write $\psi(\boldsymbol{Z}'_t\gamma)$. Since this function connects the CTPs with the covariates and the parameters it is known as a *link function*. It is assumed twice continuously differentiable and maps a subset of $\mathbb{R}^2$ bijectively onto the set $\mathbb{P} = \left\{ p_{j,t+h}(\gamma) > 0, j = 0,1, \sum_{j=0,1} p_{j,t+h}(\gamma) < 1 \right\}$. The inverse function $\psi^{-1}(\cdot)$ is also well defined. The vector of the CTPs $\boldsymbol{p}_{t+h}(\gamma)$ can thus be written as:

$$\boldsymbol{p}_{t+h}(\gamma) = \psi(\boldsymbol{Z}'_t\gamma) \quad \Leftrightarrow \quad \psi^{-1}\left[\boldsymbol{p}_{t+h}(\gamma)\right] = \boldsymbol{Z}'_t\gamma \tag{8}$$

---
[6] A constant term can also be added to each of the two columns in $\boldsymbol{Z}_t$.
[7] As noted in [10], in the case where $\boldsymbol{Z}_t$ depends on the last m observations of $x_{t+h,m}$ and any other covariates, then $x_{t+h,m}$ can be treated as a (nonhomogeneous) Markov chain of order m.

Among other link functions, an example of a convenient one is the multinomial logit link. We have that the jth element of the function (that is, the jth probability) is given by

$$p_{j,t+h}(\gamma) = \frac{\exp(\mathbf{Z}'_{j,t}\gamma)}{1+\sum_{j=0,1}\exp(\mathbf{Z}'_{j,t}\gamma)} \ , \qquad j = 0,1 \qquad (9)$$

where $\mathbf{Z}_{j,t}$ denotes the column of $\mathbf{Z}_t$ of the corresponding direction state. For the left-out "downward trend" state we have:

$$p_{-1,t+h}(\gamma) = \frac{1}{1+\sum_{j=0,1}\exp(\mathbf{Z}'_{j,t}\gamma)} \qquad (10)$$

Inference about the CTPs now depends exclusively on inference about γ. For some choice of the link function, the vector of unknown parameters can be estimated by the method of maximum likelihood (ML). The ML estimators, say $\widehat{\gamma}_n$, can then be used to estimate the CTPs. Standard errors and confidence intervals for the CTPs are also easily obtained. In [5] the authors use for estimation the more general method of partial likelihood (PL) (see also [2], [14] and [13]). The PL is similar to the usual likelihood function but it requires neither complete knowledge of the joint distribution of the covariates (as in full likelihood) nor complete covariate information throughout the period of observation (as in conditional likelihood). PL uses only what is known to the observer up to the time of actual observation. This concept fits perfectly in our set-up: the embedded direction series $x_{t+h,m}$ and the corresponding CTPs are (1) both conditioned on just the available information up to time t through $\mathcal{G}_{t,m}$ and (2) since $\psi(\mathbf{Z}'_t\gamma)$ approximates the true, but unknown cdfs, we do not have complete covariate information, that is we do not have complete information about what should be included in $\mathbf{Z}_t$. The partial log-likelihood is given by:

$$\ell_n(\gamma) = \sum_{t=m}^{n-h}\sum_{j=-1}^{1} w_{j,t}\log p_{j,t}(\gamma) \qquad (11)$$

and partial ML estimates can easily be obtained through numerical optimization of the above partial log-likelihood. Under mild assumptions, the asymptotic distribution of the partial ML estimators $\widehat{\gamma}_n$ is normal and given by:

$$\sqrt{n}(\widehat{\gamma}_n - \gamma) \to_d \mathcal{N}(\mathbf{0}, \mathbf{V}(\gamma)) \ , \ n \to \infty \qquad (12)$$

where $\mathbf{V}(\gamma)$ is the inverse of the conditional information matrix. Using the ML estimates for the parameters, we can then obtain estimates for the CTPs as $\widehat{\mathbf{p}}_{t+h}(\widehat{\gamma}_n) = \psi(\mathbf{Z}'_t\widehat{\gamma}_n)$. Standard errors and confidence intervals for the estimated CTPs can be obtained using the asymptotic distribution of the partial ML estimators $\widehat{\gamma}_n$ and the delta method, applied to the link function.

Note that the in-sample estimates of the CTPs obtained this way are similar to "fitted" values in a regression context. Alternatively, one can obtain recursive CTP estimates, iterating over all sample values $t = m, \ldots, n - h$. This second alternative might be useful since it allows the coefficient values to change as new information becomes available (this is essentially the set-up employed in [10]).

The in-sample estimates of the CTPs can be also be used in constructing an additional, simple statistic based on averaging. That is, one can consider the sample average of all in-sample CTPs, say:

$$\mathcal{P}_{j,N} = \frac{1}{N} \sum_{t=m}^{n-h} \widehat{p}_{j,t+h}(\widehat{\gamma}_n) = \frac{1}{N} \sum_{t=m}^{n-h} \psi(\mathbf{Z}_{j,t}\widehat{\gamma}_n) \, , \, j = -1, 0, 1 \qquad (13)$$

where $N = n - h - m + 1$. These can be used to concisely summarize the individual CTPs over all observations and to aid in out-of-sample prediction. Note that, since $\mathcal{P}_{j,N}$ depends on the covariate matrix $\mathbf{Z}_t$, it is expected that it would vary as one changes m and/or the number of lagged covariates of y_t or $x_{t+h,m}$ included in $\mathbf{Z}_t$. While we will not pursue the issue of the asymptotic behavior of $\mathcal{P}_{j,N}$ here, one can surmise that, since the link function is continuous and $\widehat{\gamma}_n$ is a consistent estimator of γ, some version of a law of large numbers can be applied to $\mathcal{P}_{j,N}$.

We next turn to the issue of selecting a value for m. Since the parameters of the link function and the corresponding CTPs are computed via ML, one can use any of the known model selection criteria (as the AIC in [1] or the BIC in [12]) to select m. This is an easy, data-dependent method for selecting m since one would usually change the number of lagged covariates included in $\mathbf{Z}_t$ as m changes. Thus, the number of parameters estimated would change with m which justifies the use of model selection criteria. As a reminder to the reader, most model selection criteria, say $\mathcal{M}(\widehat{\gamma}_n, m)$, (that are based on models estimated by ML) have the generic form:

$$\mathcal{M}(\widehat{\gamma}_n, m) \approx -\ell_n(\widehat{\gamma}_n) + \Pi_n(q) \qquad (14)$$

where $\Pi_n(q)$ is a "penalty" factor that is based on the number of estimated parameters (q being the dimension of γ) and possibly the sample size. For example, in the case of the AIC criterion we have that $\Pi_n(q) = 2q$ so that $\mathcal{M}(\widehat{\gamma}_n, m) = -2\ell_n(\widehat{\gamma}_n) + 2q$. In the case of the BIC criterion we have that $\Pi_n(q) = q \ln n$ so that $\mathcal{M}(\widehat{\gamma}_n, m) = -2\ell(\widehat{\gamma}_n) + q \ln n$. The optimal m would be the one minimizing $\mathcal{M}(\widehat{\gamma}_n, m)$ so that we can write $\widehat{m} = \inf_{m \geq 1} \mathcal{M}(\widehat{\gamma}_n, m)$.

Alternatively, one can use sequential likelihood ratio (LR) tests to determine the appropriate number of parameters to be included in γ. Under the same conditions leading to the asymptotic distribution of $\widehat{\gamma}_n$ in (12), we have that:

$$2\left[\ell_n(\widehat{\gamma}_n) - \ell_n(\widetilde{\gamma}_n)\right] \to_d \chi^2(J)$$

where $\tilde{\gamma}_n$ denotes the restricted estimator (with fewer parameters) and J denotes the number of extra parameters appearing in $\widehat{\gamma}_n$, the unrestricted estimator.

We can now define the out-of-sample, h-step ahead predictor for the embedded direction series $x_{t+h,m}$. We have:

Definition 3. Consider the partial ML estimators for the parameters of the link function, $\widehat{\gamma}_n$, and the in-sample corresponding estimators of the CTPs, $\widehat{p}_{j,t+h}(\widehat{\gamma}_n)$ for $j = -1, 0, 1$ and $t = m, \ldots, n-h$. The out-of-sample, h-step ahead, predictor of the embedded direction series $x_{t+h,m}$, denoted by $\mathsf{E}\left(x_{n+h,m}|\mathcal{G}_{n,m}; \widehat{\gamma}_n\right)$, is defined as:

$$\mathsf{E}\left(x_{n+h,m}|\mathcal{G}_{n,m}; \widehat{\gamma}_n\right) = \left\{ j^* \in \mathbb{S} : \widehat{p}_{j^*,n+h}(\widehat{\gamma}_n) = \sup_{j \in \mathbb{S}} \widehat{p}_{j,n+h}(\widehat{\gamma}_n) \right\} \tag{15}$$

The above predictor simply selects the direction which is assigned highest probability. Since the asymptotic distribution of the CTPs is known, one can construct a test in which the predicted direction (of highest probability) is compared to one of the remaining two directions (of lower probability). That is, we can test the null hypothesis of $H_0 : p_{j^*,n+h} = p_{j,n+h}$ for $j \neq j^*$. Also note that computation of the the predictor does not requires knowledge of the future values of the covariates in $\boldsymbol{Z}_t$: for any chosen value of the forecast horizon h, $\boldsymbol{Z}_t$ includes values only up to time t. Thus, a multi-step ($h > 1$) predictor can always be computed and, in contrast to standard prediction, we need not compute it recursively using the previous predicted values in place of the unknown future values.

4 Empirical Examples

In this section we illustrate the methodology introduced above using two sample time series from finance and economics. The first sample series consists of daily closing prices of the Standard & Poor's 500 (S&P500) market index (data from January 1995 to July 2001 for a total of 1661 observations). The second sample series consists of the quarterly growth rate of the U.S. real GDP (RGDP) (data from 1959 to 2000 for a total of 168 observations).

Our results are on in-sample average CTPs, that is $\mathcal{P}_{j,N}$, for both series. Instead for searching for the most appropriate specification (that is, searching for a choice of m and the number of lagged covariates in $\boldsymbol{Z}_t$) we opted for presenting results for various choices of both m and h, while keeping the number of lagged covariates in $\boldsymbol{Z}_t$ fixed. We have used the logit link function, given in equations (9) and (10), in all computations. For the S&P500 series we present results for three values of m, $m = \{2, 5, 20\}$ (corresponding to reference intervals of two days, a week and a month) and three values of h, $h = \{1, 5, 20\}$. For the RGDP series we present results for two values of m,

$m = \{4, 8\}$ (corresponding to a year and two years) and three values for h, $h = \{1, 4, 8\}$. For both series the covariate matrix $\boldsymbol{Z}_t$ included lagged values of only the embedded series $x_{t+h,m}$ and no lagged values of the original series y_t or a constant term. For the S&P500 series we set the lag length to $q = 5$ and for the RGDP series we set the lag length to $q = 4$. The reader should keep in mind that the results presented below are tied to these particular choices of m, h and q. A different set of average CTPs could be obtained if one uses another link function or different values for m, h and q.

In addition to the average CTPs we present, for comparison, the sample average of the direction state vector $\boldsymbol{w}_{t+h}$. These are the sample proportions for each of the three direction states. We denote them by $\widehat{\pi}_{j,N}$ and are given by:

$$\widehat{\pi}_{j,N} = \frac{1}{N} \sum_{t=m}^{n-h} w_{j,t+h} \, , \, j = -1, 0, 1$$

Note that, if the direction state series was modeled as a first-order homogeneous Markov chain, $\widehat{\pi}_{j,N}$ would have been the ML estimates for the UP $\pi_{j,t+h}$ given in (5).

Our results are summarized in Tables 1 and 2. Table 1 has the results for the S&P500 series while Table 2 has the results for the RGDP series. The format of all tables is the same: each table has three row panels corresponding to the values of the forecasting horizon h and two column panels corresponding to the average values of UPs (first column panel) and CTPs (second column panel). Each column panel has three (S&P500) or two (RGDP) columns for the corresponding values of m. The three possible directions are denoted by Up (for upward local trend), Side (for sideways movement) and Down (for downward local trend).

The results in Table 1 indicate that for the Up and Side directions the CTPs are higher than the UPs, while the opposite occurs for the Down direction. That is, the conditioning on prior information used in estimating the CTPs leads to higher average probabilities for an upward trend or a sideways movement relative to a downward trend. For any given value of m, as the forecasting horizon increases the Up direction probabilities become larger and the Side direction probabilities become smaller, both for the UPs and the CTPs. However, the Down direction probabilities either stay the same or increase for the UPs but they decrease for the CTPs, again showing the effects of conditioning. Also note that the differences in magnitude between UPs and CTPs are more pronounced for the Down direction than for the Up and Side direction. In general, as h increases the Up direction probabilities clearly dominate, indicating the long-term trend in the series. Next, note that, for any given h, as m increases the Up and Down direction probabilities decrease while the Side direction probability increases. These results are quite consistent with the trending but also fluctuating stock market. The short-run fluctuations are reflected in the larger average probabilities assigned to the

Table 1. Average UP and CTP estimates for various values of m and h. S&P 500 Market Index Daily Data 01/1995 to 07/2001.

Direction	Average UPs			Average CTPs		
	$m=2$	$m=5$	$m=20$	$m=2$	$m=5$	$m=20$
	$h=1$					
Up	0.4286	0.3134	0.2017	0.4570	0.3655	0.2387
Side	0.2495	0.4988	0.7252	0.2856	0.4994	0.7249
Down	0.3219	0.1878	0.0731	0.2575	0.1351	0.0364
	$h=5$					
Up	0.5372	0.4383	0.3177	0.5998	0.5248	0.3696
Side	0.1178	0.3154	0.5681	0.1931	0.3448	0.5783
Down	0.3450	0.2464	0.1142	0.2071	0.1304	0.0522
	$h=20$					
Up	0.6195	0.5626	0.4679	0.7005	0.6500	0.5243
Side	0.0646	0.1778	0.3921	0.1430	0.2315	0.4201
Down	0.3159	0.2596	0.1400	0.1565	0.1185	0.0556

Table 2. Average UP and CTP estimates for various values of m and h. U.S. Real GDP Quarterly Growth Data 1959-2000.

Direction	Average UPs		Average CTPs	
	$m=4$	$m=8$	$m=4$	$m=8$
	$h=1$			
Up	0.2086	0.1006	0.2830	0.1484
Side	0.5215	0.7610	0.5283	0.7677
Down	0.2699	0.1384	0.1887	0.0839
	$h=4$			
Up	0.2125	0.1026	0.2970	0.1645
Side	0.5000	0.7308	0.5342	0.7434
Down	0.2875	0.1667	0.1688	0.0921
	$h=8$			
Up	0.2500	0.1382	0.3476	0.2152
Side	0.4936	0.6908	0.5055	0.6983
Down	0.2564	0.1711	0.1469	0.0866

Side direction, as m is increasing. For example, the Side direction CTPs for $h = 1$ more than double from 28.56 percent, for $m = 1$, to 72.49 percent, for $m = 20$. The long-term trend is reflected in the increase of the Up direction average probabilities. For example, the Up direction for $m = 20$ more than double from 23.87 percent, for $h = 1$, to 52.43 percent, for $h = 20$.

The results in Table 2 are, in many respects, similar to those in Table 1. However, as the real GDP growth series is not trending, there are also some significant differences. As in Table 1, the CTPs for the Up (Down) direction are larger (smaller) than the UPs for the same direction; moreover, both the UPs and the CTPs exhibit the same monotonicity as in Table 1. This monotonicity is, however, not as strong as in the results of Table 1. The CTPs for the Side direction are practically the same to the UPs for the same direction. These results, combined with the dominance of the Side direction probabilities across m and h, are an explicit indication for the absence of long-term trend in this series. Note that, for all table entries, the Side direction probabilities are 50 percent and up, while the largest Up direction probability is only 34.76 percent.

5 Concluding Remarks

In this paper we proposed a method for predicting the direction of future observations of a real-valued time series. Direction was defined as the relative position of the future observations with respect to a predetermined interval of past coordinates. Using a two-step procedure that consisted of a state-embedding step and an estimation step (the second step being based on inference results about nonstationary categorical time series) we defined a probability-based predictor for direction. The usefulness of this methodology can only be assessed by further applications using more time series and by comparing its results with other predictive time series models.

References

1. H. Akaike: 'Information Theory and an Extension of the Maximum Likelihood Principle'. In: *2nd International Symposium on Information Theory*, ed. by B.N. Petrov and F. Csaki (Akademiai Kiado, Budapest, 1973)
2. D.R. Cox: Biometrika **62**, 441 (1975)
3. F.X. Diebold, J.H. Lee, G.C. Weinbach, 'Regime Switching with Time Varying Probabilities'. In: *Nonstationary Time Series and Cointegration*, ed. by C. Hargreaves (Oxford University Press, Oxford 1994)
4. L. Fahrmeir, H. Kaufmann: J. Time Series Analys. **8**, 147 (1987)
5. K. Fokianos and B. Kedem: J. Mult. Analys. **67**, 277 (1998)
6. W. Greene: *Econometric Analysis*, 4th ed., (Prentice Hall, Upper Saddle River 2000)
7. J.D. Hamilton: Econometrica **57**, 357 (1989)
8. J.D. Hamilton: J. Econometrics **45**, 39 (1990)

9. J.D. Hamilton: J. Bus. Econ. Statist. **9**, 27 (1991)
10. H. Kaufmann: Ann. Statist. **15**, 79 (1987)
11. P. McCullagh, J.A. Nelder: *Generalized Linear Models*, 2nd. ed. (Chapman and Hall, London 1989)
12. F. Schwarz: Ann. Statist. **6**, 461 (1978)
13. E. Slud: Scand. J. Statist. **19**, 97 (1992)
14. W.H. Wong: Ann. Statist. **14**, 88 (1986)

On the Variability of Timing in a Spatially Continuous System with Heterogeneous Connectivity

Viktor K. Jirsa

The processing of "relevant" or "meaningful" information has been related to the occurrence of macroscopic phase transitions in the system (see Jirsa (2004) [1] for a discussion). Information processing in a spatially distributed system is a function of the connectivity and the intrinsic dynamics of the elements constituting the system. The macroscopic phase transitions are influenced in particular by heterogeneous connections which connect far-distant sites. This type of connectivity is found commonly in biological systems such as the cortex. However, even before phase transitions occur, their signatures are present in a reduced variability of the timing patterns of the connected sites. Here we discuss how, and to what extent, the introduction of additional connectivity reduces the variability of timing patterns, even though the system does not perform a macroscopic phase transition.

1 Introduction

Macroscopic pattern formation and self-organization phenomena have been studied systematically in physical and chemical systems (see [2,3] for reviews). As a consequence, most methods and mathematical models describing pattern formation have been developed to solve problems posed by physics and chemistry. Many examples can be found in physics, e.g. hydrodynamics, laser, and chemistry, e.g. Belousov Zhabotinsky reaction The connection topology of the underlying material substrate is exclusively homogeneous in the sense of space invariance. To the extent that inhomogeneous properties are considered, they are typically introduced as spatially varying parameters or inputs to the homogeneous system (see e.g. [4] for a recent example). Systems with a spatially invariant connection topology allow for a differential partial differential description of their dynamics. Only later, biological and other systems were viewed from the dynamics system and pattern formation perspective posing different and exciting new questions. Neuronal networks, biological tissue and geographic spread of epidemics (see [15] for an introduction) are just a few examples. Such systems display a *heterogeneous* connection topology, that is their connectivity is spatially variant. In such cases, a description of the dynamics as a partial differential equation results in high order differentials and a multitude of delays raising the complexity enormously. An integral description, however, allows for spatiotemporal pattern formation and provides a simpler representation and control of the inhomogeneous connection topology by its integral kernel. An initial treatment of an inhomogeneous connection topology in a dynamic system has been made by the introduction of instantaneous long-range connections between areas in a discretely coupled chain of oscillators [5]. In previous work [7,12] we have reported an integral formulation of a spatially continuous dynamic system with a heterogeneous connection topology and propagation delays along its connections. Experimentally, these systems are probed by detectors at specific locations. These detectors register the on-going activity simultaneously at each site. After data collection, the analysis of the timing and its variability between individual sites provides information on how sites communicate with each other and how they are connected. A reduced variability in the timing of events between different locations is typically understood to be a consequence of the presence of a coupling and thus enhanced information exchange. However, the question remains open how different couplings or pathways influence the timing variability. The emergence of a spatiotemporal pattern will naturally reduce the timing variability. In previous work [7,12], we showed that the introduction of additional pathways may induce macroscopic pattern formation implying a reduced timing variability. In the present study, we investigate the contributions of homogeneous and heterogeneous couplings below, but close to the threshold of macroscopic pattern formation, however in the presence of noise. In particular we study the communication between individual sites from the modeling perspective, that is we define a spatiotemporal system with a homogeneous connectivity and embedd an additional pathway connecting two sites A and B. Then we study the timing of activation and its variability at the individual sites in dependence of the presence or non-presence of a direct pathway between A and B. In Sect. 2 we briefly review the formalism [12] which allows such a system's analytic treatment of pattern formation by means of a mode decomposition. In Sect. 3 the paradigmatic example of a two-point connection embedded in a continuous homogeneously connected medium illustrates the destabilization mechanism of a heterogeneous connection. The destabilization leads to a non-equilibrium phase transition and causes macroscopic pattern formation. In Sect. 4 we show how the variability of the timing between individual areas is affected by different types of connectivity.

2   Spatiotemporal Dynamics and Integral Equations

We define the spatiotemporal dynamics of a scalar field $\psi(x,t)$ with space $x \in \mathcal{R}^n$ and time $t \in \mathcal{R}$ as a nonlinear retarded integral equation of the form

$$\psi(x,t) = \int_A dX f(x,X) S(\psi(X,T) + I(X,T)) \tag{1}$$

where $f(x,X)$ describes a general connectivity function and S a nonlinear function of ψ at space-point X and a time-point $T = t - |x - X|/v$ delayed by the propagation time over the distance $|x - X|$. A denotes the surface area of the medium and v the constant signal velocity. $I(t)$ is the input to the field $\psi(x,t)$. Variations of this type of integral equation have been widely used in theoretical neuroscience [8–12] to describe the dynamics of neural activity. But in all these cases, as well as generally in physical systems, translational variance $f(x,X) = f(|x - X|)$ has been employed. Here we assume $f(x,X)$ to be a general function of x and X. We decompose the field $\psi(x,t)$ into spatial modes $g_n(x)$ and complex time dependent amplitudes $\psi_n(t)$ such as

$$\psi(x,t) = \sum_{n=-\infty}^{\infty} g_n(x)\psi_n(t) \qquad (2)$$

The choice of the spatial basis functions will depend on the surface A and its boundary conditions, but also on practical considerations about the type of connectivity $f(x,X)$ and inputs I to the system. An adjoint basis is given by

$$\int_A dx\, \bar{g}_m(x) g_n(x) = \delta_{mn} \qquad (3)$$

where δ_{mn} is the Kronecker symbol. We choose a polynomial representation of the nonlinear function S in (7) such as

$$S(X,T) = a_0 + a_1 \psi(X,T) + a_2 \psi 2(X,T) + ... = \sum_{m=0}^{\infty} a_m \psi^m(X,T) \qquad (4)$$

where $a_m \in \mathcal{R}$ and $I(X,T)$ is not considered (no restriction of generality). By projection of (7) on a spatial basis function $\bar{g}_q(x)$ and restricting its dimension N to be finite, we obtain a set of N coupled integral equations

$$\psi_q(t) = \int_A dx\, \bar{g}_q(x) \int_A dX\, f(x,X) \sum_{m=0}^{\infty} a_m \left(\sum_{n=-N}^{N} g_n(X)\psi_n(T) \right)^m \qquad (5)$$

Note that T still depends on the distance $|x - X|$. We perform the following trick: exchange space and time integration by introducing a time integral and a δ-function $\delta(t - \tau - v^{-1}|x - X|)$ in (5); then perform the integration over dX preserving $t - \tau \geq 0$ for causality. The resulting equations are

$$\Psi(t) = v \int_{-\infty}^{t} d\tau\, (\Gamma_0(t-\tau) + \Gamma_1(t-\tau)\Psi(\tau) + \Gamma_2(t-\tau)\Psi(\tau)\Psi(\tau) + ...) \qquad (6)$$
$$= v \int_{-\infty}^{t} d\tau\, \Gamma_0(t-\tau) + L^t \Psi(t) + N^t(\Psi(t))$$

where vector notation has been used:

$$\Psi(t) = (\cdots \psi_q(t) \cdots)^T \qquad (7)$$

and

$$\Gamma_0(t-\tau) = (\cdots \Gamma_q(t-\tau) \cdots)^T,\ \Gamma_1(t-\tau) = (\cdots \Gamma_{qn}(t-\tau)\cdots),\cdots \qquad (8)$$

Formally L^t and N^t represent the linear and nonlinear temporal evolution operators. The tensor matrix elements are

$$\Gamma_q(t-\tau) = a_0(\gamma_q^-(t-\tau) + \gamma_q^+(t-\tau))$$
$$\Gamma_{qn}(t-\tau) = a_1(\gamma_{qn}^-(t-\tau) + \gamma_{qn}^+(t-\tau)) \qquad (9)$$
$$\vdots$$

where

$$\gamma_q^\pm(t-\tau) = \int_A dx\, \bar{g}_q(x) f(x, x \pm v(t-\tau))$$
$$\gamma_{qn}^\pm(t-\tau) = \int_A dx\, \bar{g}_q(x) f(x, x \pm v(t-\tau)) g_n(x \pm v(t-\tau)) \qquad (10)$$
$$\vdots$$

Our particular interest is in macroscopic phase transitions from one mode to another, i.e. the destabilization of a stationary spatiotemporal state. We consider a stationary solution Ψ_0 of (6) and its small deviations $\epsilon(t)$ resulting in $\Psi(t) = \Psi_0 + \epsilon(t)$. Then (6) may be written as

$$\begin{aligned}\epsilon(t) &= L^t \Psi(t) + N^t(\Psi(t)) - L^t \Psi_0 + N^t(\Psi_0) \\ &= \tilde{L}^t \epsilon(t) + \tilde{N}^t(\epsilon(t))\end{aligned} \qquad (11)$$

where $\tilde{N}^t(\epsilon(t))$ is of the order $|\epsilon|^2$. The general solution of the linear parts of (11) is

$$\epsilon(t) = \sum_{n=1}^{N} \left(\sum_{m=0}^{M-1} c_{mn} t^m \right) e^{\lambda_n t} O_n \qquad (12)$$

where c_{mn} is a constant, λ_n the eigenvalue, M its multiplicity and O_n the right-hand eigenvector. The eigenvalue problem of (11) is defined as

$$\det(e^{-\lambda t} L^t(e^{-\lambda \tau}) - I) = 0 \qquad (13)$$

and provides λ_n. The identity matrix is I. Decomposing $\epsilon(t) = \sum_{i=1}^{N} \xi_i(t) O_i$ and defining the left-hand eigenvectors $\bar{O}_k$ of the linear problem in (11) we obtain by projecting (11) on $\bar{O}_k$

$$\xi_k(t) = \int_{-\infty}^{t} \lambda_k \xi_k(\tau) d\tau + \bar{O}_k \tilde{N}^t \left(\sum_{i=1}^{N} \xi_i(\tau) O_i \right) \qquad (14)$$

The nonlinear contributions can be rewritten such that

$$\xi_k(t) = \int_{-\infty}^{t} d\tau \left(\lambda_k \xi_k(\tau) + \sum_{rs=1}^{N} \bar{\Gamma}_{2krs}(t-\tau) \xi_r(\tau) \xi_s(\tau) + \text{higher orders} \right) \qquad (15)$$

with

$$\bar{\Gamma}_{2krs}(t-\tau) = \sum_{jmn=1}^{N} \bar{O}_k^j \Gamma_{2jmn}(t-\tau) O_r^m O_s^n \qquad (16)$$

where the superscript of the eigenvectors denotes their elements. Here the entire complexity of the connection topology is contained in the tensor matrices $\bar{\Gamma}$ which can be explicitly expressed as integrals of the connectivity $f(x,X)$ and the spatial basis functions $\bar{g}_n(x), g_m(x)$. The linear stability of an eigenmode $\xi_k(t)$ is determined by its eigenvalue λ_k. Thus we are able to express the entire dynamics of (7) in terms of its eigenmodes under appropriate choice of $\bar{g}_n(x), g_m(x)$ in the vicinity of a phase transition.

3 Influence of Connectivity: A Two-Point Connection

Let us now consider the example of a one-dimensional homogeneous medium with a heterogeneous two-point connection between locations x_1 and x_2. The connectivity function $f(x,X)$ shall be given by

$$f(x,X) = f(|x-X|) + f_{12}(x,X) + f_{21}(x,X) \qquad (17)$$

where f_{12} is the link from x_2 to x_1 and f_{21} the link in the opposite direction as illustrated in Fig. 1. The distance between the inhomogeneous contributions of connectivity is $d = |x - X|$ which serves as our control parameter.

The dynamics of our example is given by

$$\psi(x,t) = a \int_A dX f(x,X) S(\psi(X,T) - \bar{\psi}(X,T)) \qquad (18)$$

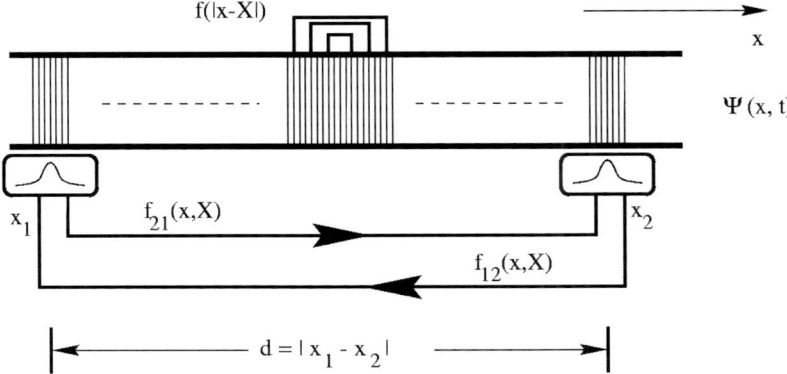

Fig. 1. Two-point connection. The homogeneous connection topology is illustrated within a one-dimensional continuous medium whose activity is described by $\psi(x,t)$. A projection from x_1 to x_2 introduces a heterogeneity into the connectivity.

with $T = t - |x - X|/v$ and $x, X \in \mathcal{R}$. $\bar{\psi}(x,t) = \int_A dX\, \psi(X,t)$ is subtracted to reduce spatially uniform saturation effects [10]. Periodic boundaries are imposed. The connectivities are specified as

$$f(|x-X|) = f_0 e^{-|x-X|/\sigma} \qquad f_{ij}(x,X) = f_{ij}\delta(x-x_i)\delta(X-x_j)\ i \neq j \quad (19)$$

where $a, \sigma, f_{12}, f_{21}$ are constant parameters. We choose the spatial basis system to be spanned by the trigonometric functions $\sin nkx, \cos mkx$ with $n, m \in \mathcal{Z}$ and $k = 2\pi/L$ where L is the length of the one-dimensional closed loop. $x_1 = 0$ is not varied thereby resulting in pinning of the spatial modes around $x = 0$. The nonlinear function S in (18) is assumed to be sigmoidal [10]. We study the stability of the origin $\Psi_0 = 0$ and expand S around its deflection point, $S[n] \approx \alpha n - 4/3\alpha^3 n3 \pm \cdots$. We consider spatial basis function $g_m(x) = \cos mkx$ to 2nd order in m and truncate the expansion of the sigmoid after the 3rd order studying small amplitude dynamics. Application of (6)–(16) provides the following eigenvalue problem

$$v \int_{-\infty}^{t} d\tau\, e^{-im\omega(t-\tau)} \Gamma_{mm}(t-\tau) e^{-\lambda_m(t-\tau)} = 1 \quad (20)$$

where ω is a constant, m refers to the spatial mode number and

$$\Gamma_{mm}(t-\tau) = \frac{a\alpha}{\sigma} e^{-\omega_0(t-\tau)} \cos mkv(t-\tau) + d_1\delta(|x_1 - x_2| - v(t-\tau)) \quad (21)$$

where

$$d_1 = a\alpha/L \left(\frac{f_{12}f_{21}}{f_{12}+f_{21}}\right) \cos mkx_1 \cos mkx_2 \quad (22)$$

Inserting (15) in (11) provides us with a transcendental equation for $z = \lambda_m + im\omega$

$$\left(z2 + 2\omega_0 z + \omega_0 2 + m^2 k^2 v2\right)\left(1 - d_1 e^{zd/v}\right) - (z+\omega_0)\omega_0 a\alpha = 0 \quad (23)$$

The real part λ_m of the eigenvalue z of the m-th mode and its frequency ω can be determined graphically from (12) as a function of the control parameter d. The destabilization of a particular state m depends on d_1 and d. For $d_1, d = 0$ the purely homogeneous system is obtained in which the origin is the only stable state for sufficiently small α. The introduction of a heterogeneous projection $f_{ij}(x,X)$ may cause a desynchronization of the connected areas and thus a destabilization of the respective spatial mode resulting in a phase transition. The desynchronization properties are determined by the tensor matrices in (10) which describe the interplay of the connectivity and distribution of areas represented as spatial modes. This interplay prescribes the timing relationship of activity between areas via the time delay and thus determines its desynchronizing properties.

4 Variability of the Timing of Distant Sites

We study the the timing and its variability in dependence of the presence of a two-point connection. The connectivity matrix is defined by (17), (19), and illustrated in Fig. 2. In addition to the influence of the connectivity, given in (7), we allow for an intrinsic dynamics at each site, that is independent of the neighboring active elements. Here we choose the intrinsic dynamics to be an excitable medium, for instance of FitzHugh-Nagumo (FHN) type [13,14]. Then the activation $\psi(x,t)$ is described by two variables $\psi(x,t) = (u(x,t), v(x,t))$ for which only the first component, the excitatory u, has connections, the other component, the inhibitory v, is local. Then the system reads

$$\psi(x, t + \Delta t) \approx \cdots \dot{\psi}(x,t)\Delta t + \psi(x,t) \qquad (24\text{a})$$

$$= F(\psi(x,t)) + \int_A dX f(x, X) S(\psi(X, T)) \qquad (24\text{b})$$

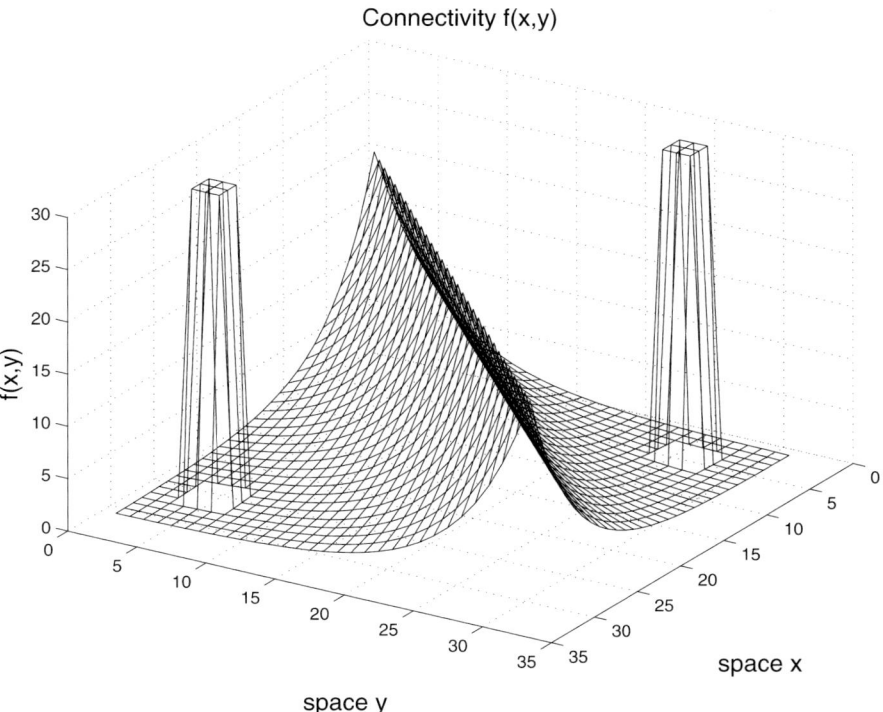

Fig. 2. Connectivity matrix $f(x,y)$ with homogeneous contributions and the bilateral two-point connection. The latter appears as the two singularities f_{12} and f_{21} in the connectivity matrix.

with $F(\psi(x,t))$ as the intrinsic FHN dynamics and the additional Δt expanded into a Taylor series assuming $\Delta t \ll 1$. The explicit equations are

$$\dot{u} = c_1(u - u3/3 - v + I_1) + \int_A dX f(x,X) S(u - \bar{u})(X,T) \tag{25a}$$

$$\dot{v} = c_2(-v + \frac{5}{4}u + I_2) \tag{25b}$$

The parameters are $c_1 = 10, c_2 = 0.8, I_2 = 1.5$. I_1 may vary in space and time, but will chosen here to excite one particular site in the medium constantly, $I_2 = 0.2$, and being zero otherwise. The connectivity parameters are $f_0 = 20, f_{12} = f_{21} = 30, \sigma = 0.5$. The time T in the connectivity integral is $T = t - |x - X|/v$. For a formal analysis of the intrinsic FHN dynamics we refer, for instance, to [16]. In principle, there is a separatrix in phase space for which the FHN system shows properties of an excitable medium: Below a threshold value of u the system relaxes to a fixed point, but above this value the system gets excited, emits a large amplitude spike and returns eventually to the fixed point. The inclusion of additive noise in space and time turns equations (25a, 25b) into a stochastic system and allows for a random emission of spikes. The present parameter space is chosen such that the origin, that is $u(x,t) = 0 \; \forall x$, is stable. However, the proximity to the threshold, when the origin gets destabilized, is expected to cause a spatially non-uniform distribution of events in the stochastic system.

We focus on the following question: Given three sites A, B and C in the active medium (see Fig. 3), how does the inclusion of an additional two-point connection from A to B render the variability of the occurrence of spike timings for a given noise level? The site C shall be located somewhere far from A, B and we will study the variability of the time difference of spikes between A and B, as well as between A and C. In Fig. 3 we illustrate the spatiotemporal dynamics of the activity $u(x,t)$ while a bilateral pathway connects sites A and B. Just from visual inspection, it is evident that a clustering of spikes occurs at these sites over time. Spikes occur more frequently with a fixed time lag between A and B, rather than between A and C. Our aim is to quantify this observation. We will compare four situations: 1) homogeneous connectivity only; 2) homogeneous connectivity and a one-way projection from A to B; 3) homogeneous connectivity and a bilateral pathway between A and B, that is feed forward and feed backward; 4) bilateral pathway between A and B only, that is all the active units in between are disconnected. For each of these four scenarios we study the difference $\Delta t = \min(|t_A - t_{B,C}|)$ between the time of occurrence of a spike at sites A and B or C, respectively. The distribution of the timing difference Δt between sites provides information on the amount of communication exchanged between areas in the noisy environment. Our interest is to identify which mechanism may be used to obtain an increased degree of synchronization, that is a reduced variability in the timing difference.

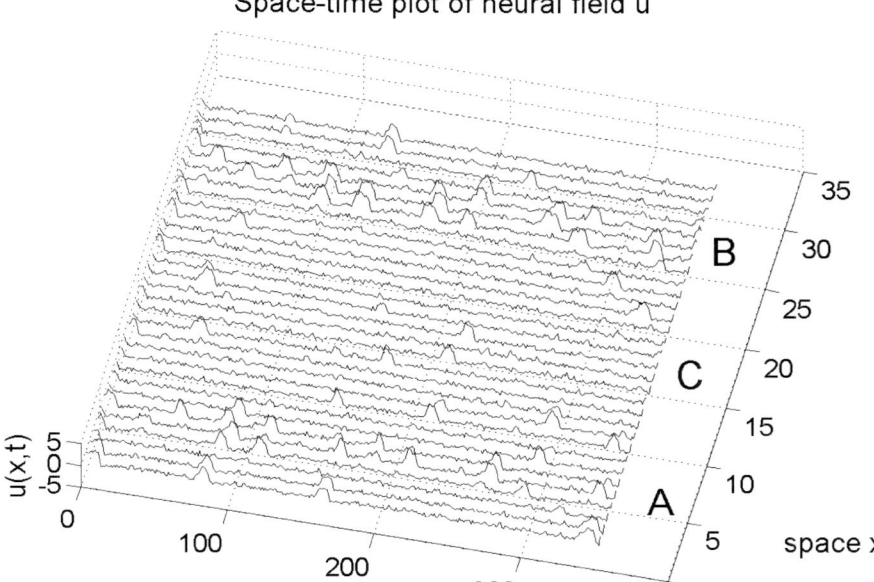

Fig. 3. Dynamics of the activity of $u(x,t)$ over space and time. The connectivity has a homogeneous component and a bilateral pathway between sites A and B. Note the enhanced clustering of spike occurrences at these sites in comparison to site C.

4.1 Homogeneous Connectivity Only

If no heterogeneous pathways are present, then the connectivity is translationally invariant. The connectivity still has a spatial structure, in this case excitatory connections decaying with increasing distance. The histograms in Fig. 4 show the timing distribution between sites A, B and sites A, C. The number of events has been restricted to $N = 675$ for both present and all following cases. The timing for A, B shows a clustering with broad wings around 50 time units ($dt = 0.025$ on an arbitrary scale), the timing for A, C shows a much broader distribution, even though there is also a clustering around 50 time units observed. Note that the time period of a linear wave model, that is distance between A, B divided by velocity, falls into a similar regime implying that wave propagation mechanisms may contribute to the clustering in the distribution. Note also that the intrinsic time scale of the FHN model is in the same regime, that is $T = 80$ time units $= 2$.

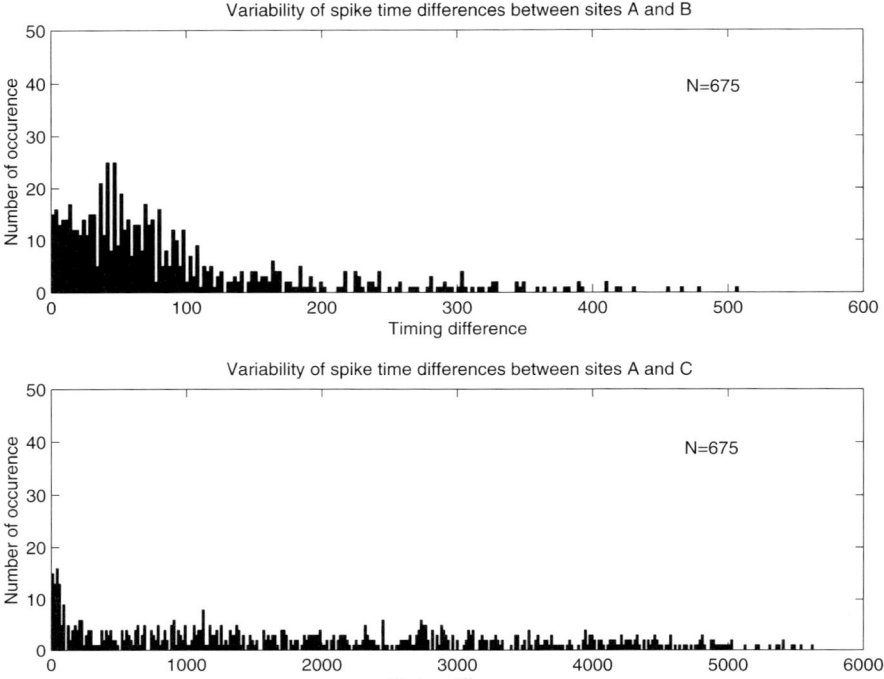

Fig. 4. Homogeneous projections are present only ($f_0 = 20, f_{12} = f_{21} = 0$). The distributions show a modest clustering of timing differences around its intrinsic firing rates.

4.2 Homogeneous Connectivity and Projection from A to B

In addition to the previous situation, a direct pathway is introduced from A to B providing an additional strengthening of the communication between these areas, but only uni-directionally. Surprisingly, (see Fig. 5), the variability in the timing between these areas does actually not change qualitatively compared to the purely homogeneous connectivity. No significant changes of the timing distribution can be observed between A, C either.

4.3 Homogeneous Connectivity and Bilateral Pathway Between A and B

The introduction of a bilateral fiber track, i.e. feed forward and feed backward communication, enhances the robustness of the timing of spike firings significantly (Fig. 6). Higher harmonics of the dominant timing difference around 50 time units are also clearly present. The timing distribution for the sites A, C becomes flat. This appears to be a consequence of the reduced

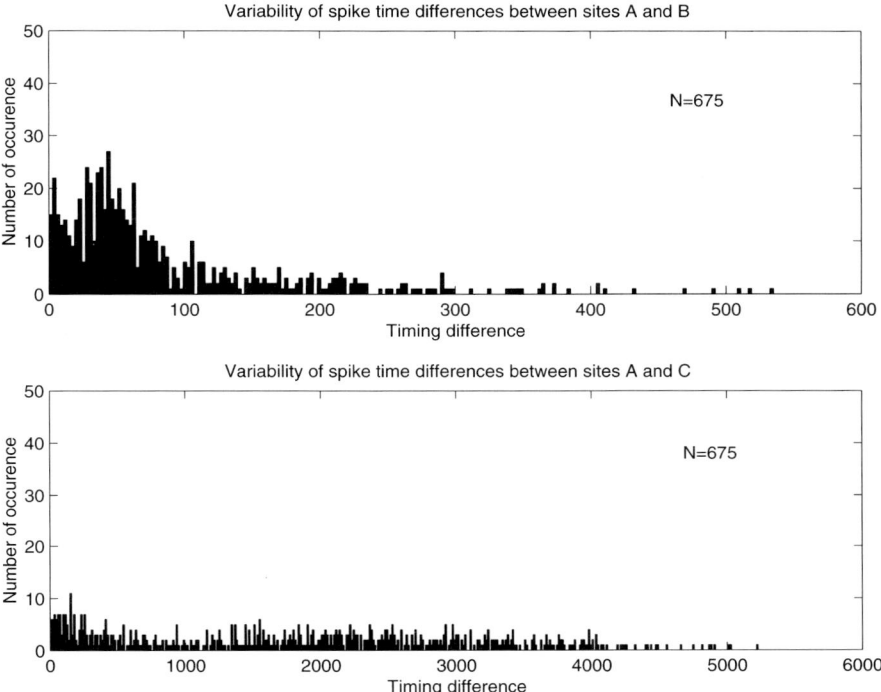

Fig. 5. Homogeneous connectivity and a uni-directional projection from A to B ($f_0 = 20, f_{12} = 30, f_{21} = 0$).

variability in the timing of A, B which also may be interpreted as the signature of the proximity of the system to the point at which spatial patterns emerge oscillating over time with the sites of A and B in anti-phase. In the purely deterministic system, that is without noise, the origin is still stable. Note that global inhibition is present in the current formulation of the model, hence the spatially uniform pattern of organization is suppressed and higher order patterns, such as the present, may be observed.

4.4 Heterogeneous Pathways Only

Finally we consider the isolated pathway between A, B only, i.e. the homogeneous local connectivity is suppressed and all active units except A, B act independently (Fig. 7). It is notable that a broad clustering is observed around the same time point as with the homogeneous connections in the previous case, but with a higher variability. We hypothesize that the additional mechanism of traveling wave propagation through the active medium contributes to a sharpening of the timing between distant sites which are directly connected with heterogeneous pathways. This result is surprising, in

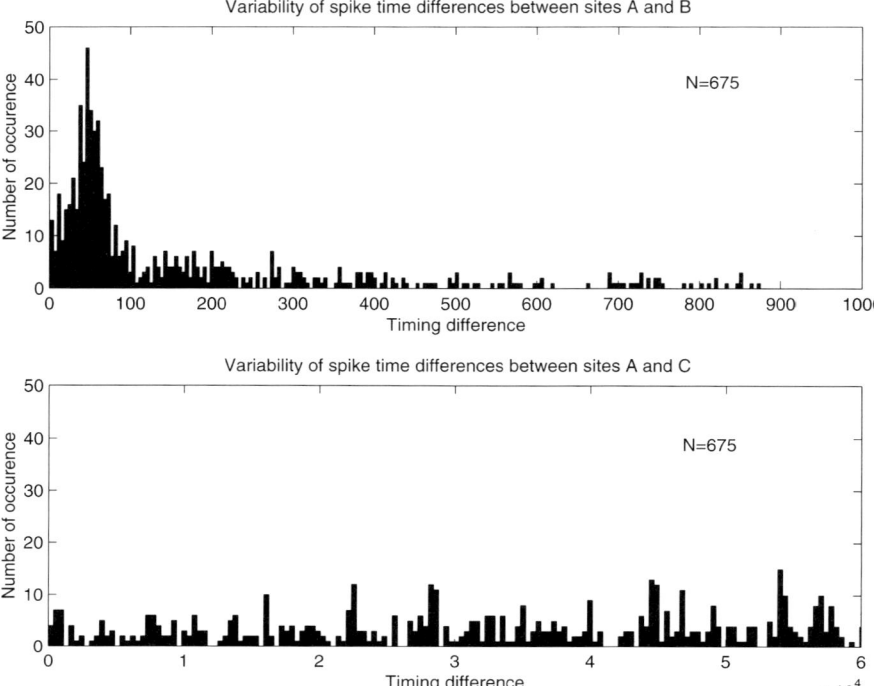

Fig. 6. Homogeneous connectivity with the additional bilateral projection between A and B ($f_0 = 20, f_{12} = f_{21} = 30$).

particular because both individual cases, heterogeneous projections only and homogeneous connectivity only, provide a broader distribution of timings independently. The variability of the timings between the sites A, C remains uniform as expected.

5 Conclusions

We described macroscopic coherent pattern formation in a spatially continuous system by means of an integral equation to capture effects caused by a heterogeneous connection topology. The connectivity serves as an intrinsic topological control parameter which systematically controls spatiotemporal bifurcations. We discussed how the introduction of an additional pathway connecting two distant sites may cause a phase transition and hence macroscopic pattern formation. Close to this threshold, we introduced noise into the system and studied the variability of the timings of events at different locations. It turns out that there is a parameter space for which the introduction of a unilateral pathway, from one site to the other, does not significantly alter

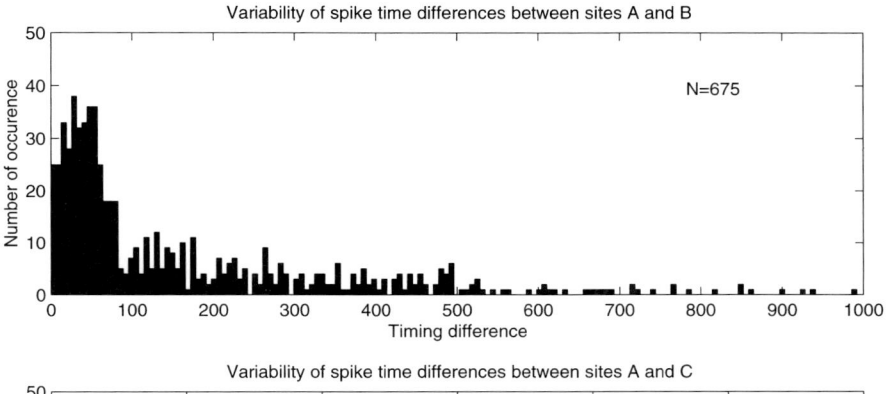

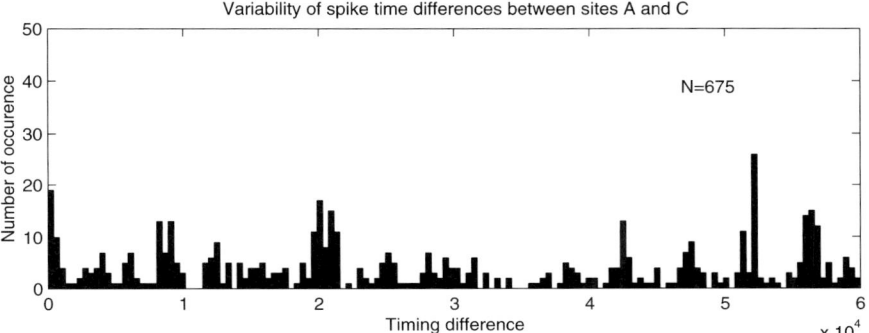

Fig. 7. Bilateral projections between A, C only ($f_0 = 0, f_{12} = f_{21} = 30$). All intermediate active units are independent.

the timing precision of events at the connected sites. However, the bilateral connection causes a significant reduction of the timing variability.

References

1. V.K. Jirsa: Intern. Journ Bif. Chaos **14**, 2 (2004)
2. H. Haken: *Synergetics. An Introduction* 3rd ed. (Springer, Berlin 1983)
3. M.C. Cross, P.C. Hohenberg: Rev. Mod. Phys. **65**, 851 (1993)
4. M. Hendrey, E. Ott, T.M. Antonsen Jr.: Phys. Rev. Lett. **82**, 859 (1999)
5. G.B. Ermentrout, N. Kopell: SIAM J. Appl. Math. **54**, 478 (1994)
6. V.K. Jirsa, J.A.S. Kelso: Phys. Rev E **62**, 8462 (2000)
7. V.K. Jirsa: Prog. Theo. Phys. Suppl. **139**, 128 (2000)
8. S. Amari: Biol. Cybern. **27**, 77 (1977)
9. P.L. Nunez: *Neocortical Dynamics and Human EEG Rhythms*, (Oxford University Press, Oxford 1995)
10. V.K. Jirsa, H. Haken: Phys. Rev. Lett. **77**, 960 (1996)
11. J.J. Wright, D.T.J. Liley: Behav. Brain. Sci. **19**, 285 (1996)

12. P.A. Robinson, C.J. Rennie, J.J. Wright: Phys. Rev. E **56**, 826 (1997)
13. R. FitzHugh: Biophys. J. **1**, 445 (1961)
14. J. Nagumo, S. Arimoto, S. Yoshizawa: Proc. IRE **50**, 2061 (1962)
15. J.D. Murray: *Mathematical Biology* (Springer, Berlin Heidelberg New York 1993)
16. H.R. Wilson: *Spikes, Decisions and Actions*. (Oxford University Press, Oxford 1999)

First Passage Time Problem: A Fokker-Planck Approach

Mingzhou Ding and Govindan Rangarajan

This chapter reviews the first passage time problem for one-dimensional stochastic processes and presents closed-form solutions for the underlying distribution function. Using the Fokker-Planck approach the case of Brownian motion with drift is solved in the diffusive limit. This technique is then generalized to obtain exact solutions in the case of anomalous diffusion, corresponding to a continuous time random walk.

1 Introduction

Let $X(t)$ be an one dimensional stochastic process. Consider the time when the process first crosses a threshold. This time T is obviously a random variable and is called the First Passage Time (FPT) [1]. An important problem is to find the probability density function (pdf) of T. This is known as the first passage time problem and has a long history [2–9].

The FPT problem finds applications in many areas of science and engineering [10–13]. A sampling of these applications is listed below:

- probability theory (study of Wiener process, fractional Brownian motion etc.)
- statistical physics (study of anomalous diffusion)
- neuroscience (analysis of neuron firing models)
- civil and mechanical engineering (analysis of structural failure)
- chemical physics (study of noise assisted potential barrier crossings)
- hydrology (optimal design of dams)
- financial mathematics (analysis of circuit breakers)
- imaging (study of image blurring due to hand jitter)

We now consider one application of the FPT problem in some detail. Consider a neuron which receives an input current $x(t)$. We wish to study the output spike train generated by the neuron. For simplicity, we restrict ourselves to the simplest model of neuron firing – the perfect or nonleaky Integrate and Fire (IF) model first introduced by Lapicque [12]. In this model, the membrane potential $V(t)$ of the neuron is obtained by integrating the input current. When $V(t)$ reaches a threshold value V_0, an action potential is generated and $V(t)$ is reset instantaneously to its resting value (assumed here to be zero). This discrete generation of the action potential leads to an output spike train. The time interval between two successive spikes is called the interspike interval (ISI). Since these intervals are known to codify information, it is important to characterize their distribution.

For most neurons, the output ISI is found to be a random variable [14] and hence we need to find its pdf. To find this, we first need a mechanism to generate stochastic spike trains. They can be generated by making the input current $x(t)$ a stochastic process. Therefore $V(t)$ which is obtained by integrating this input current is now a random walk. Assume that the neuron has generated an output spike at time t_i. It generates the next spike after a time interval T when the random walk $\{V(t), t \geq t_i\}$ starting from $V(t) = 0$ at $t = t_i$ first crosses the threshold value V_0. Therefore it is clear that T is a first passage time and a random variable. Its pdf is nothing but the FPT distribution. Thus the pdf of the output ISI's is nothing but the distribution of FPT's obtained as above.

Given the ubiquitous role played by the FPT distribution in various applications, it is natural to derive the FPT distribution for different types of stochastic processes. In particular, we will first investigate Brownian motion with drift in the diffusive limit. Then we consider continuous time random walks which give rise to anomalous diffusions where mean squared displacement varies as t^γ, $\gamma \neq 1$ for large t. We consider both subdiffusive processes ($0 < \gamma < 1$) and superdiffusive processes ($1 < \gamma < 2$). We employ the Fokker-Planck approach [10] (and its generalizations) in this review. It should be noted that in some cases, probabilistic arguments can be used to derive the FPT distribution in a much simpler fashion. But this approach is not always applicable. For pedagogical reasons and for the sake of consistency we will use the Fokker-Planck approach even in cases where simpler probabilistic derivations are possible.

2 FPT Distribution for Brownian Motion

We first study the first passage time problem for the simplest case of Brownian motion (Wiener process). This will illustrate the manner in which the Fokker-Planck equation enters the picture and how it facilitates the solution of the FPT problem.

We start with a simple random walk. A step Y is taken for every τ units of time with the following probabilities:

$$Y = \begin{cases} l, & \text{with probability } 0.5; \\ -l, & \text{with probability } 0.5. \end{cases} \quad (1)$$

Consider a new random variable X_n defined as

$$X_n = \sum_{i=1}^{n} Y_i, \quad (2)$$

with $X_0 = 0$. It is obvious that X_n gives the position of the random walker at time $t = n\tau$. Assume that the steps Y_i are mutually independent. Then the mean squared displacement $\langle X_n^2 \rangle$ is given by

$$\langle X_n^2 \rangle = \sum_{i=1}^{n} \langle Y_i^2 \rangle + \sum_{i \neq j} \langle Y_i Y_j \rangle. \tag{3}$$

The second term is zero since the Y_i's are mutually independent. Hence

$$\langle X_n^2 \rangle = \sum_{i=1}^{n} l^2 = nl^2. \tag{4}$$

The above equation can be written as

$$\langle X_n^2 \rangle = nl^2 = \frac{l^2}{\tau}(n\tau) = \frac{l^2}{\tau}t. \tag{5}$$

The random walk described above has discontinuous jumps. It is often easier to deal with continuous quantities. We can make the sample paths continuous by taking the so-called diffusive limit: $\tau \to 0$, $l \to 0$ such that

$$\frac{l^2}{\tau} = 2D. \tag{6}$$

Here D is called the diffusion constant. In this diffusive limit, we have

$$\langle X^2(t) \rangle = 2Dt. \tag{7}$$

The reader will recognize this as the equation that characterizes Brownian motion (or Wiener process). More formally, $X(t)$ is a Wiener process if

- $X(t)$ is a Gaussian process,
- $X(0) = 0$, $\langle X(t) \rangle = 0$,
- $\langle (X(t) - X(s))^2 \rangle = 2D|t-s| \quad \forall t, s$.

Further, $X_\mu(t)$ is a Wiener process (or Brownian motion) with drift μ if

$$X_\mu(t) = \mu t + X(t). \tag{8}$$

Brownian motion is an example of a regular or ordinary diffusion. All processes that belong to this class are characterized by the following relation:

$$\langle X^2(t) \rangle \sim t \text{ for large } t. \tag{9}$$

Thus in regular diffusive processes other than Brownian motion, the proportionality of the mean squared displacement with time is satisfied only for large times.

In the analysis above, we have directly dealt with the stochastic process X_n (or $X(t)$ in the diffusive limit). It turns out that it is easier to deal with the pdf for the stochastic process than the process itself for many applications (including the calculation of the FPT distribution). As we show below, the pdf for the simple random walk described above satisfies a partial differential equation known as the Fokker-Planck equation in the diffusive limit.

Let $W(ml, (n+1)\tau)$ denote the probability that the random walker is at position ml at time $(n+1)\tau$ (i.e. after $n+1$ time steps). But the walker could have reached this position only by either being at position $(m-1)l$ at time $n\tau$ and jumping right (with probability 0.5) or being at position $(m+1)l$ at time $n\tau$ and jumping left (again with probability 0.5). Hence we have from simple probability considerations:

$$W(ml, (n+1)\tau) = \frac{1}{2}[W((m+1)l, n\tau) + W((m-1)l, n\tau)]. \tag{10}$$

If τ is small, we have

$$W(ml, (n+1)\tau) \approx W(ml, t) + \tau \frac{\partial}{\partial t} W(ml, t), \tag{11}$$

where $t = n\tau$. Hence, we get (for τ small)

$$\tau \frac{\partial}{\partial t} W(ml, t) = \frac{1}{2}[W((m+1)l, t) + W((m-1)l, t) - 2W(ml, t)]. \tag{12}$$

But for l small, we have

$$[W((m+1)l, t) + W((m-1)l, t) - 2W(ml, t)] \approx l^2 \frac{\partial^2}{\partial x^2} W(x, t), \tag{13}$$

where $x = ml$. Thus, we have

$$\frac{\partial}{\partial t} W(x, t) \approx \frac{l^2}{2\tau} \frac{\partial^2}{\partial x^2} W(x, t). \tag{14}$$

Taking the diffusive limit described earlier, we finally get

$$\frac{\partial}{\partial t} W(x, t) = D \frac{\partial^2}{\partial x^2} W(x, t). \tag{15}$$

This is nothing but the diffusion equation for $W(x, t)$ and is known as the Fokker-Planck Equation (FPE). The natural boundary conditions for this equation are that $W(x, t) = 0$ at $x = \pm\infty$. Since the random walker is assumed to start at the origin, we also have the initial condition $W(x, 0) = \delta(x)$.

We can easily verify that the above Fokker-Planck equation describes the evolution of the pdf for Brownian motion. First we compute $\langle X \rangle$. Multiplying the FPE given in Eq. (15) by x and integrating from $x = -\infty$ to $x = \infty$ we get

$$\frac{\partial}{\partial t} \langle X \rangle = -D \int_{-\infty}^{\infty} dx \, \frac{\partial W(x, t)}{\partial x}, \tag{16}$$

where we have performed integration by parts and assumed that $x\frac{\partial W(x,t)}{\partial x} = 0$ at $x = \pm\infty$. Evaluating the remaining integral using the natural boundary conditions, we get

$$\frac{\partial}{\partial t}\langle X \rangle = 0. \tag{17}$$

This gives $\langle X \rangle = 0$ which is the expected result.

Next we compute $\langle X^2 \rangle$. Multiplying the FPE in Eq. (15) by x^2 and integrating we get

$$\frac{\partial}{\partial t}\langle X^2 \rangle = -2D \int_{-\infty}^{\infty} dx\, x \frac{\partial W(x,t)}{\partial x}, \tag{18}$$

where we have again used integration by parts and assumed that $x^2 \frac{\partial W(x,t)}{\partial x} = 0$ at $x = \pm\infty$. Evaluating the remaining integral using the the boundary conditions and the normalization condition $\int_{-\infty}^{\infty} dx\, W(x,t) = 1$ we obtain

$$\langle X^2 \rangle = 2Dt. \tag{19}$$

This shows that the FPE in Eq. (15) does describe the usual Brownian motion.

For Brownian motion with drift, the corresponding FPE is:

$$\frac{\partial}{\partial t}W(x,t) = \mu\frac{\partial}{\partial x}W(x,t) + D\frac{\partial^2}{\partial x^2}W(x,t). \tag{20}$$

Next we obtain the FPT distribution for Brownian motion with drift starting from the Fokker-Planck equation. Consider a stochastic process $X(t)$ with $X(0) = 0$. The first passage time (FPT) T to the point $X = a > 0$ is defined as [1]

$$T = \inf\{t : X(t) = a\}. \tag{21}$$

We would like to obtain the probability density function for T for Brownian motion.

Since Brownian motion (in the diffusive limit) is described by Fokker-Planck equations, the problem of obtaining the FPT density function can be recast as a boundary value problem with absorbing boundaries [10]. In our case, to obtain the FPT density function, we first need to solve Eq. (20) with absorbing boundaries at $x = -\infty$ and $x = a$, where a is the predetermined level of crossing, with the initial condition $W(x,0) = \delta(x)$ [10]. An equivalent formulation, due to symmetry, is to solve Eq. (20) with the following boundary and initial conditions:

$$W(0,t) = 0, \quad W(\infty,t) = 0, \quad W(x,0) = \delta(x-a), \tag{22}$$

where $x = a$ is the new starting point of the Brownian motion, containing the initial concentration of the distribution. The equivalence is easily seen by making the change of variables $x \to a - x$ in Eq. (20). This latter formulation makes the subsequent derivation less cumbersome.

Once we solve for $W(x,t)$, the first passage time density $f(t)$ can be determined as follows. From simple probability considerations, the probability $P(T > t)$ that the first passage time exceeds a given time t is nothing but

$$P(T > t) = \int_0^\infty dx \, W(x,t). \tag{23}$$

Therefore,

$$P(T \le t) = 1 - \int_0^\infty dx \, W(x,t). \tag{24}$$

Hence the FPT density $f(t)$ is given by

$$f(t) = \frac{d}{dt} P(T \le t) = -\frac{d}{dt} \int_0^\infty dx \, W(x,t). \tag{25}$$

First we solve for $W(x,t)$ using the given boundary and initial conditions. We solve the FPE using the method of separation of variables [15]. Let $W(x,t) = X(x)T(t)$. Substituting in Eq. (20) we obtain

$$X(x) \frac{dT(t)}{dt} = T(t) \left[\mu X'(x) + D X''(x) \right], \tag{26}$$

where the primes denote the derivatives with respect to x. Separating out the variables and introducing the separation constant λ we get

$$D X''(x) + \mu X'(x) = -\lambda X(x), \tag{27}$$

and

$$\frac{dT(t)}{dt} = -\lambda T(t). \tag{28}$$

First we solve the simple Eq. (28). We obtain

$$T(t) = \exp\left[-\lambda t\right]. \tag{29}$$

Next consider Eq. (27). The solution of this equation satisfying the boundary conditions is given by

$$X(x) = \exp[-\mu(x-a)/2D] \frac{\sin[x\sqrt{\lambda/D - \mu^2/4D^2}]}{2\sqrt{D\lambda - \mu^2/4}}, \quad \lambda \ge \mu^2/4D. \tag{30}$$

Thus we have a continuous spectrum for λ. Combining the solutions for $X(x)$ and $T(t)$, $W(x,t)$ is given by the following integral over λ:

$$W(x,t) = \frac{2}{\pi} \int_{\mu^2/4D}^\infty d\lambda A(\lambda) \exp[-\mu(x-a)/2D] \frac{\sin[x\sqrt{\lambda/D - \mu^2/4D^2}]}{2\sqrt{D\lambda - \mu^2/4}}$$
$$\times \exp\left[-\lambda t\right]. \tag{31}$$

The coefficient $A(\lambda)$ is fixed by the initial condition $(W(x,0) = \delta(x-a))$ and we get

$$W(x,t) = \frac{2}{\pi} \int_{\mu^2/4D}^{\infty} d\lambda \exp[-\mu(x-a)/2D] \sin[a\sqrt{\lambda/D - \mu^2/4D^2}]$$
$$\times \frac{\sin[x\sqrt{\lambda/D - \mu^2/4D^2}]}{2\sqrt{D\lambda - \mu^2/4}} \exp[-\lambda t]. \quad (32)$$

Letting $\lambda' = \sqrt{\lambda/D - \mu^2/4D^2}$ we obtain

$$W(x,t) = \frac{2}{\pi} \int_0^{\infty} d\lambda' \sin \lambda' a \sin \lambda' x \exp[-\mu(x-a)/2D]$$
$$\times \exp\left[-(D\lambda'^2 + \mu^2/4D)t\right]. \quad (33)$$

Using standard trigonometric identities and dropping the primes, the above equation can be rewritten as

$$W(x,t) = \frac{1}{\pi} \int_0^{\infty} d\lambda \exp[-\mu(x-a)/2D] \exp\left[-(D\lambda^2 + \mu^2/4D)t\right]$$
$$\times [\cos \lambda(x-a) - \cos \lambda(x+a)]. \quad (34)$$

Taking the Laplace transform with respect to time we get

$$q(x,s) = \frac{1}{\pi} \int_0^{\infty} d\lambda \exp[-\mu(x-a)/2D] \frac{1}{s + \mu^2/4D + D\lambda^2}$$
$$\times [\cos \lambda(x-a) - \cos \lambda(x+a)], \quad (35)$$

where $q(x,s)$ is the Laplace transform of $W(x,t)$. Here we have used the fact that the Laplace transform of $\exp(-Bt)$ is $1/(s+B)$ [16].

We can now perform the integration over λ by using the following result [17]

$$\int_0^{\infty} d\lambda \frac{\cos \lambda x}{\alpha^2 + \lambda^2} = \frac{\pi}{2\alpha} e^{-\alpha|x|}. \quad (36)$$

Thus we obtain

$$q(x,s) = \frac{1}{2\sqrt{D}} \frac{\exp[-\mu(x-a)/2D]}{\sqrt{s + \mu^2/4D}} \exp[-\sqrt{s + \mu^2/4D}\,|x-a|/\sqrt{D}]$$
$$- \frac{1}{2\sqrt{D}} \frac{\exp[-\mu(x-a)/2D]}{\sqrt{s + \mu^2/4D}} \exp[-\sqrt{s + \mu^2/4D}\,(x+a)/\sqrt{D}]. \quad (37)$$

To obtain the Laplace transform $F(s)$ of the FPT density function $f(t)$, we take the Laplace transform of Eq. (25) to get

$$F(s) = -s \int_0^{\infty} dx\, q(x,s) + \int_0^{\infty} dx\, W(x,0). \quad (38)$$

Here we have used the fact that Laplace transform of $dW(x,t)/dt$ is given by [17] $sq(x,s) - W(x,0)$. Since $W(x,0) = \delta(x-a)$, we obtain

$$F(s) = 1 - s \int_0^\infty dx\, q(x,s). \tag{39}$$

Substituting for $q(x,s)$ from Eq. (37), we get

$$\begin{aligned}
F(s) = 1 &- \frac{1}{2\sqrt{D}} \frac{1}{\sqrt{s + \mu^2/4D}} \\
&\times \int_0^\infty dx\, \exp[-\mu(x-a)/2D] \exp[-\sqrt{s+\mu^2/4D}\, |x-a|/\sqrt{D}] \\
&+ \frac{1}{2\sqrt{D}} \frac{1}{\sqrt{s + \mu^2/4D}} \\
&\times \int_0^\infty dx\, \exp[-\mu(x-a)/2D] \exp[-\sqrt{s+\mu^2/4D}\, (x+a)/\sqrt{D}].
\end{aligned} \tag{40}$$

Upon evaluating the integrals we obtain

$$F(s) = \exp\left[-a\left(-\mu + \sqrt{\mu^2 + 4Ds}\right)/2D\right]. \tag{41}$$

Performing the inverse Laplace transform [17] we finally get the desired FPT distribution for a Brownian motion with drift:

$$f(t) = \frac{a}{\sqrt{4\pi D t^3}} \exp\left[-\frac{(a-\mu t)^2}{4Dt}\right], \quad a > 0, \; t > 0. \tag{42}$$

This is the famous inverse Gaussian distribution. This distribution when $\mu = 0$ was first derived by Bachelier [2] and the $\mu \neq 0$ case was first derived by Schrödinger and Smoluchowski [3,4].

3 FPT Distribution for Continuous Time Random Walks

Consider a one dimensional continuous time random walk [18,19] described by the following Langevin equation:

$$\frac{dX}{dt} = \sum_{i=1}^\infty Y_i\, \delta(t - t_i). \tag{43}$$

Here the random walker starts at $x = 0$ at time $t_0 = 0$. Subsequently, the random walker waits at a given location x_i for time $t_i - t_{i-1}$ before taking a jump Y_i which could depend on the waiting time. The waiting time $u > 0$ and the jump size y ($-\infty < y < \infty$) are drawn from the joint probability density function $\phi(y,u)$. The waiting time distribution $\psi(u)$ is given by

$$\psi(u) = \int_{-\infty}^\infty dy\, \phi(y,u). \tag{44}$$

First Passage Time Problem 39

The process is non Markovian if $\psi(u)$ is a non-exponential distribution since the probability for the next jump to occur depends on how long the the random walker has been waiting since the previous jump. But CTRW is non Markovian in a special way since it does not depend on the history of the process prior to the previous jump.

We now relate [18] the probability distribution $W(x,t)$ for the CTRW to $\phi(y,u)$ and $\psi(u)$. The probability density $\eta(x,t)$ of the random walker just arriving at x in the time interval t to $t+dt$ is

$$\eta(x,t) = \int_{-\infty}^{\infty} dx' \int_0^t \eta(x',\tau)\phi(x-x',t-\tau)d\tau + \delta(t)\delta(x). \tag{45}$$

Thus $\eta(x,t)$ is obtained by summing over all x' and τ the probability $\eta(x',\tau)$ of being exactly at x' at time τ multiplied by the probability $\phi(x-x',t-\tau)$ of jumping a distance $x-x'$ in time $t-\tau$ to exactly arrive at x at time t. The second term $\delta(x)\delta(t)$ is just the initial condition. At $x=0$, $t=0$, we have $\eta(0,0)=1$ i.e. the walker starts at $x=0$ at time $t=0$ with probability 1.

The probability $W(x,t)$ of the random walker being at x at time t is obtained by summing over all τ' the probability $\eta(x,t-\tau')$ of exactly arriving at x at time $t-\tau'$ and then waiting without jumping up to time t. The probability of not jumping from time $t-\tau'$ to t is given by $\xi(\tau')$ where $\xi(t)$ is the so-called survival probability:

$$\xi(t) = 1 - \int_0^t \psi(\tau)d\tau. \tag{46}$$

Hence we have

$$W(x,t) = \int_0^t \eta(x,t-\tau')\xi(\tau')d\tau'. \tag{47}$$

From the equation for $\eta(x,t)$ we have

$$\eta(x,t-\tau') = \int_{-\infty}^{\infty} dx' \int_0^{t-\tau'} \eta(x',\tau)\phi(x-x',t-\tau'-\tau)d\tau \\ + \delta(t-\tau')\delta(x). \tag{48}$$

Substituting this expression in $W(x,t)$ we get

$$W(x,t) = \int_{-\infty}^{\infty} dx' \int_0^t d\tau' \int_0^{t-\tau'} \eta(x',\tau)\phi(x-x',t-\tau'-\tau)\xi(\tau')d\tau \\ + \xi(t)\delta(x). \tag{49}$$

Letting $\tau'' = \tau + \tau'$ we obtain

$$W(x,t) = \int_{-\infty}^{\infty} dx' \int_0^t d\tau' \int_{\tau'}^t \eta(x',\tau''-\tau')\phi(x-x',t-\tau'')\xi(\tau')d\tau'' \\ + \xi(t)\delta(x). \tag{50}$$

Changing the order of integration, the above expression can be rewritten as

$$W(x,t) = \int_{-\infty}^{\infty} dx' \int_0^t d\tau'' \left[\int_0^{\tau''} \eta(x', \tau'' - \tau')\xi(\tau')d\tau' \right] \phi(x - x', t - \tau'') \\ + \xi(t)\delta(x). \qquad (51)$$

But the term within brackets is $W(x', \tau'')$. Thus

$$W(x,t) = \int_{-\infty}^{\infty} dx' \int_0^t d\tau'' W(x', \tau'')\phi(x - x', t - \tau'') + \xi(t)\delta(x). \qquad (52)$$

Taking the Fourier-Laplace transform and denoting the transforms of $W(x,t)$ and $\phi(y,u)$ by $\tilde{W}(k,s)$ and $\tilde{\phi}(k,s)$ respectively (where k is the Fourier transform of the space variable and s the Laplace transform of the time variable) we get [18]

$$\tilde{W}(k,s) = \tilde{W}(k,s)\tilde{\phi}(k,s) + \tilde{\xi}(s). \qquad (53)$$

Here $\tilde{\xi}(s)$ is the Laplace transform of $\xi(t)$. It can be related to the Laplace transform $\tilde{\psi}(s)$ of the waiting time distribution $\psi(u)$ as follows [cf. Eq. (46)]:

$$\tilde{\xi}(s) = \frac{1}{s} - \frac{\tilde{\psi}(s)}{s}. \qquad (54)$$

Here we have used the fact that the Laplace transforms of 1 and $\int_0^\tau \psi(u)du$ are $1/s$ and $\tilde{\psi}(s)/s$ respectively. Substituting the above expression in the equation for $\tilde{W}(k,s)$ and solving for $\tilde{W}(k,s)$ we finally obtain:

$$\tilde{W}(k,s) = \frac{1}{s} \frac{1 - \tilde{\psi}(s)}{1 - \tilde{\phi}(k,s)}. \qquad (55)$$

It can be further shown that, depending on the specific form of $\phi(y,u)$, the CTRW can produce anomalous diffusion as well as ordinary diffusion [18,20]. For example, consider

$$\phi(y,u) = \frac{1}{\sqrt{2\pi\sigma^2}} \exp[-y^2/2\sigma^2] \frac{(\alpha - 1)/\tau}{(1 + u/\tau)^\alpha}, \qquad (56)$$

where y and u are decoupled with y being a Gaussian variable with zero mean. Here the parameters σ and τ can be thought of as giving characteristic step size and waiting time for the random walk. For $1 < \alpha < 2$, the corresponding CTRW gives subdiffusion where $\langle X^2(t) \rangle \sim t^\gamma$ for large t with $\gamma = \alpha - 1$ between 0 and 1 [38]. We call this Lévy type anomalous diffusion since the waiting time distribution $\psi(u)$ given by [cf. Eqs. (56) and (44)]

$$\psi(u) = \frac{(\alpha - 1)/\tau}{(1 + u/\tau)^\alpha} \qquad (57)$$

is a Lévy type distribution [38]. For $\alpha \geq 2$, one gets ordinary diffusion with $\gamma = 1$. Similarly, for a CTRW characterized by [18]

$$\phi(y, u) = \frac{1}{2}\delta(u/\tau - |y|/\sigma)\frac{(\beta - 1)/\tau}{(1 + u/\tau)^\beta}, \tag{58}$$

where $2 < \beta < 3$ and $\delta(\cdot)$ is the Dirac delta function, we obtain a Lévy type superdiffusive process with $\gamma = \beta - 1$.

Above we had described CTRW processes that gives rise to Lévy type anomalous diffusion. However, it is difficult to derive analytical results directly from the process. As mentioned earlier, it is more convenient to work in the general framework of Fokker-Planck equations [10]. One can go from the CTRW process to a fractional Fokker-Planck equation (FFPE) [22] by taking the generalized diffusive limit. This limit is analogous to the regular diffusive limit that we considered earlier to derive the regular Fokker-Planck equation from a random walk with $\langle X^2(t)\rangle \sim t$ for large t. In the regular diffusive limit, we took $\sigma, \tau \to 0$ such that σ^2/τ is maintained a constant. For a CTRW where $\langle X^2(t)\rangle \sim t^\gamma$ $(0 < \gamma < 2)$, the generalized diffusive limit is obtained by taking the limit $\sigma, \tau \to 0$ such that σ^2/τ^γ is maintained a constant.

Thus to obtain a FFPE from the CTRW process, we need to take the limit $\sigma, \tau \to 0$. We will take this limit for the Fourier-Laplace transform $\tilde{W}(k, s)$ of $W(x, t)$ and then invert the transform to obtain the FFPE. First consider the subdiffusive CTRW characterized by Eq. (56). The Laplace transform $\tilde{\psi}(s)$ of $\psi(u)$ given in Eq. (57) is [17]

$$\tilde{\psi}(s) = (\alpha - 1)(\tau s)^{\alpha - 1}\Gamma(1 - \alpha, \tau s)e^{\tau s}. \tag{59}$$

The Fourier-Laplace transform $\tilde{\phi}(k, s)$ of $\phi(y, u)$ in Eq. (56) is given by [17]

$$\tilde{\phi}(k, s) = \exp(-\sigma^2 k^2/2)\tilde{\psi}(s). \tag{60}$$

To obtain an expression for $W(k, s)$ in the limit $\tau \to 0$, we first consider $\tilde{\psi}(s)$. We have [23]

$$\Gamma(1 - \alpha, \tau s) = \Gamma(1 - \alpha) - \sum_{n=0}^{\infty}\frac{(-1)^n(\tau s)^{1-\alpha+n}}{n!(1 - \alpha + n)}. \tag{61}$$

As $\tau \to 0$,

$$\Gamma(1 - \alpha, \tau s) \approx -\frac{\Gamma(2 - \alpha)}{\alpha - 1} + \frac{(\tau s)^{1-\alpha}}{\alpha - 1} + \frac{(\tau s)^{2-\alpha}}{2 - \alpha}. \tag{62}$$

Therefore $\tilde{\psi}(s)$ as $\tau \to 0$ is given by

$$\tilde{\psi}(s) \approx 1 - \Gamma(2 - \alpha)(\tau s)^{\alpha-1}, \quad 1 < \alpha < 2, \tag{63}$$
$$\approx 1 - (2\alpha - 3)\tau s/(\alpha - 2), \quad \alpha > 2. \tag{64}$$

Substituting this in the Eq. (55) we have [cf. Eq. (60)]

$$\tilde{W}(k,s) \approx \frac{1}{s} \frac{\Gamma(1-\gamma)(\tau s)^\gamma}{1 - [1 - \Gamma(1-\gamma)(\tau s)^\gamma]\exp(-\sigma^2 k^2/2)}. \qquad (65)$$

Here we have used $\gamma = \alpha - 1$ ($0 < \gamma < 1$) instead of α since it is the physically relevant quantity. Now we take the further limit $\sigma \to 0$ such that $\sigma^2/2\Gamma(1-\gamma)\tau^\gamma = K$ is a constant. K is called the generalized diffusion constant. We then obtain

$$\tilde{W}(k,s) = \frac{1}{s + Kk^2 s^{1-\gamma}}. \qquad (66)$$

This can be rewritten as

$$\tilde{W}(k,s) - \frac{1}{s} = -Kk^2 s^{-\gamma} \tilde{W}(k,s). \qquad (67)$$

To take the inverse Fourier-Laplace transform of the above equation, we need the inverse Laplace transform of $s^{-\gamma}\tilde{W}(k,s)$. This is given by the Riemann-Liouville fractional integral $_0D_t^{-\gamma}W(k,t)$ which is defined as [24,25]

$$_0D_t^{-\gamma}W(k,t) = \frac{1}{\Gamma(\gamma)} \int_0^t dt' \, (t-t')^{\gamma-1} W(k,t'), \quad \gamma > 0. \qquad (68)$$

This result enables us to take the inverse Fourier-Laplace transform of Eq. (67) giving [22]

$$W(x,t) - W(x,0) = K \, _0D_t^{-\gamma} \frac{\partial^2}{\partial x^2} W(x,t), \quad 0 < \gamma < 1. \qquad (69)$$

Here we have incorporated the initial condition $W(x,0) = \delta(x)$. One obtains the same equation for a subdiffusive process but with a different K.

To obtain the FPT distribution for the above process, we follow a procedure identical to that we used for analyzing the regular Fokker-Planck equation in the previous section. We first solve for $W(x,t)$ (the boundary and initial conditions are the same as for the FPE case). Then we use Eq. (25) to obtain the FPT distribution.

Let $W(x,t) = X(x)T(t)$. Substituting in Eq. (69) we obtain

$$X(x)T(t) - X(x) = \left[_0D_t^{-\gamma} T(t) \right] [KX''(x)], \qquad (70)$$

where the primes denote the derivatives with respect to x. Separating out the variables and introducing the separation constant λ we get

$$KX''(x) = -\lambda X(x), \qquad (71)$$

and

$$T(t) - 1 = -\lambda \, _0D_t^{-\gamma} T(t). \qquad (72)$$

First we solve Eq. (72). Taking its Laplace transform we obtain

$$T(s) - \frac{1}{s} = \frac{\lambda}{s^\gamma} T(s). \tag{73}$$

Here we have used the fact that the Laplace transform of $_0D_t^{-\gamma}T(t)$ is given by $T(s)/s^\gamma$. Solving for $T(s)$ we get

$$T(s) = \frac{1}{s - \lambda s^{1-\gamma}}. \tag{74}$$

Taking the inverse Laplace transform [16] we finally obtain

$$T(t) = E_\gamma\left[-\lambda t^\gamma\right], \tag{75}$$

where $E_\gamma(z)$ is the Mittag-Leffler function [16] with the following power series expansion:

$$E_\gamma(z) = \sum_{n=0}^{\infty} \frac{z^n}{\Gamma(1+\gamma n)}. \tag{76}$$

Note that $E_\gamma(z)$ reduces to the regular exponential function when $\gamma = 1$.

Next consider Eq. (71). The solution of this equation satisfying the boundary conditions is given by

$$X(x) = \frac{\sin[x\sqrt{\lambda/K}]}{2\sqrt{K\lambda}}, \quad \lambda \geq 0. \tag{77}$$

Thus we have a continuous spectrum for λ. Combining the solutions for $X(x)$ and $T(t)$, $W(x,t)$ is given by the following integral over λ:

$$W(x,t) = \frac{2}{\pi} \int_0^\infty d\lambda A(\lambda) \frac{\sin[x\sqrt{\lambda/K}]}{2\sqrt{K\lambda}} E_\gamma\left[-\lambda t^\gamma\right]. \tag{78}$$

The coefficient $A(\lambda)$ is fixed by the initial condition ($W(x,0) = \delta(x-a)$) and we get

$$W(x,t) = \frac{2}{\pi} \int_0^\infty d\lambda \sin[a\sqrt{\lambda/K}] \frac{\sin[x\sqrt{\lambda/K}]}{2\sqrt{K\lambda}} E_\gamma\left[-\lambda t^\gamma\right]. \tag{79}$$

Letting $k = \sqrt{\lambda/K}$ we obtain

$$W(x,t) = \frac{2}{\pi} \int_0^\infty dk \sin[ka]\sin[kx] E_\gamma\left[-k^2 K t^\gamma\right]. \tag{80}$$

Taking the Laplace transform and performing the integral over k we get

$$q(x,s) = \frac{s^{\gamma/2-1}}{2\sqrt{K}} \left[\exp(-s^{\gamma/2}|x-a|/\sqrt{K}) - \exp(-s^{\gamma/2}(x+a)/\sqrt{K})\right].$$

The inverse Laplace transform of $s^{\gamma/2-1}\exp(-|x|s^{\gamma/2})$ is known [26]. We get

$$p(x,t) = \frac{1}{2(Kt^\gamma)^{1/2}} H_{1,1}^{1,0}\left(\frac{|x-a|}{(Kt^\gamma)^{1/2}}\bigg|\begin{matrix}(1-\gamma/2,\gamma/2)\\(0,1)\end{matrix}\right) \quad (81)$$
$$-\frac{1}{2(Kt^\gamma)^{1/2}} H_{1,1}^{1,0}\left(\frac{x+a}{(Kt^\gamma)^{1/2}}\bigg|\begin{matrix}(1-\gamma/2,\gamma/2)\\(0,1)\end{matrix}\right).$$

Here, the Fox or H-function [27,28] has the following alternating power series expansion:

$$H_{p,q}^{m,n}\left(z\bigg|\begin{matrix}(a_j,A_j)_{j=1,\ldots,p}\\(b_j,B_j)_{j=1,\ldots,q}\end{matrix}\right) = \sum_{l=1}^{m}\sum_{k=0}^{\infty}\frac{(-1)^k z^{s_{lk}}}{k! B_l} \quad (82)$$
$$\times \frac{\prod_{j=1,j\neq l}^{m}\Gamma(b_j-B_j s_{lk})\prod_{r=1}^{n}\Gamma(1-a_r+A_r s_{lk})}{\prod_{u=m+1}^{q}\Gamma(1-b_u+B_u s_{lk})\prod_{v=n+1}^{p}\Gamma(a_v-A_v s_{lk})},$$

where $s_{lk} = (b_l+k)/B_l$ and an empty product is interpreted as unity. Further, m, n, p, q are nonnegative integers such that $0 \leq n \leq p$, $1 \leq m \leq q$; A_j, B_j are positive numbers; a_j, b_j can be complex numbers. For further discussions of the H-function, see Mathai [28].

Substituting Eq. (81) into Eq. (25) we have

$$f(t) = -\frac{d}{dt}\left[\frac{1}{2(Kt^\gamma)^{1/2}}\int_0^\infty dx\, H_{1,1}^{1,0}\left(\frac{|x-a|}{(Kt^\gamma)^{1/2}}\bigg|\begin{matrix}(1-\gamma/2,\gamma/2)\\(0,1)\end{matrix}\right)\right]$$
$$+\frac{d}{dt}\left[\frac{1}{2(Kt^\gamma)^{1/2}}\int_0^\infty dx\, H_{1,1}^{1,0}\left(\frac{x+a}{(Kt^\gamma)^{1/2}}\bigg|\begin{matrix}(1-\gamma/2,\gamma/2)\\(0,1)\end{matrix}\right)\right].$$

Defining $z = (x-a)/(Kt^\gamma)^{1/2}$, $z' = (x+a)/(Kt^\gamma)^{1/2}$, we obtain

$$f(t) = -\frac{d}{dt}\int_{-a/(Kt^\gamma)^{1/2}}^\infty dz\, H_{1,1}^{1,0}\left(|z|\bigg|\begin{matrix}(1-\gamma/2,\gamma/2)\\(0,1)\end{matrix}\right)$$
$$+\frac{d}{dt}\int_{a/(Kt^\gamma)^{1/2}}^\infty dz'\, H_{1,1}^{1,0}\left(z'\bigg|\begin{matrix}(1-\gamma/2,\gamma/2)\\(0,1)\end{matrix}\right)$$
$$= \frac{a\gamma}{2K^{1/2}t^{(2+\gamma)/2}} H_{1,1}^{1,0}\left(\frac{a}{(Kt^\gamma)^{1/2}}\bigg|\begin{matrix}(1-\gamma/2,\gamma/2)\\(0,1)\end{matrix}\right).$$

Thus using the fractional Fokker-Planck approach and H-functions, we are able to obtain an exact form for the FPT distribution for Lévy-type anomalous diffusion with zero drift. When $\gamma = 1$, the above expression reduces to the inverse Gaussian distribution with $\mu = 0$. Thus FPT distribution for Brownian motion is contained as a special case of this more general result. Expressions for the FPT distribution (or its Laplace transform) have been derived [38,29,30] in other cases also.

4 Summary

In this paper, we highlighted the important role played by the first passage time distribution in different areas. We explained in detail how the Fokker-Planck approach can be used to find the FPT distribution for ordinary Brownian motion. We then extended the approach using fractional Fokker-Planck equation to find the FPT distribution for anomalous diffusion.

Acknowledgements

MD's work was supported by US Office of Naval Research and National Science Foundation. GR was supported by the Homi Bhabha Fellowship. GR is also associated with the Jawaharlal Nehru Centre for Advanced Scientific Research, Bangalore, India as an honorary faculty member.

References

1. G.R. Grimmet and D.R. Stirzaker: *Probability and Random Processes* (Oxford University Press, New York 1994)
2. L. Bachelier: Annales des Sciences de l'Ecole Superieure **17**, 21 (1900)
3. E. Schrödinger: Physikalische Zeitschrift **16**, 289 (1915)
4. M.V. Smoluchowski: Physikalische Zeitschrift **16**, 318 (1915)
5. H.A. Kramers: Physica **7**, 284 (1940)
6. D.A. Darling, A.J.F. Siegert: Ann. Math. Stat. **24**, 624 (1953)
7. E.W. Montroll, K.E. Shuler: *Adv. Chem. Phys.* **1**, 361 (1958)
8. R. Landauer, J.A. Swanson: Phys. Rev. **21**, 1668 (1961)
9. G.H. Weiss: In *Stochastic Processes in Chemical Physics* ed. by I. Oppenheim, K.E. Schuler, G.K. Weiss (MIT Press, Cambridge, Masss. 1977)
10. H. Risken: *The Fokker-Planck Equation* (Springer-Verlag, Berlin 1989)
11. C.W. Gardiner: *Handbook of Stochastic Methods* (Springer Verlag, Berlin 1997)
12. H.C. Tuckwell: *Introduction to Theoretical Neurobiology*, Vol. 1 & 2 (Cambridge University Press, Cambridge 1988)
13. Y.K. Lin, G.Q. Cai: *Probabilistic Structural Dynamics* (McGraw-Hill, New York 1995)
14. H.C. Tuckwell: *Stochastic Processes in Neurosciences* (SIAM, Philadelphia 1989)
15. P.M. Morse, H. Feshbach: *Methods of Theoretical Physics* (McGraw-Hill, New York 1953)
16. A. Erdelyi: *Higher Transcendental Functions*, Vol. 3 (McGraw-Hill, New York 1955)
17. A. Erdelyi: *Tables of Integral Transforms* (McGraw-Hill, New York 1954)
18. M.F. Shlesinger, J. Klafter, Y.M. Wong: J. Stat. Phys. **27**, 499 (1982); J. Klafter, A. Blumen, M.F. Shlesinger: Phys. Rev. A **35**, 3081 (1987)

19. A. Blumen, J. Klafter, G. Zumofen: In *Optical Spectroscopy of Glasses* ed. by I. Zschokke (Reidel, Dordrecht 1986); S. Havlin, D. Ben-Avraham: Adv. Phys. **36**, 695 (1987); J.-P. Bouchaud, A. Georges: Phys. Rep. **195**, 12 (1990); G. Zumofen, J. Klafter, A. Blumen: J. Stat. Phys. **65**, 991 (1991); *Fractals in Biology and Medicine,* ed. by T.F. Nonnenmacher, G.A. Losa, E.R. Weibl (Birkhauser, Basel 1993); H.-P. Muller, R. Kimmich, J. Weis: Phys. Rev. E **54**, 5278 (1996); F. Amblard, A.C. Maggs, B. Yurke, A.N. Pargellis, S. Leibler: Phys. Rev. Lett. **77**, 4470 (1996); A. Klemm, H.-P. Muller, R. Kimmich: Phys. Rev. E **55**, 4413 (1997); B.J. West, P. Grigolini, R. Metzler, T.F. Nonnenmacher: Phys. Rev. E **55**, 99(1997); Q. Gu, A. Schiff, S. Grebner, R. Schwartz: Phys. Rev. Lett. **55**, 4413 (1997); E.R. Weeks, H.L. Swinney: Phys. Rev. E **57**, 4915 (1998)
20. X.-J. Wang, Phys. Rev. A **45**, 8407 (1992)
21. G. Rangaran, M. Ding: Phys. Rev. E **62**, 120 (2000)
22. R. Metzler, J. Klafter, I.M. Sokolov: Phys. Rev. E **58**, 1621 (1998); R. Metzler, E. Barkai, J. Klafter: Phys. Rev. Lett. **82**, 3563 (1999)
23. I.S. Gradshteyn, I.M. Ryzhik: *Tables of Integrals, Series, and Products* (Academic, New York 1965)
24. K.B. Oldham, J. Spanier: *The Fractional Calculus* (Academic, New York 1974)
25. K.S. Miller, B. Ross: *An Introduction to the Fractional Calculus and Fractional Differential Equations* (Wiley, New York 1993)
26. I. Podlubny: *Fractional Differential Equations* (Academic Press, San Diego 1999)
27. C. Fox: Trans. Am. Math. Soc. **98**, 395 (1961)
28. A.M. Mathai, R.K. Saxena: *The H-function with Applications in Statistics and Other Disciplines* (Wiley Eastern, New Delhi 1978)
29. G. Rangaran, M. Ding: Phys. Lett. A **273**, 322 (2000)
30. G. Rangaran, M. Ding: Fractals **8**, 139 (2000)

First- and Last-Passage Algorithms in Diffusion Monte Carlo

James A. Given, Chi-Ok Hwang, and Michael Mascagni

This chapter provides a review of a new method of addressing problems in diffusion Monte Carlo: the Green's function first-passage method (GFFP). In particular, we address four new strands of thought and their interaction with the GFFP method: the use of angle-averaging methods to reduce vector or tensor Laplace equations to scalar Laplace equations; the use of the simulation-tabulation (ST) method to dramatically expand the range of the GFFP method; the use of the Feynman-Kac formula, combined with GFFP to actually perform path integrals, one patch at a time; and the development of last-passage diffusion methods; these drastically improve the efficiency of diffusion Monte Carlo methods. All of these techniques are described in detail, with specific examples.

1 Introduction

Many researchers have used diffusion Monte Carlo methods to calculate the bulk properties of porous and composite media. Basic examples of such properties include: the electrical or thermal conductivity [1–4] or shear modulus of structural composites; the permeability of porous media [5]; the electrostatic contribution to the free energy of a biomolecule in solution; and the mutual capacitance matrix describing interaction of micro-components in a transistor matrix on a microchip. Porous and composite media have basic geometric similarities: they involve samples of bulk matter that are composed of small patches of two (or more) pure phases. Both the bulk material properties of each pure phase and the statistics of the mixture, *i.e.*, its correlation functions, are assumed to be known. This information can be used to determine the bulk properties of the multi-phase medium.

The two classes of problems also share a deeper, mathematical foundation: they involve the solution of elliptic or parabolic partial differential equations in domains that contain a large amount of surface area, *i.e.*, interface area, at which boundary conditions must be imposed. Standard finite-element or boundary element methods require long computation times in these cases, especially when high accuracy is required. Considerable cost is associated with discretizing complicated interfaces. These problems can be efficiently solved by diffusion Monte Carlo techniques: the problem in question is modeled as

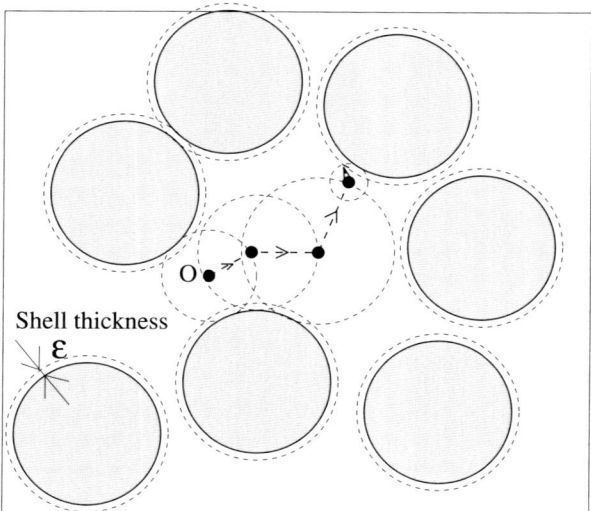

Fig. 1. A two-dimensional schematic representation of a Brownian trajectory using WOS algorithm. If the diffusing particle reaches the ϵ-layer, it is taken to be absorbed

an (in general) anisotropic, biased diffusion problem. Many methods, at this step, employ a discrete representation in either space or time of the underlying Brownian motion; as we show, the availability of Green's functions for the continuum problem makes this unnecessary.

Here we describe a new approach to such problems, the Green's function first-passage (GFFP) method. It is a synthesis of advances developed by this group, and those developed elsewhere; of ideas from pure mathematics and those from applied mathematics. In particular, the GFFP method involves:

- Using the angle-averaging method to reduce problems based on vector or tensor Laplace equations to problems based on scalar Laplace equations, *i.e.*, on biased diffusion equations.
- Defining the solution to the problem in question in terms of sources and sinks of diffusing particles. For example, an electrostatic problem is cast as an effort to calculate the surface charge density on all interfaces. Once this is done, voltages, or other quantities defined as weighted averages over the surface charge density, can be calculated efficiently by using, *e. g.*, the fast multipole method.
- Describing the calculation to be conducted as the simulation of a large number of Brownian trajectories. These may either begin at charge sources or sinks, as in last-passage algorithms [6], or terminate at them, as in first-passage algorithms [7].

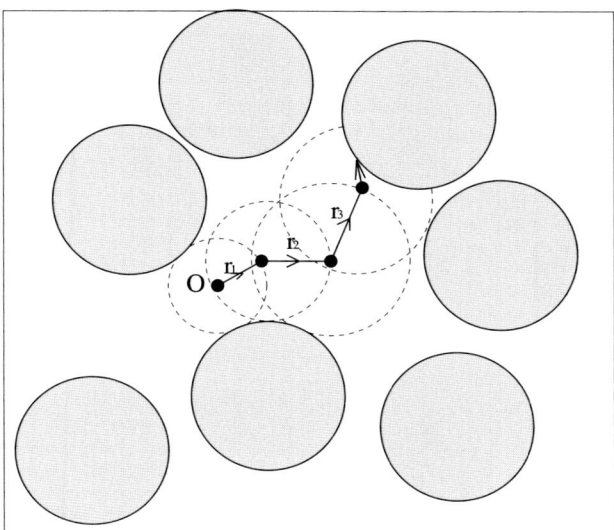

Fig. 2. A two-dimensional schematic representation of a Brownian trajectory using both the WOS algorithm (r_1 to r_3) and the GFFP algorithm (the final step). The solid circles are FP boundaries and absorbing. We use a δ-boundary layer as a criterion such that WOS is used outside the δ-boundary layer and GFFP in the δ-boundary layer

- Modeling the interface between phases as locally smooth, *i.e.*, as locally consisting of patches that are flat, spherical, cylindrical, or otherwise understood in terms of their Laplacian Green's function.
- Modeling the free diffusion of particles in such an environment by using a first-passage (FP) strategy. We divide the trajectory of a Brownian particle into a series of jumps, each one taking the Brownian particle from the center of a FP volume to a point on the FP surface. FP surfaces far from absorbing boundaries are spheres (this approach replicates the "walk on spheres" (WOS) algorithm [8–10]; see Fig. 1). But near absorbing boundaries, FP surfaces can be more complicated. They can include portions of an absorbing boundary. Acceptable FP surfaces, at this stage of analysis, are those for which a quasi-analytic Green's function exists for the corresponding Dirichlet problem. Such Green's functions (actually the normalized distribution functions corresponding to them) can be tabulated for each set of values of the dimensionless geometric parameters they depend on. This tabulation can then be closely approximated by a spline or other interpolatory fit, which in turn allows rapid and accurate sampling of the FP position during a Monte Carlo simulation (see Fig. 2). It has been shown that this method is substantially more efficient computationally in applications for which high accuracy is required [11].

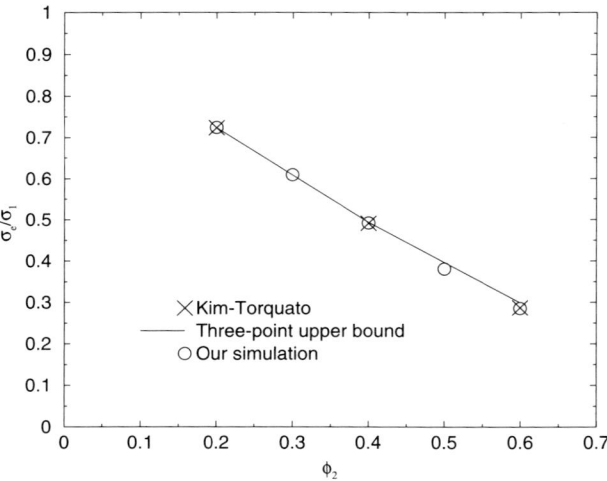

Fig. 3. Scaled effective conductivity σ_e/σ_1 of equilibrium distributions of nonoverlapping insulating spheres in a matrix of conductivity σ_1 with $\epsilon = 0$ and $\delta = 0.1a$

As an example, in Fig. 3 we show the effective conductivity of a two-phase medium, consisting of an ensemble of nonoverlapping, insulating spherical inclusions dispersed randomly in a matrix phase of finite conductivity σ_1. We compare the CPU time of the GFFP algorithm with that of the WOS algorithm in Figs. 4 and 5. In the WOS algorithm, CPU time depends on the ϵ-shell thickness while in the GFFP algorithm it depends on δ-boundary layer. The ϵ-shell around the target is used to establish convergence in the WOS method, such that any Brownian particle inside it is taken to be absorbed. Also, we use a δ-boundary layer as a criterion such that WOS is used outside the δ-boundary layer and GFFP in the δ-boundary layer, because GFFP is more efficient as the Brownian particle approaches the boundary. Here, $\epsilon = 10^{-1}$ in the WOS method approximately corresponds to the optimal case of GFFP.

Algorithms developed from the GFFP method already provide the most efficient algorithms known for certain important classes of problems, including the electrostatic capacitance of an arbitrary object. For example, the most accurate value for the capacitance of the unit cube is $C = 0.660675(5)$. For comparison, the most accurate value for this quantity yet obtained from boundary element methods is uncertain in the third digit, due to the logarithmic convergence involved in applying these methods to surfaces with edges and corners.

But much more is possible, using the diffusion Monte Carlo methodology. We combine it with:

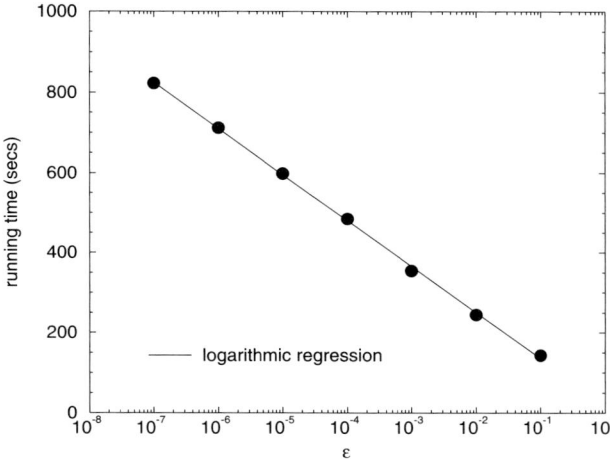

Fig. 4. CPU time required to calculate the effective conductivity of a system of non-overlapping, insulating spherical inclusions dispersed randomly in a conducting matrix with sink volume fraction $\phi_2 = 0.2$. Here, we used the WOS method with mean diffusion path length $X^2/a^2 = 100$. Times here were measured on a 500 MHz Pentium III work station running Linux over 10^4 Brownian trajectories. The simulations show the expected relation for WOS: CPU time proportional to $\ln(\epsilon)$)

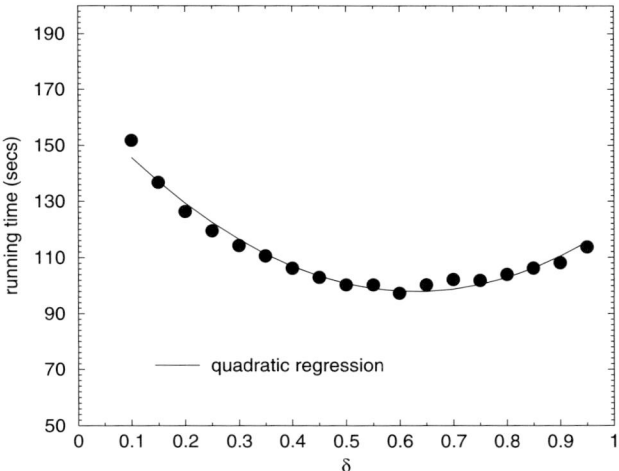

Fig. 5. CPU time required to calculate the effective conductivity of a system of non-overlapping, insulating spherical inclusions dispersed randomly in a conducting matrix with sink volume fraction $\phi_2 = 0.2$. Here, we used the GFFP method with mean diffusion path length $X^2/a^2 = 100$. Times here were measured on a 500 MHz Pentium III work station running Linux over 10^4 Brownian trajectories. This figure shows that an optimal δ is around 0.65

- optimal applied mathematics methods; these include the simulation-tabulation (ST) method [12], and the efficient generation of quasi-random numbers;
- important developments in probability theory; these include both the last-passage methods, and methods based on the Feynman-Kac formula [13–15].

Combining all of these methods, allows the treatment of classes of important problems, including the linearized Poisson-Boltzmann equation [16,17].

This paper is organized as follows: in Sect. 2, we describe the angle-averaging approximations that allow the reduction of a problem based on a vector or tensor Laplace equation, to a problem based on a scalar Laplace equation, $i.e.$, to a diffusion problem. In Sect. 3, we describe the simulation-tabulation (ST) method, that allows extension of the GFFP method to problems in which quasi-analytic Green's functions are not available. In Sect. 4, we discuss the Feynman-Kac method, which allows the use of Monte Carlo diffusion methods to solve a more general class of elliptic boundary value problems. Our emphasis here is on the interaction between the Feynman-Kac methods and the other methods presented here. In Sect. 5, we describe two classes of last-passage algorithms, $i.e.$, Monte Carlo diffusion algorithms in which diffusing particles "initiate" at the point at which they are absorbed, and diffuse "backwards in time." In Sect. 6 we give our conclusions and suggestions for further study.

2 The Angle-Averaging Method

In this section, we describe the angle-averaging method, which allows one to approximate a problem based on a vector or tensor Laplace equation, with a problem based on a scalar Laplace equation. The latter can then be solved using diffusion Monte Carlo methods. The first application of the angle-averaging method was by Hubbard and Douglas [18–20], who gave the following approximation for the translational hydrodynamic friction, f, of an arbitrary object:

$$f = 4\pi\eta C, \tag{1}$$

where η is the fluid viscosity and C the electrical capacitance of an ideal conductor having the same size and shape as the object.

The present authors recently generalized this result to give an algorithm for the permeability of a packed bed, or other porous medium. As an example, in Fig. 6 we present simulation results of packed beds composed of polydispersed overlapping, randomly placed, impenetrable spherical inclusions. The inclusion sphere radii are chosen at random from the values 1.5, 3.5, 5.5, and 7.5 with equal probability. We compare our results with the available numerical solutions of the Stokes equation, [21].

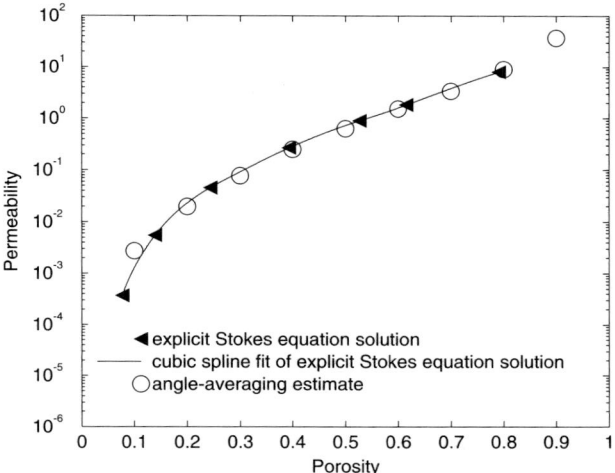

Fig. 6. Permeability, k, versus porosity for a porous medium consisting of a polydispersed mixture of randomly overlapping impermeable spheres. The sphere radii are chosen to have the four values $a = \{1.5, 3.5, 5.5, 7.5\}$ with equal probability

We have also developed an efficient first-passage implementation of this relation. A generalization to the case of a packed bed has direct applications to the properties of suspensions.

The angle-averaging method also provides an approximate relation between the hydrodynamic viscosity of an object and the electrostatic polarizability of an object of the same shape [22,23].

3 The Simulation-Tabulation (ST) Method

In this section, we explain the ST method, and how to use it to extend the GFFP method to classes of problems for which the Green's function is not available in quasi-analytic form. A basic example is the class of problems involving either mixed or reflecting, *i.e.* Neumann, boundary conditions. This class of problems includes the calculation of the conductivity of a composite medium involving insulating inclusions dispersed in a conducting matrix.

The many-body problem is reduced to the solution of a Laplace or Poisson equation with complicated boundary conditions, *i. e.* it is solved by taking mathematical expectations over Brownian motion trajectories. Since Brownian motion is the microscopic manifestation of diffusion, one can also reinterpret these as diffusion problems [7]. Thus, we need methods to efficiently generate Brownian trajectories in complicated domains. We do this with the help of a useful result from probabilistic potential theory [24,13]. Namely, we need the fact that the first-passage probability for Brownian motion in a region is *equivalent* to the surface Green's function for the Laplacian in that same region.

While this equivalence is a well-known fact to probabilists and many experts in Monte Carlo, we feel it useful for the rest of our presentation to give an elementary proof of this fact. First consider the Dirichlet problem for the Laplace equation (see Fig. 7),

$$\Delta u(\mathbf{x}) = 0, \quad \mathbf{x} \in \Omega, \quad u(\mathbf{x}) = f(\mathbf{x}), \quad \mathbf{x} \in \partial\Omega. \tag{2}$$

The solution at point $\mathbf{x}$ can be represented probabilistically as the average over all the boundary values of Brownian motion, $X^{\mathbf{x}}(\tau_{\partial\Omega})$, starting at $\mathbf{x}$ where the Brownian particle first strikes the boundary. The time, $\tau_{\partial\Omega}$, when the Brownian particle first strikes the boundary is called the first-passage time, and the place where the Brownian particle first strikes the boundary, $X^{\mathbf{x}}(\tau_{\partial\Omega})$, is called the first-passage location. Thus we claim that the probabilistic solution, $u_p(\mathbf{x})$, to (2) is given by

$$u_p(\mathbf{x}) = E[f(X^{\mathbf{x}}(\tau_{\partial\Omega}))]. \tag{3}$$

The proof that this is the case is simple. Place a sphere centered at $\mathbf{x}$ completely lying within Ω. Clearly the particle will have to hit this sphere before hitting $\partial\Omega$. The probability distribution of the first-passage location, $\mathbf{z}$, on the sphere is clearly uniform due to the isotropic nature of Brownian motion. Now we continue the Brownian particle from $\mathbf{z}$ until it hits the boundary. Here $X^{\mathbf{x}}(\tau_{\partial\Omega})$ is the first passage location on the boundary, $\partial\Omega$ (see Fig. 7). Averaging over the first-passage boundary values of Brownian paths started at $\mathbf{z}$ gives us $u_p(\mathbf{z})$. Since each trajectory starting at $\mathbf{x}$ that hits $\partial\Omega$ must first hit the sphere with uniform probability, $u_p(\mathbf{x})$ must be the mean of the values of $u_p(\mathbf{z})$ over the sphere. Thus $u_p(\mathbf{x})$ has the mean-value property and is harmonic, *i. e.* it obeys the Laplace equation, [25]. If we then think of moving the starting point for our Brownian particles to the boundary, we clearly will,

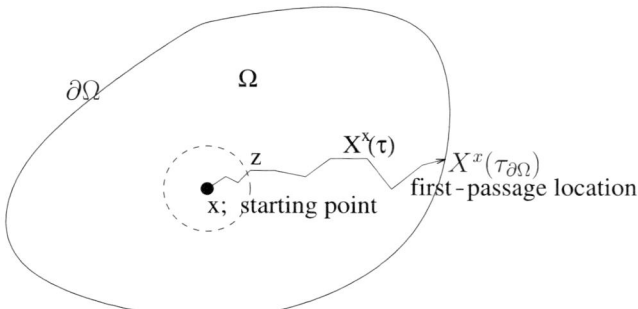

Fig. 7. This figure shows a Brownian motion which starts at $\mathbf{x}$ and terminates at $X^x(\tau_{\partial\Omega})$ on the boundary $\partial\Omega$ while passing through $\mathbf{z}$, a point on a sphere centered at $\mathbf{x}$

in the limit, have the first passage location coincide with the limit of $\mathbf{x}$ on the boundary. This argues that, in addition, $u_p(\mathbf{x})$ has the correct boundary values, and so it is the unique solution to (2).

We now use an interpretation of this fact to prove the equivalence mentioned above. Equation (3) can be interpreted as an average of the boundary values, $f(\mathbf{x})$, $\mathbf{x} \in \partial\Omega$ over $\partial\Omega$. The weighting in this average is the first-passage probability $p(\mathbf{x}, \mathbf{y})$ of a Brownian particle starting at $\mathbf{x}$ hitting the boundary first at $y = X^{\mathbf{x}}(\tau_{\partial\Omega}) \in \partial\Omega$. Thus we can represent $u(\mathbf{x})$ as an integral over the boundary, $\partial\Omega$, via

$$u(\mathbf{x}) = \int_{\partial\Omega} p(\mathbf{x}, \mathbf{y}) f(\mathbf{y}) dy. \tag{4}$$

However, there is another representation of the solution of the Dirichlet problem for the Laplace equation in terms of an integral over the boundary. This is provided by means of the Green's function, $G(\mathbf{x}, \mathbf{y})$, [26]

$$u(\mathbf{x}) = \int_{\partial\Omega} \frac{\partial G(\mathbf{x}, \mathbf{y})}{\partial \mathbf{n}} f(\mathbf{y}) dy. \tag{5}$$

The normal derivative of the Green's function on $\partial\Omega$ is what we refer to as the "surface Green's function" for the domain Ω. Thus the surface Green's function for a domain, Ω, must be *equivalent* to the first-passage probability distribution for that same domain: $p(\mathbf{x}, \mathbf{y}) = \partial G(\mathbf{x}, \mathbf{y})/\partial \mathbf{n}$. With this fact, our strategy becomes the use of surface Green's functions to act as probability distributions to move Brownian particles quickly through their trajectories while maintaining their exact distribution properties.

The ST method greatly extends the GFFP method by allowing its application to domains for which we have no analytic representation of the Green's function. Perhaps the most basic example is the escape of a diffusing particle from a reflecting, *i.e.*, non-absorbing sphere. This is important as many FP domains involving either reflecting, or mixed, boundary conditions provide examples of this type.

The ST method is implemented as follows: for each set of values of the geometric parameters that characterize a particular FP surface, one performs a large number of simulations. For each simulation, the FP position is noted and the dimensionless parameters that characterize it are binned. The normed average of these binned values is partially integrated to give the distribution function for first-passage. This quantity is then tabulated, and a high-precision interpolatory fit is applied to it. This procedure, though numerically intensive, need only be carried out once for each FP geometry. The result, a tiny dataset consisting of the values of the resulting interpolation parameters, can then be used, rapidly and efficiently, to sample the FP position for this absorbing FP surface. This is a bootstrap methodology: the simulation phase of an initial ST application uses only WOS; subsequent applications are more efficient because each uses the results of the previous ST tabulations.

The ST method can be used to sample the FP position, *i.e.*, the absorption position, for Dirichlet Laplace problems in which the FP surface can be characterized by either one or two dimensionless parameters. This last limitation is purely computational; a tabulation of a problem of this kind that uses three parameters will be a natural supercomputer project once it is motivated.

The ST method has been applied to calculate the electrical conductivity of a composite material composed of non-overlapping, non-conducting spherical inclusions randomly dispersed in a conducting matrix. This is a specific case of a problem first studied in detail by Kim and Torquato. Our results (see Figs. 3- 5) agree with theirs in detail, although our computation times are shorter.

Second, the ST method is not limited to obtaining the FP position of a diffusing particle. Calculations of electrical conductivity require knowledge of the FP time. But any quantity may be sampled using the ST method. For example, the Feynman-Kac formulation of the linearized Poisson-Boltzmann equation requires sampling the exponential of the FP time. We discuss this next.

4 The Feynman-Kac Method

In this section, we explain the Feynman-Kac method for solving the Schrödinger equation, and other elliptic partial differential equations. We discuss the interaction between this method and methods already detailed in this review.

The Feynman-Kac formulation of the Dirichlet problem for the classical Schrödinger equation:

$$\frac{1}{2}\Delta\phi + V(x)\phi = 0, \qquad \mathbf{x} \in \Omega \tag{6}$$

$$\phi(\mathbf{x}) = \Psi(\mathbf{x}), \qquad \mathbf{x} \in \partial\Omega \tag{7}$$

provides a formula for the value of the field ϕ at a point x, namely [13,14]:

$$\phi(\mathbf{x}) = E\left[\Psi(X(\tau_{\partial\Omega}))\exp\left\{\int_0^{\tau_{\partial\Omega}} V(X(\tau_{\partial\Omega}))d\tau\right\}\right]. \tag{8}$$

where $\tau_{\partial\Omega} = \{\tau : X(\tau) \in \partial\Omega\}$ is the FP time and $X(\tau_{\partial\Omega})$ is the FP point of X, the Brownian motion started at **x**.

An elementary example of the Schrödinger equation, but one of extensive importance in molecular biology, is the linearized Poisson-Boltzmann equation [16]:

$$\Delta\phi - \kappa^2\phi = 0, \qquad \mathbf{x} \in \Omega \tag{9}$$

$$\phi(\mathbf{x}) = \Psi(\mathbf{x}), \qquad \mathbf{x} \in \partial\Omega \tag{10}$$

Here ϕ gives the electrical potential in the neighborhood of a biomolecule immersed in an electrolyte; κ is the Debye constant of the electrolyte. The Feynman-Kac formulation, applied to this problem, gives the formula for ϕ:

$$\phi(\mathbf{x}) = E\left[\Psi(X(\tau_{\partial\Omega}))\exp\left\{-\int_0^{\tau_{\partial\Omega}} \kappa^2 d\tau\right\}\right], \tag{11}$$

where $\tau_{\partial\Omega} = \{\tau : X(\tau) \in \partial\Omega\}$ is the FP time and $X(\tau_{\partial\Omega})$ is the FP point. For a spherical FP surface, the κ^2-term, interpreted as a standard decay rate, produces the decimation probability [27,17]:

$$E[\exp(-\kappa^2 t)] = \frac{d\kappa}{\sinh d\kappa}, \tag{12}$$

Here d is the FP sphere radius. The FK formulation given above reproduces the formula. But, it is much more general. For a more complicated FP surface, the LHS of the above equation can still be determined, using the ST method, even though analytic results are no longer available. Using the entire first-passage time probability distribution, we illustrate its computational equivalence to the decimation probability via the parallel plates problem used in previous research [16,17]. The results are given in Fig. 8.

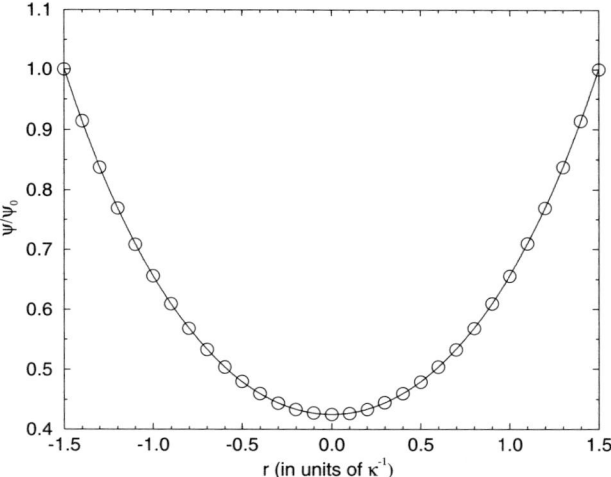

Fig. 8. The electric potential in an electrolyte between two infinite charged parallel flat plates. The solid line is the analytic solution and the circles are the simulation results with 10^6 random walks using an absorption layer thickness of $\delta = 10^{-4}$. Here, r is the distance from the mid-point of the plates and κ the inverse Debye length

5 Last Passage Methods for Diffusion Monte Carlo

In this section, we develop the concept of last-passage diffusion and explain its importance to the realm of Monte Carlo diffusion problems. Because this concept will be novel to most readers of this review, we will explain both the motivations for the concept and its two rather different realizations in practice.

The methodology for solving diffusion Monte Carlo problems that we have described in previous sections of this review is optimal for a large class of problems in which diffusing particles initiate outside a complex material domain and terminate on portions of its surface.

It is a fact, well known in pure mathematics, but apparently not in the realm of applied mathematics, that many diffusion Monte Carlo problems can be adequately described by using both 'first-passage diffusion', and also 'last-passage diffusion' methods [6]. The latter involves diffusing particles that initiate on or near their absorption points, and diffuse "backwards in time." Here we develop the first of two basic last-passage algorithms: the external-origin last-passage (EOLP) algorithms. These were developed by our group.

The charge density $\sigma(x)$ at a point, x, on the surface of an absorbing object is given by the equation:

$$\sigma(x) = \frac{1}{4\pi} \int_{y \in \partial \Omega_x} d^2 y\, G(x,y) P_{y \to \infty}. \tag{13}$$

Here the surface, $\partial \Omega_x$, is that part of a sphere with x as center, that is, outside the absorbing surface and the factor $P_{y \to \infty}$ is the probability that a diffusing particle, initiating at point $y \in \partial \Omega_x$, diffuses to infinity without returning to the absorbing surface. The function $G(x,y)$ is defined by:

$$G(x,y) = \frac{d}{d\delta_\epsilon}\bigg|_{\delta_\epsilon = 0} g(x,y). \tag{14}$$

Here $g(x,y)$ is the Laplace Green's function on the surface, $\partial \Omega_x$ with source point x at a distance δ_ϵ from the absorbing surface.

To derive Eq. (13), first consider the function $V(x)$ that gives the probability that a diffusing particle initiating at a point x near to the surface of an absorbing object, touches, *i.e.* makes first passage at the surface of the object in finite time (see Fig. 9). This is a harmonic function; it is unity on the surface of the object, and zero at infinity. Thus, by uniqueness of solutions to the Laplace equation, it is identical to the voltage surrounding the object when it is at voltage unity with respect to infinity. By the Gauss theorem, the charge density, $\sigma(x)$, at a point on the surface is given by

$$\sigma(x) = -\frac{1}{4\pi} \frac{d}{d\delta_\epsilon}\bigg|_{\delta_\epsilon = 0} V(x). \tag{15}$$

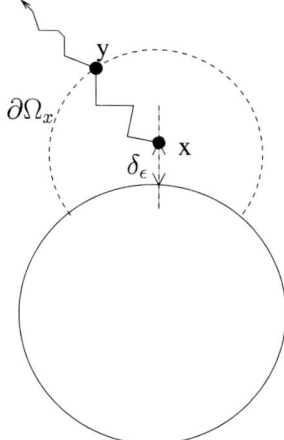

Fig. 9. The voltage $V(x)$ near a conducting sphere, at the point x, is given in diffusion language by the probability that a diffusing particle starting at point x will diffuse away to infinity without hitting the sphere. In order to do so, it must first reach a FP surface, $\partial\Omega_x$, drawn around point x and then proceed to diffuse far away without returning

Representing $V(x)$ as in Fig. 9, and realizing that only the δ_ϵ-dependence of the probability density for the first step is relevant (because it is proportional to δ_ϵ), gives Eq. (13).

The function $G(x,y)$ is a point-dipole Green's function. To see this, note that taking the δ_ϵ-derivative and setting δ_ϵ to zero in Eq. (15) has the same effect as taking the dipole limit: allowing both the magnitude Q of the source at point x, and that of its image charge, to grow without limit as $\delta_\epsilon \to 0$, while keeping the quantity $Q^2\delta_\epsilon$ finite. Placing a point dipole at point x provides a source for all of the diffusing particle trajectories that originate at the point x leave and never return. The probability of escape from an absorbing surface is rigorously zero; this Green's function samples only the measure-zero subset of trajectories that succeed. Figure 10 shows a simple case in which this claim can be easily verified by inspection. This formula (and this Green's function) were developed to provide a local formula for charge density, *i.e.*, a formula that could be used regardless of other nearby charges and conductors.

Both the calculation of the capacitance of a non-smooth object and the two other classes of problems mentioned above can also be treated with the other class of last-passage methods, the integral-origin last-passage (IOLP) methods. The equilibrium charge distribution $\sigma(x)$ on an absorbing object is given by:

$$\sigma(x) = 2\pi |x - z| L(x, z), \tag{16}$$

where $L(x, z)$ is the last passage distribution. Here diffusing particles initiate at a point z interior to an absorbing object. They diffuse, ignoring the absorbing object, until they eventually diffuse away to infinity. At this time, the point, x, of last contact, *i.e.*, last-passage, with the absorbing object is determined. We do this using a generalization of the dipole Green's function defined in Eq. (14); this is detailed in a forthcoming publication.

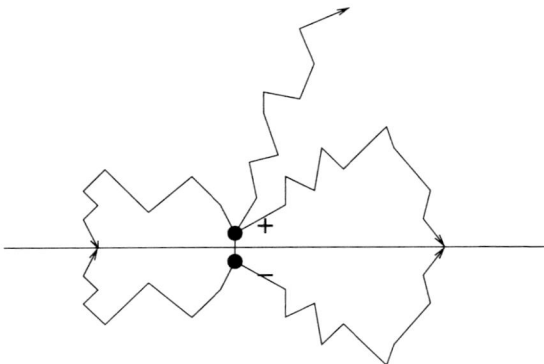

Fig. 10. The Green's function for a point dipole oriented normal to an absorbing surface is a generating function for diffusing particle trajectories that leave the absorbing surface and never return. The effect of trajectories that leave and do return is zero; they cancel out in pairs

None can yet describe the relative advantages of these two sets of last-passage methods; this research is now in progress.

There are at least three classes of diffusion Monte Carlo problems for which last-passage algorithms are optimal:

- Charge distribution on a conducting object with edges and corners. In such problems, a large fraction of the charge will collect very close to the edges and corners, *i.e.*, on a very small subset that is readily identified in advance. Thus last-passage algorithms are appropriate.
- Problems in which a large fraction of the absorption takes place on a very small fraction of the surface, because of the imposed boundary conditions. The basic example here is the problem of diffusion-limited absorption of a ligand molecule at a small absorbing site on a macromolecule. If the absorbing site is small enough, it must become optimal computationally for the diffusing particles to initiate on the absorbing site rather than to initiate on an external launch sphere and 'search for' the absorbing site. The Solc-Stockmayer model of protein-ligand binding is perhaps the best-studied model of this process [12,28].
- Problems in which more than one conducting object is present, at close proximity, and at different voltages. In these cases (modern micro-electronics provides many examples), one seeks to calculate not a capacitance but an entire capacitance matrix. Here, no launch surface for diffusing particles can be defined; so first passage algorithms are not a possibility.

We will discuss examples of the first class of applications; examples of the two other classes will be published separately.

A basic well-studied example of the first class is that of a conducting cube. If first-passage methods are used to study this problem, importance sampling will occur, *i.e.*, the correct surface charge distribution will be obtained only if a large launch sphere is used. This will not be optimal in a computational sense. optimality in a computational sense. In the last-passage algorithm, the capacitance of the unit cube is defined to be the integral, over the surface of the cube of the surface charge density $\sigma(x)$ as given by Eq. (11); it is defined as a double integral. Importance sampling is readily imposed on the outside integral, *i.e.*, the integral over σ by using the measure:

$$x = (1 - \eta^{3/2}) \tag{17}$$

$$y = (1 - \eta^{3/2}). \tag{18}$$

Here, (x, y) is the sampling point on the sampling area, $(0, 1) \times (0, 1)$, in x-y plane and the η's are independent random numbers uniformly distributed in $(0, 1)$. If this measure is used, almost all points at which the charge density must be sampled will be close to the edge of the cube. The statistics of the inner integral, *i.e.*, the integral that gives $\sigma(x)$, will be very poor because the probability $P_{y \to \infty}$ will be very close to zero. An important method of overcoming this problem is the method of the edge distribution.

For any edge on a conducting surface the charge distribution $\sigma(x, \delta_e)$ on a curve parallel to the edge, but separated from it by distance δ_e, with δ_e small, is given by:

$$\sigma(x, \delta_e) = \delta_e^{\pi/\alpha - 1} \sigma_e(x). \tag{19}$$

Here $\sigma_e(x)$ is what we term the edge distribution. α is the angle between the two intersecting surfaces, here $\alpha = 3\pi/2$. The edge distribution has a natural probabilistic interpretation: it is the (rescaled) probability density that a diffusing particle makes last passage on the edge point x. This distribution can be calculated either by simulation (see Fig. 10) or by application of the general formula from Eq. (11). The point is that this one-dimensional distribution need be calculated only once for each edge on each absorbing object in a problem. An extension of Eq. (13) for $\sigma(x)$ gives a formula for the edge distribution:

$$\sigma_e(x) = \frac{1}{4\pi} \lim_{\delta_e \to 0} \delta_e^{1 - \pi/\alpha} \int_{y \in \partial \Omega_e} d^2 y G(x, y) P_{y \to \infty}. \tag{20}$$

Here $\partial \Omega_e$ is a cylindrical surface that intersects the pair of absorbing surfaces meeting at angle α (see Fig. 11). The other quantities have already been defined.

For the function $G(x, y)$, we start with the potential inside a grounded cylindrical box defined by the surfaces $z = 0$, $z = L$, and $\rho = a$ with a unit point charge located at the point (ρ', ϕ', z') given by [29]:

$$\Phi(\rho, \phi, z) = \frac{4}{L} \sum_{m=-\infty}^{\infty} \sum_{n=1}^{\infty} e^{im(\phi - \phi')} \sin\left(\frac{n\pi z}{L}\right) \sin\left(\frac{n\pi z'}{L}\right) \frac{I_m\left(\frac{n\pi \rho_<}{L}\right)}{I_m\left(\frac{n\pi a}{L}\right)}$$
$$\times \left[I_m\left(\frac{n\pi a}{L}\right) K_m\left(\frac{n\pi \rho_>}{L}\right) - K_m\left(\frac{n\pi a}{L}\right) I_m\left(\frac{n\pi \rho_>}{L}\right) \right]. \tag{21}$$

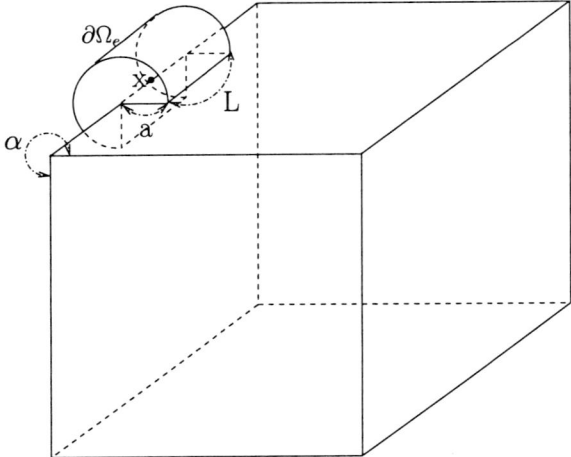

Fig. 11. A three-quarter cylinder of radius a and length L on the edge of a cube is shown

Here, I_m and K_m are modified Bessel functions and

$$\rho_< = \min(\rho', \rho), \tag{22}$$
$$\rho_> = \max(\rho', \rho). \tag{23}$$

Modifying Eq. (21) to satisfy the boundary conditions of a three-quarter cylinder (see Fig. 12), the potential inside the three-quarter cylinder when $\rho' < \rho$ is obtained as

$$\Phi_c(\rho, \phi, z) = \frac{4}{L} \sum_{n=1}^{\infty} \sin\left(\frac{2}{3}\phi\right) \sin\left(\frac{2}{3}\phi'\right) \sin\left(\frac{n\pi z}{L}\right) \sin\left(\frac{n\pi z'}{L}\right) \frac{I_{2/3}(\frac{n\pi \rho'}{L})}{I_{2/3}(\frac{n\pi a}{L})}$$
$$\times \left[I_{2/3}\left(\frac{n\pi a}{L}\right) K_{2/3}\left(\frac{n\pi \rho}{L}\right) - K_{2/3}\left(\frac{n\pi a}{L}\right) I_{2/3}\left(\frac{n\pi \rho}{L}\right) \right]. \tag{24}$$

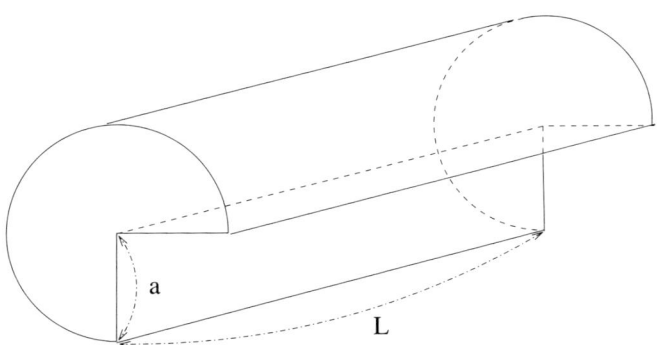

Fig. 12. A three-quarter cylinder of radius a and length L is shown

Hence, $G(\phi, z)_{\rho=a}$ for the side of the three-quarter cylinder is given by

$$G(\phi, z)_{\rho=a} = \frac{1}{\Gamma(5/3)2^{2/3}} \frac{4}{9\pi La} \sum_{n=1}^{\infty} \sin\left(\frac{2}{3}\phi\right) \sin\left(\frac{n\pi z}{L}\right) \sin\left(\frac{n\pi z'}{L}\right)$$
$$\times \left(\frac{n\pi}{L}\right)^{2/3} \frac{1}{I_{2/3}\left(\frac{n\pi a}{L}\right)} \quad (25)$$

and $G(\rho, \phi)_{z=0}$ for the lower cap of the three-quarter cylinder

$$G(\rho, \phi)_{z=0} = \frac{1}{\Gamma(5/3)2^{2/3}} \frac{4}{9\pi L} \sum_{n=1}^{\infty} \sin\left(\frac{2}{3}\phi\right) \left(\frac{n\pi}{L}\right)^{5/3} \sin\left(\frac{n\pi z'}{L}\right)$$
$$\times \frac{1}{I_{2/3}\left(\frac{n\pi a}{L}\right)} \left[I_{2/3}\left(\frac{n\pi a}{L}\right) K_{2/3}\left(\frac{n\pi \rho}{L}\right) - K_{2/3}\left(\frac{n\pi a}{L}\right) I_{2/3}\left(\frac{n\pi \rho}{L}\right)\right]. \quad (26)$$

Thus the edge distribution of a cube (see Fig. 13), $\sigma_e(x)$ is obtained as

$$\sigma_e(x) = 2 \int_0^a \int_0^{3\pi/2} G(\rho, \phi)_{z=0} P_{y \to \infty} \rho d\phi d\rho$$
$$+ a \int_0^L \int_0^{3\pi/2} G(\phi, z)_{\rho=a} P_{y \to \infty} d\phi dz, \quad (27)$$

where $P_{y \to \infty}$ is the probability of going to infinity from the point y on the side or caps of the chopped cylinder.

For convenience, a chopped cylinder of $L = a = 0.02$ is used and the edge calculations are done at the center of the chopped cylinder on a unit cube

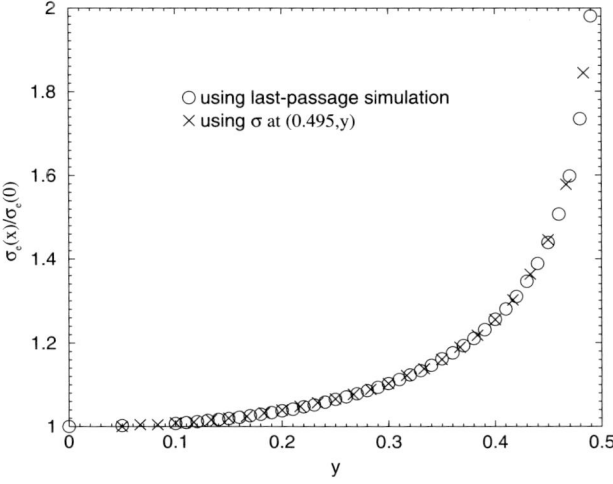

Fig. 13. The edge distribution of a unit cube calculated using Eq. (11) and using Eq. (20)

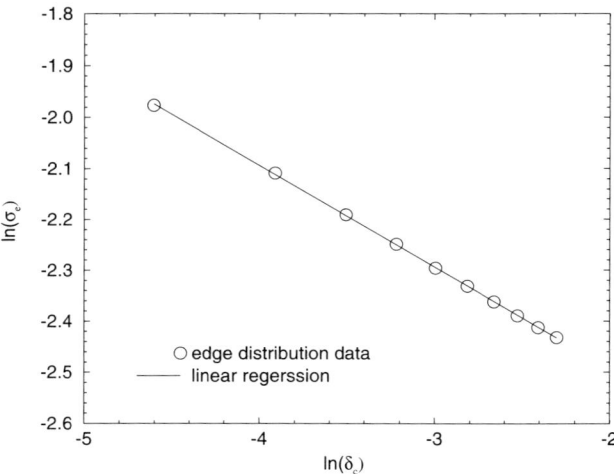

Fig. 14. Asymptotic values near the corner of the edge distribution for a unit cube: δ_c is the distance from the corner and the slope is approximately -0.20

moving the chopped cylinder along the edge of the cube (see Fig. 11). Due to the symmetry of the cube, the edge distribution is obtained in the range of $(0, 0.49)$ with 0.01 step. The result is shown in Fig. 13. However, because the charge singularity at a corner is stronger that that at an edge, the edge distribution near the corner diverges. To use the edge distribution for the fast calculation of the capacitance of a cube, we calculate the asymptotic behavior near the corner. The exponent of the edge distribution near the corner is approximately -0.20, that is, $\sigma_e \sim \delta_c^{-1/5}$ (see Fig. 14).

The edge distribution can be calculated for a conducting object, provided it consists of a segment of a line or an arc of a circle, using generalizations of the methods discussed here. These distributions can then be used to calculate the surface charge density on the conducting object using a rapid importance sampling algorithm. Such algorithms are rapid because they tell us in advance the detailed charge distribution near edges and corners.

We believe that this last passage method is the fastest method to date for solving this basic set of problems. Computing the capacitance of a cube has been considered to be "one of the major unsolved problems of electrostatic theory" [20].

6 Conclusions and Suggestions for Further Study

In this paper, we review the results so far obtained from applying the set of Monte Carlo diffusion methods we have developed and assembled. The results already exhibited demonstrate that a number of classes of important problems can be solved far more efficiently using these methods.

The potential of this class of methods is yet to be tapped. Here we note just two examples of important extensions:

- Solution of the linearized Poisson-Boltzmann equation in general requires solution of problems with dielectric boundaries, *i.e.*, nontrivial values of the dielectric constant on both sides of the interface. Green's functions for this purpose are available; they can be tabulated.
- Calculation of the mutual capacitance matrix for a system of conductors in close proximity. Last passage methods allow calculation of this quantity using diffusion Monte Carlo. It remains to be seen which of these methods is most efficient.

Acknowledgements

We give special thanks to John David Jackson, Joe Hubbard, Kai Lai Chung, Karl K. Sabelfeld, and Henry P. McKean for their useful discussions.

References

1. A. Haji-Sheikh, E.M. Sparrow: J. Heat Transfer **89**, 121 (1967)
2. I.C. Kim, S. Torquato: J. Appl. Phys. **69**, 2280 (1991)
3. I.C. Kim, S. Torquato: J. Appl. Phys. **71**, 2727 (1992)
4. S. Torquato, I.C. Kim, D. Cule: J. Appl. Phys. **85**, 1560 (1999)
5. C.-O. Hwang, J.A. Given, M. Mascagni: Phys. Fluids A **12**, 1699 (2000)
6. K.L. Chung: *Green, Brown, and Probability* (World Scientific, Singapore 1995)
7. J.A. Given, J.B. Hubbard, J.F. Douglas: J. Chem. Phys. **106**, 3721 (1997)
8. M.E. Müller: Ann. Math. Stat. **27**, 569 (1956)
9. T.E. Booth: J. Comput. Phys. **39**, 396 (1981)
10. S. Torquato, I.C. Kim: Appl. Phys. Lett. **55**, 1847 (1989)
11. C.-O. Hwang, J.A. Given, M. Mascagni: Monte Carlo Meth. Appl. **7**, 213 (2001)
12. C.-O. Hwang, J.A. Given, M. Mascagni: J. Comput. Phys. **174**, 925 (2001)
13. M. Freidlin: *Functional Integration and Partial Differential Equations.* (Princeton University Press, Princeton 1985)
14. K.L. Chung, Z. Zhao: *From Brownian Motion to Schrödinger's Equation.* (Springer Verlag, Berlin 1995)
15. K.K. Sabelfeld: Math. Comput. Modell. **23**, 111 (1996)
16. R. Ettelaie: J. Chem. Phys. **103**, 3657 (1995)
17. C.-O. Hwang, M. Mascagni: Appl. Phys. Lett. **78**, 787 (2001)
18. J.B. Hubbard, J.F. Douglas: Phys. Rev. E **47**, 2983 (1993)
19. J.F. Douglas, H.X. Zhou, J.B. Hubbard: Phys. Rev. E **49**, 5319 (1994)
20. H.X. Zhou, A. Szabo, J.F. Douglas, J.B. Hubbard: J. Chem. Phys. **100**, 3821 (1994)
21. N.S. Martys, S. Torquato, D.P. Bentz: Phys. Rev. E **50**, 403 (1994)
22. H.X. Zhou: Biophys. J. **69**, 2286 (1995)
23. H.X. Zhou: Biophys. J. **69**, 2298 (1995)
24. S.C. Port, C.J. Stone: *Brownian Motion and Classical Potential Theory* (Academic Press, New York 1978)

25. F. John: *Partial Differential Equations.* (Springer-Verlag, Berlin, Heidelberg, and New York 1982)
26. E.C. Zachmanoglou, D.W. Thoe: *Introduction to Partial Differential Equations with Applications.* (Dover, New York 1976)
27. B.S. Elepov, G.A. Mihailov: Soviet Math. Dokl. **14**, 1276 (1973)
28. H.X. Zhou: J. Chem. Phys. **108**, 8139 (1998)
29. J.D. Jackson: *Classical Electrodynamics.* (John Wiley and Sons, New York 1975)

Part II

Econophysics

An Updated Review of the LLS Stock Market Model: Complex Market Ecology, Power Laws in Wealth Distribution and Market Returns

Sorin Solomon and Moshe Levy

Introduced in 1994, the Levy-Levy-Solomon (LLS) model developed into a standard tool to study in realistic detail the emergence of complex dynamics of stock markets with heterogeneous quasi-rational partially informed investors. We review the main features of this model and several of its extensions. We study the effects of investor heterogeneity and show that predation, competition, or symbiosis may occur between different investor populations. Many properties observed in actual markets appear as natural consequences of the LLS dynamics: truncated Lévy distribution of short-term returns, excess volatility, a "U-shape" pattern of return autocorrelation, and a positive correlation between volume and absolute returns. We also comment on the emergence of power law tails in the distribution of investor wealth and its relation to the efficient market hypothesis.

1 Introduction to the Levy-Levy-Solomon (LLS) Model

LLS is a microscopic simulation model of the stock market. Its details and some generalizations of it can be found in [2]. In the present account we introduce the basic ideas of the LLS model and its main results. The model involves a large number of virtual investors characterized each by current wealth, portfolio structure, expectations and risk-taking preferences. These personal characteristics come into play in each investor's decision making process as described below.

We consider a market with only two investment options: a bond and a stock, which can be thought of as a proxy for the market index (see [3] for an extension to the case of multiple stocks). The bond is assumed to be a riskless asset yielding a constant return r_f. The bond is exogenous and investors can buy as much of it as they wish at the given rate.

The stock is a risky asset. The total return on the stock, H_t, is composed of two elements:
(i) Capital gain (loss): If an investor holds a stock, any rise (fall) in the market price of the stock contributes to an increase (decrease) in the investor's wealth.
(ii) Dividends: The company earns income and distributes dividends. Dividends reflect economic fundamentals and are assumed to follow a stochastic process $\bar{D}_t = D_{t-1}(1+\bar{g})$, where $\bar{D}_t$ is the dividend distributed at period t and $\bar{g}$ is a stochastic growth rate.

The total return on the stock, denoted by $\bar{H}_t$ is given by:

$$\bar{H}_t = \frac{\bar{P}_t + \bar{D}_t}{P_{t-1}}, \tag{1}$$

where $\bar{P}_t$ is the price of the stock at time t.

Each investor i is confronted with a decision where the outcome is uncertain: which is the optimal fraction x_i of her wealth to invest in stock? According to the standard theory of investment each investor is characterized by a utility function of wealth, $U(w)$, that reflects her personal risk-taking preference. Investor i chooses her optimal investment proportion in the stock such as to maximize her expected utility $EU(w_i)$. The optimal investment proportion depends, of course, on the expectation regarding the stock's return.

In the basic LLS model investors estimate the ex-ante stock return distribution by employing the distribution of historical returns. An investor who looks at the past k returns in forming her estimation of the ex-ante return distribution believes:

$$\text{Prob}(\bar{H}_{t+1} = H_{t-j}) = \frac{1}{k} \quad \text{for} \quad j = 1, \cdots, k. \tag{2}$$

Thus, this investor's expected utility is given by:

$$EU(w_i(t+1)) = \frac{1}{k} \sum_{j=1}^{k} U[w_i(t)(1-x_i)(1+r_f) + w_i(t)x_i H_{t-j}]. \tag{3}$$

Investor i chooses her investment proportion in the stock, x_i, such as to maximize her expected utility given above. The extrapolation range k may differ between various investors, and it will be the main parameter inducing market inhomogeneity later on in this chapter. Here we assume the logarithmic utility function, i.e. $U(w) = \ln(w)$, which is a special case of the more general power utility functions considered in [1]. To account for the many unknown factors influencing decision-making, such as liquidity constraints or deviations from rationality, we add a small random variable (or idiosyncratic "noise") to the optimal proportion x_i.

At fixed time intervals each investor revises the composition of her portfolio and submits a new demand schedule. The aggregate of these demands determines the new stock price by the market clearance condition. To be specific, once each investor decides on the proportion of her wealth x_i that she wishes to hold in stocks, one can derive the number of stocks $N(i, p_h)$ she wishes to hold corresponding to each hypothetical stock price p_h. Since the total number of shares in the market, N, is fixed, there is a particular value of the price p for which

$$\sum_i N(i, p) = N, \tag{4}$$

where the sum is over all investors. This value p is the new market equilibrium price. (In [4] it is shown that the demand curves are downward sloping and that the equilibrium price indeed exists and is unique.) Upon updating accordingly the traders' portfolios, wealth and list of last k stock returns, one is ready for the next market iteration. This process is repeated for each time step, and the market price, volume, and the wealth, and portfolio composition of each investor are recorded at each step throughout the run.

Notice that power utility functions in LLS imply constant relative risk aversion. This means that given a set of beliefs, the investor's demand for the stock is proportional to her wealth. The empirical evidence lends support for this class of utility functions [20–22]. Thus, in the LLS model the wealth dynamics play a key role, and is intertwined with the price dynamics. We analyze the wealth distribution generated by the model and show that it in turns determines the distribution of stock returns. Other approaches in the literature (for example, [24]) assume constant *absolute* risk aversion, which implies that the dollar amount invested in the stock is independent of the investor's wealth. While this approach is analytically more tractable, it disconnects between the price dynamics and the wealth dynamics.

The basic version of the LLS model does not include learning or migration of investors towards strategies which have shown past success. These effects are modeled in [24,25,27], for example, and also in later versions of LLS. However, even in the basic model with no migration we observe "ecological" market dynamics – investors with successful strategies become wealthier, and their impact on the market grows.

Various extensions of the basic LLS model include adding "fundamentalists" who employ the discounted dividend stream model to derive the stock's fundamental value [1], and investigating the model with investors characterized by Prospect Theory preferences [1,23]. Reference [26] studies the effect of introducing investors with buy-and-hold policies. The effect of the number of investors on the price dynamics in LLS is examined in [3].

2 Crashes, Booms and Cycles

The LLS model provides already at the level of a quite homogeneous trader population a convincing description of the emergence of cycles of booms and crashes in the stock market. In a market with one species of investors all having a homogeneous memory (extrapolation) range spanning the last k returns of the stock, the stock price alternates regularly between two very different price levels. The explanation for this behavior is as follows.

Assume that the rate of return H_t on the stock at a time t is higher than the oldest remembered return (H_{t-k}). The addition of H_t and the elimination of H_{t-k} creates then a new distribution of past returns that is better than the previous one. Since the LLS extrapolating investors use the past k returns to estimate the distribution of the next period's return, they will be led to be more optimistic and increase their investments in the stock. This, in turn, will cause the stock price to rise, which will generate an even higher return.

This positive feedback loop stops only when investors reach the maximum investment proportion (i.e. $X(i) = 100\%$: we do not allow borrowing or short selling), and can no longer increase their investment proportion in the stock. The dividend contribution to the returns is small compared with this high price at this stage. In the absence of noise the returns on the stock at this plateau converge to a constant growth rate which is just slightly higher than the riskless interest rate (see [4]). In other words, in the absence of noise the price remains almost constant, growing only because of the interest paid on the bond (more money entering the system and being invested in the stock).

As there is some noise in the system, generated by the small random deviations from the optimal investment proportions, the price fluctuates a little around the asymptotic high level. These fluctuations generate some negative returns (on a downward fluctuation) and some high returns (when the price goes back up). One might suspect that a large downward fluctuation might trigger a reverse positive feedback effect, with investors' expectations becoming lower, investment proportions decreasing, the price dropping, generating further negative returns, and so on: a crash. This can happen during the "plateau" period but only after the previous sharp price boom, which generated an extremely high return, is forgotten. And, indeed, this is exactly what happens. Since it takes k steps to forget the boom, the high price plateaus are a bit longer than the extrapolation span (k days to forget the boom + $O(1)$ more days until a large enough negative fluctuation occurs).

The crash generates a disastrous return and, until it is forgotten, investment proportions and hence the price remain very low. When the price is low, the dividend yield becomes significant again and the returns on the stock are relatively high (compared with the bond). Once the crash has been forgotten, all the returns that are remembered are therefore high, and the price jumps back up. Thus, the low price plateaus are k steps long. This completes one cycle, which is repeated throughout the run. This (quasi-)periodicity is best viewed in the Fourier transform of the price time evolution (Fig. 1) as a series of narrow peaks around the frequency $2k + O(1)$ and its harmonics. (Note however that the dynamics is not perfectly periodic and therefore in spite of its simplicity, according to some mathematical criteria it may fall into the "complex" category. In the present paper we reserve however the term "complex" for dynamics that is truly complicated to the degree that it does not admit simple verbal or mathematical description or understanding.)

The homogeneous stock market described above exhibits booms and crashes. However, the homogeneity of investors leads to unrealistic periodicity. As shown below, when there is more than one investor species the dynamics becomes much more complex and realistic.

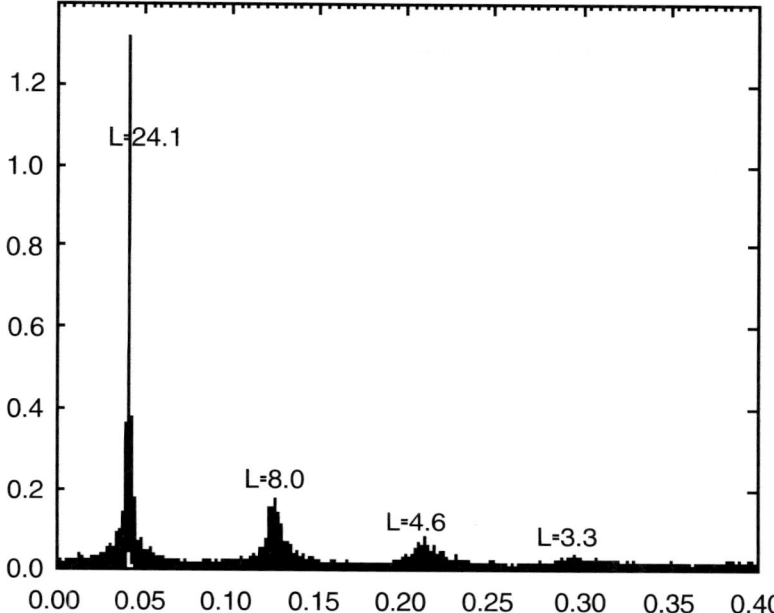

Fig. 1. The Fourier transform of the price in a market with one investor species with extrapolation range $k = 10$. The market contained 10000 traders that had initially equal wealth invested half in stock and half in bonds

3 Predation, Competition and Symbiosis Between Trader Species

In the preceding section it was explained that a homogenous population of traders that extrapolate the last k returns leads to cycles of booms and crashes of period $2k + O(1)$. When there are two species with extrapolation ranges k_1 and respectively k_2, we observe sharp irregular transitions between eras where one species dominates (cycles of period $2k_1 + O(1)$) and market eras where the other species dominates (cycles of period $2k_2 + O(1)$). When the number of trader species is three, there are dramatic qualitative changes: generically, the dynamics becomes complex. We show that complexity is an intrinsic property of the stock market. This suggests an alternative explanation to the widely accepted but empirically debatable random walk hypothesis. We discuss below some of the market ecologies possible with only two species of traders.

3.1 Market Ecologies with Two Trader Species

When there are two trader species with different extrapolation spans it turns out that the nature of the dynamics is determined by the ratio of the extrapolation spans of the two species. In [9] we performed a qualitative theoretical analysis of this phenomenon and supported it by microscopic simulations. We showed that in market eras in which one species (of extrapolation range k_0) dictates the dynamics (i.e. boom-crash cycles have periods of length $2k_0 + O(1)$) the second species (with extrapolation range k) has generically the following performance:

A : If $k_0 < k < 2k_0$, then k is performing very poorly (looses money);
B : If $2nk_0 < k < (2n+1)k_0$ (with n a natural number), then k is doing relatively well;
C : If $(2n+1)k_0 < k < 2nk_0, n > 1$, then k does better than in A but worse than in B;
D : If $k < k_0$, then k is doing well.

These facts turned out sufficient to understand the three main cases that a two-species ecology can display as will now be discussed.

Case 1: Predator-Prey Dynamics

If one considers one species with an extrapolation range $k_1 = 10$ and a second species with an extrapolation range $k_2 = 14$ it turns out that the resulting

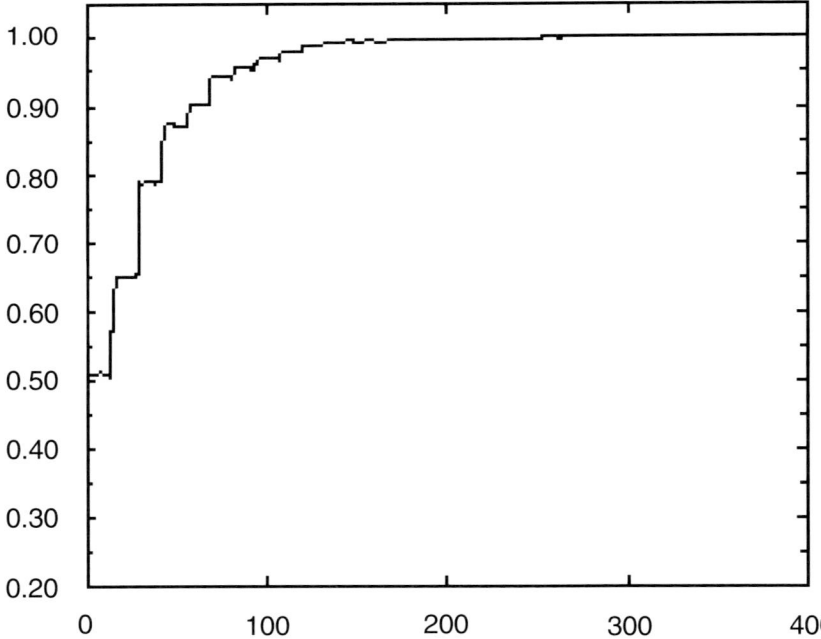

Fig. 2. Fraction of the wealth that the species $k_1 = 10$ possesses in Case 1. The traders in the market belonged to two species consisting each of 5000 traders. Each trader owed at the beginning 5000 dollars in cash and 5000 shares (worth each 1.4 dollars)

ecology dynamics is a predator-prey one. In fact, the LLS market dynamics leads in this case to the extinction (total impoverishment) of the k_1 species: after some time the entire wealth on the market belongs to the species $k_1 = 10$ (Fig. 2). As a consequence, the market price presents clear cycles of booms and crashes of periodicity clustered around $24 = 2 \times 10 + O(1)$.

This is easily understood since according to the property A above the $k_2 = 14$ population is performing poorly when the k_1 dictates the market periodicity while the population 10 is performing well according to property D in the hypothetical periods when $k_2 = 14$ dictates the market periodicity. This is only an example of a large class of parameters that lead to predator-prey systems and which may result in the total extinction of one of the species.

Case 2: Competitive Species

If one chooses $k_1 = 10$, $k_2 = 26$, the species with extrapolation range 26 gains during the periods when the species $k_1 = 10$ dominates (property B) but species $k_1 = 10$ gains when the species $k_2 = 26$ dominates (property D). It is therefore reasonable that one species can not dominate the other indefinitely. Indeed, a look at the fraction of the wealth held by the species with extrapolation range $k_1 = 10$ reveals alternating eras of dominance (Fig. 3).

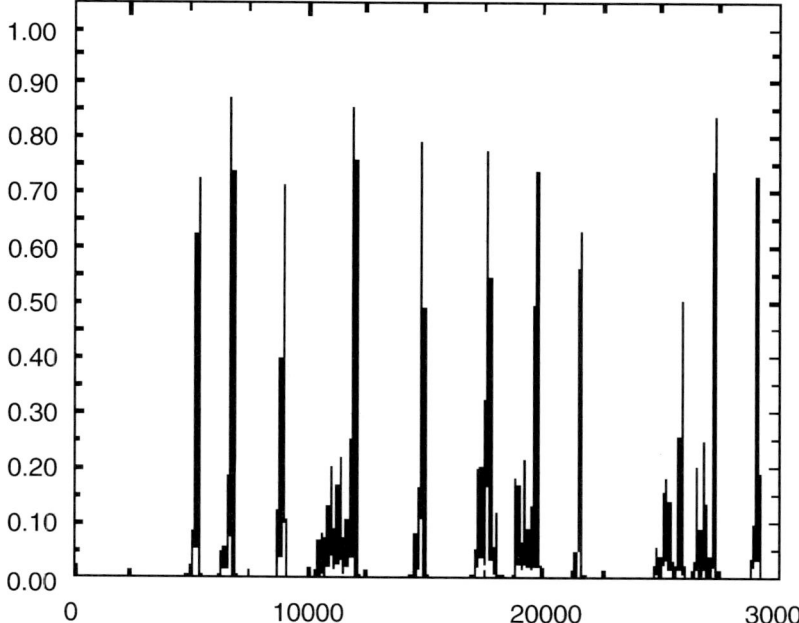

Fig. 3. Fraction of the wealth that the species $k_1 = 10$ possesses in Case 2. The initial conditions were similar to Fig. 2

This is also reflected in the alternations between price cycles (~ 56) corresponding to $k_2 = 26$ and price cycles (~ 24) corresponding to $k_1 = 10$. Clearly this alternation between the 2 species corresponds to a classical competitive ecology, in which two competing species take turns in dominating the ecology. Note however that most of the time it is the population k_2 which dominates the wealth. This seems to be a generic tendency in the long runs limit.

Case 3: Symbiotic Species

In the case $k_1 = 10$, $k_2 = 36$, similarly to the $10 - 26$ market, the investors with extrapolation range $k_2 = 36$ are doing better than those with extrapolation range $k_1 = 10$ when $k_1 = 10$ dictates the dynamics (cf. property C). On the other hand $k_1 = 10$ are doing better when the species $k_2 = 36$ dictates the dynamics (cf. D). Hence, we may speculate that again we will find alternating eras of dominance. Figure 4 shows that this is not the case. The difference between this case and the $10 - 26$ case is that here the market remains stuck in a "metastable" state: the extrapolation range 36 population never gains enough wealth to dictate long cycles. Thus, the system remains in a state of *symbiosis* throughout the run: the price cycles correspond to the

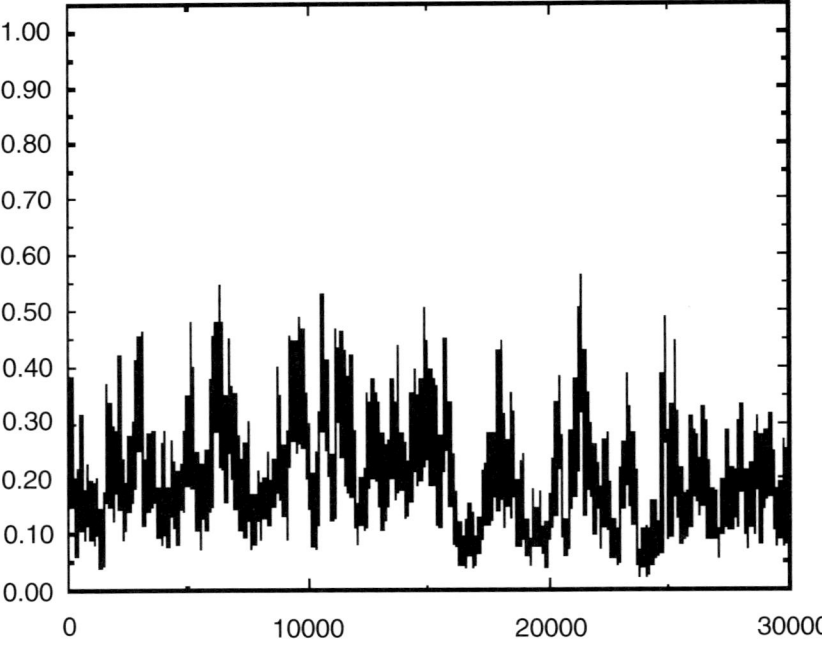

Fig. 4. Fraction of the wealth that the species $k_1 = 10$ possesses in Case 3

short species extrapolation range span $k_1 = 10$ while $70 - 80\%$ of the wealth stays with the long extrapolation span species k_2.

For very long k_2 extrapolation ranges, the share of the total wealth detained by k_2 can be even larger (approaching unity).

In conclusion, [9] has uncovered a quite lively ecology of the traders populations in the LLS model and

"observed phenomena ranging from complete dominance of one population to alternating eras of domination and to symbiosis $\cdots$. Our results suggest that complexity is an intrinsic property of the stock market. The dynamic and complex behavior of the market need not be explained as an effect of external random information. It is a natural property of the market, emerging from the strong nonlinear interaction between the different investor subgroups of the market $\cdots$."

The main source of endogenous dynamics in the LLS model turns out to be the feedback between the market price fluctuations and the wealth of the investors belonging to various species:

- On the one hand the wealth of the investors determines their influence on the price changes (at the short range): e.g. the richest determine the periodicity of the boom-crash cycles.
- On the other hand, the variations in the price determine changes in the distribution of wealth, which iterated over longer time intervals, result in changes in the market price cycle periodicity regime.

The entire cycle of rise and fall of a given species can be schematically described as follows.

The species has by chance a (momentary) winning strategy → Investors belonging to the species gain wealth → Overall wealth of the investors belonging to the species increases → Orders of investors belonging to the species become large → Investors' actions influence the market price adversely (self-defeating) → Trading of investors belonging to the species becomes inefficient → Investors lose money → Investors belonging to the species become poor → Species wealth and market impact decrease → Other species with different strategies become winners → Cycle re-starts (with the new winning strategies).

A few comments are in order.

1. The concept of a winning strategy is only a temporary one as it depends crucially on the state of the market: by its very success at a certain moment, a strategy prepares the seeds of its failure in the future.
2. The biological and cognitive analogies are useful but their limits should be understood:
   - In biology, the species selection mechanism is based on the disappearance of the unsuccessful individuals.
   - In the learning adaptive agents' case, the individuals may discard loosing strategies for new ones.

In the LLS market framework, while it is possible to include the above effects, they are not necessary: the strategies selection takes place automatically by their carriers (traders belonging to the species) losing or gaining: for the market to be efficient, no *a priori* intelligence nor explicit criteria for the evaluation and comparison of market performance are required: just the natural (Adam Smith's "invisible hand") market mechanisms. (See [24,25,27] for models with explicit learning and investor migration.)

3. While the adverse influence on the market price implied by the large orders coming from rich agents automatically leads to self-defeat (except for rich agents which follow a buy-and-hold strategy and therefore do not influence (adversely) the market), the mere lack of market influence due to poverty does not guarantee a winning strategy. It is necessary therefore that there are enough strategies and enough agents in the market for insuring that every investment "niche" is exploited, which is key to the notion of market efficiency, as described below.

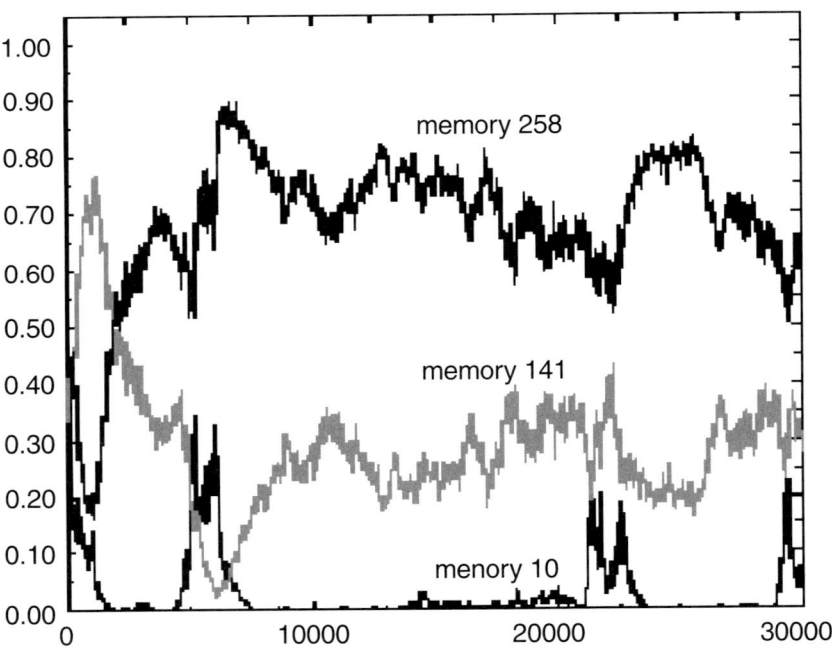

Fig. 5. The species wealths in a market with three species of extrapolation ranges of respectively 10, 141 and 256 days. Initially the three species possessed equal wealth distributed equally between stock and bond. Each species consisted of 1000 traders

3.2 Three Investor Species

One might suspect that the three species dynamics is a natural extension of the two species dynamics. Instead of alternating between two cycle lengths the system may just alternate between the three possible states of dominance. Figures 5–6 shows that this is not at all the case. These figures depict a typical part of the dynamics of a three species market, with extrapolation spans of 10, 141 and 256 respectively. With the introduction of a third species the system has undergone a qualitative change: there is no specific cycle length describing the time series. Instead, we see a mixture of different time scales: the system has become complex. Prediction becomes very difficult, and in this sense the market is much more realistic. Figure 5 shows the power struggle between the three species while Fig. 6 depicts the Fourier transform of the price evolution during this run.

Although the dynamics is complex, it is clear from Figs. 5 and 6 that there is an underlying structure, which perhaps may be analyzed by the properties A, B, C, D and their generalizations. For instance, it would appear from Fig. 5 that 141 and 256 take turns in dominating while 10 has a chance to a non-vanishing wealth share only occasionally in the transition intervals

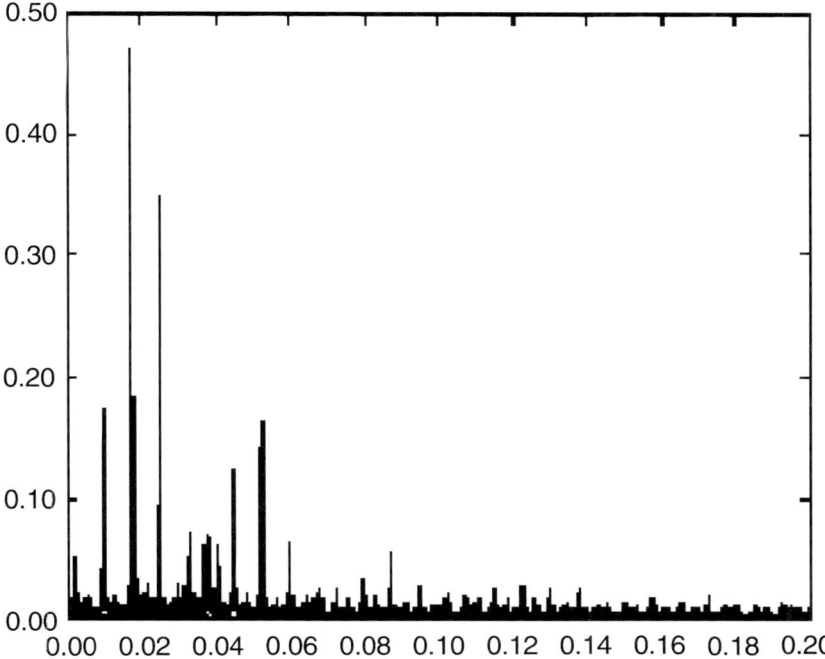

Fig. 6. Fourier transform of the stock price time evolution in the market described in Fig. 5

between 141 and 256 dominated eras. The dynamics generated by only three investor species can be extremely complex, even without any external random influences. This is similar to the conclusion reached in [24], although the frameworks of the two models are quite different. When there are more than three investors species the LLS dynamics become even more complex, and it generates many of the phenomena observed in real financial markets, as described in the next section.

4 LLS with Many Species: Realistic Dynamics of Market Returns

Our numerical experiments within the LLS framework have found that already a small number of trader species (characterized by different extrapolation ranges k) leads qualitatively to many of the empirically observed market phenomena.

In reality, we would expect not just a few trader types, but rather an entire spectrum of investors. When the full spectrum of different trader species (fundamentalists and various other types – see [1] for the detailed operational definition) is considered it turns out that "more is different" [5]: the price dynamics becomes realistic – booms and crashes are not periodic or predictable, and they are also less frequent and dramatic. At the same time, we still obtain many of the usual market anomalies described by the experimental studies.

We discuss in the following subsections a few such realistic features.

4.1 Return Autocorrelations: Momentum and Mean-Reversion

In the heterogeneous population LLS model trends are generated by the same positive feedback mechanism that generated cycles in the homogeneous case (Sect. 2): high (low) returns tend to make the extrapolating investors more (less) aggressive, this generates more high (low) returns, etc.

The difference between the two cases is that in the heterogeneous case there is a very complicated interaction between all the different investor species and as a result there are no distinct regular cycles but rather, smoother and more irregular trends. There is no single cycle length – the dynamics is a combination of many different cycles corresponding to the many extrapolation ranges k. This makes the autocorrelation pattern also smoother and more continuous. The return autocorrelations in the heterogeneous LLS model conform to the empirical findings: In the short run the autocorrelation is positive – this is the empirically documented phenomenon known as momentum: high returns during a trading quarter tend to be followed by more high returns in the following months, (and low returns tend to be followed by more low returns). In the longer run the autocorrelation is negative (after a few years of boom, one usually experiences a few "dry" years), which is known as mean-reversion. For even longer lags the autocorrelation eventually tends to zero [1]. The short run momentum, longer run mean-reversion, and eventual diminishing autocorrelation create the general "U-shape" found in empirical studies [7,28,29].

4.2 Excess Volatility

In markets with a large fundamentalist population (see [1] for their detailed operative definition in the LLS model), the price level is generally determined by the fundamental value of the stock. However, the market extrapolating investors occasionally induce temporary departures of the price away from the fundamental value. These temporary departures from the fundamental value make the price more volatile than the fundamental value.

Following Shiller's [8] methodology we measured the standard deviations of the detrended price and fundamental value. Averaging over 100 independent simulations, we found a standard deviation of 19.2% for the fundamental value, and 27.1% for the price, which is an excess volatility of 41%.

4.3 Heavy Trading Volume

In an LLS market with both fundamentalists and market extrapolating investors (over various k ranges), shares change hands continuously between the various groups. When a "boom" starts, the extrapolating investors observe higher ex-post returns and become more optimistic, while the fundamentalists view the stock as becoming overpriced and become more pessimistic. Thus, at this stage the market extrapolators buy most of the shares from the fundamentalists.

When the stock crashes, the opposite is true: the extrapolators are very pessimistic, but the fundamentalists buy the stock once it falls below the fundamental value. Thus, there is substantial trading volume in this market. The average trading volume in a typical LLS simulation was about 1,000 shares per period, or about 10% of the total outstanding shares.

4.4 Volume is Positively Correlated with Absolute Returns

The typical scenario in an LLS run is that when a positive trend is induced by the extrapolating investors, the opinions of the fundamentalists and the extrapolating investors change in opposite directions:

- the extrapolating investors see a trend of rising prices as a positive indication about the future return distribution, while
- the fundamentalists believe that the higher the price level is (the more overpriced the stock is), the harder it will eventually fall.

The exact opposite holds for a trend of falling prices. Thus, price trends are typically interpreted differently by the two investor types, and therefore induce heavy trading volume. The more pronounced the trend (large price changes), the more likely it is to lead to heavy volume.

In order to verify this relationship quantitatively we regressed volume V_t on the absolute returns for 100 independent simulations. We ran the regressions:

$$V_t = a + b|H_t| + \epsilon_t. \tag{5}$$

We found an average value of 870 for b with an average t-value of 5.0. Similar results were obtained for time lagged-returns.

The fact that the model leads to the phenomena of excess volatility, a U-shape return autocorrelation pattern, heavy volume, and correlation between volume and returns, all of which are empirically observed, is very encouraging. The parameters of the model, e.g. the number of investors of each type, can be calibrated to fit the empirical data quantitatively. Alternatively, one can endogenize these parameters by introducing learning and migration of investors to successful strategies. Both approaches are currently under investigation.

5 The Emergence of Pareto's Law in LLS

The efficient market hypothesis and the Pareto law are some of the most striking and basic concepts in economic thinking. Our analysis shows that these two fundamental concepts are closely related to one another.

More than a hundred years ago, Pareto [14] discovered that the number of individuals with wealth (or incomes) with a certain value w is proportional to $w^{-1-\alpha}$. This later became known as the Pareto Law, and was empirically verified for different countries and different time periods [30,31]. The LLS model treats the individual investor's wealth as a crucial quantity, and it views its feedback relation to the market dynamics as the main source driving the endogenous dynamics of the market.

It turns out that in the conditions in which the participants in the market do not have a systematic advantage one over the other (as expected in an efficient market), a dynamics of the LLS type leads to a Pareto law. The actual value of the exponent α depends on the particular parameters used in the model. Mainly, as explained below, α is determined by the minimal wealth amount of wealth required in order to participate in the market. If individuals participate in the market only if their wealth exceeds a certain minimal threshold value, given by a certain fraction c of the current average wealth, then, for a wide range of conditions we obtain $\alpha = 1/(1-c)$ [12]. This is confirmed in Fig. 7 which plots the wealth distribution in the LLS model with $k = 3$ and $c = 0.2$.

An insight for the deep relationship between market efficiency and the Pareto law can be obtained by considering an analogy between financial mar-

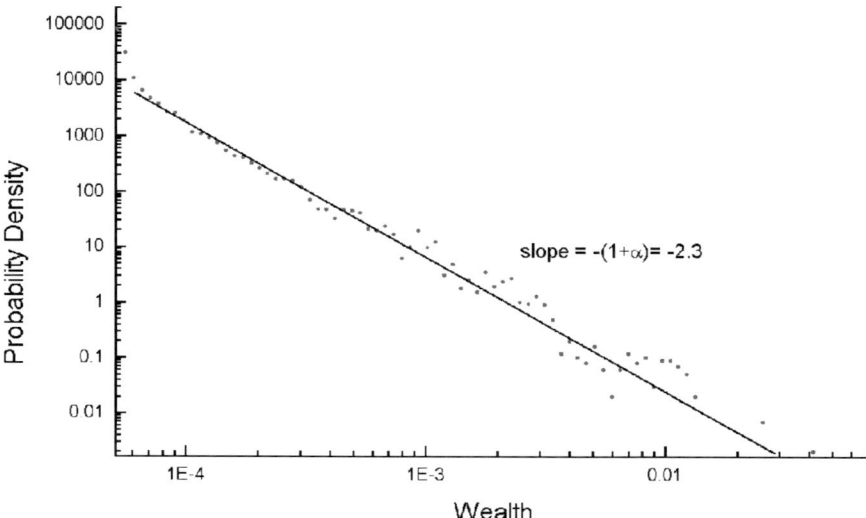

Fig. 7. The wealth distribution of the investors in an LLS model with a poverty line of $c = 20\%$ of the average wealth. On a double logarithmic scale one obtains a straight line with slope 2.2 corresponding to an α of 1.2. The market consisted of 10,000 traders and the measurement was performed as a "snapshot" after 1,000,000 "thermalization" market steps. Initially all the traders had equal wealth ($1000) equally distributed between bond and stock

ket dynamics and statistical mechanics. As the wealth dynamics in financial markets is multiplicative by nature, one can think of the following very simplified dynamics as capturing the essential feature of wealth accumulation through capital investment:

$$\bar{w}_i(t+1) = w_i(t)\bar{\lambda}_i(t), \qquad (6)$$

where $\bar{\lambda}_i(t)$ is a stochastic total return. In an efficient market we would not expect to have investors who obtain abnormal returns, i.e. draw λ's from distributions superior to those of the other investors in the market. It turns out that a system with the same distribution of λ for all investors is formally equivalent to a statistical mechanical (additive) system in thermal equilibrium, when expressed in terms of the variables $u_i(t) \equiv \ln w_i(t)/\bar{w}(t)$, and is mapped into a system of particles diffusing in an energy potential field u with a ground level $u_0 = \ln c$. In thermal equilibrium, all such systems (independent on the details of the interactions between their particles) have a universal probability density distribution discovered by Boltzmann more than 100 years ago:

$$P(u) \sim \exp(-\alpha u). \qquad (7)$$

When re-expressed in terms of the original w_i variables, this gives a Pareto power law distribution:

$$P(w) \sim w^{-1-\alpha}, \tag{8}$$

(see [10,12]). Thus, the Pareto wealth distribution is an almost inevitable result of stochastic wealth accumulation processes via capital investment, and indeed it is observed in the LLS model (see Fig. 7).

6 Market Efficiency, Pareto Law and Thermal Equilibrium

The formal equivalence between the non-stationary systems of interacting w_i's and the equilibrium statistical mechanics systems governed by the universal Boltzmann distribution has far reaching implications: it relates the Pareto distribution to the efficient market hypothesis: In order to obtain a Pareto power law wealth distribution it is necessary and sufficient that the returns of all the strategies practiced in the market are stochastically the same, i.e. there are no investors that can obtain "abnormal" returns.

Therefore, the presence of a Pareto wealth distribution is an indication of the market efficiency in analogy to the Boltzmann distribution whose presence is an indication of thermal equilibrium. Indeed physical systems which are *not* in thermal equilibrium (e.g. are forced by some external field – say by laser pumping) do *not* fulfill the Boltzmann law. Similarly, markets that are not efficient (e.g. when some groups of investors reap abnormal returns systematically) do not yield power laws (see Fig. 8). Efficient markets and power laws are the short time and long time sides of the same coin/phenomenon.

This analogy is consistent with the interpretation of market efficiency as analog to the Second law of Thermodynamics as follows.
First:

- One can extract energy (only) from systems that are not in thermal equilibrium.
- One can extract abnormal returns (only) from markets that are not efficient.

Also:

- By extracting energy from a non-equilibrium thermal system one gets it closer to an equilibrium one.
- By extracting abnormal returns from a non-efficient market one brings it closer to an efficient one.

And:

- In the process of approaching thermal equilibrium, one also approaches the Boltzmann energy distribution.

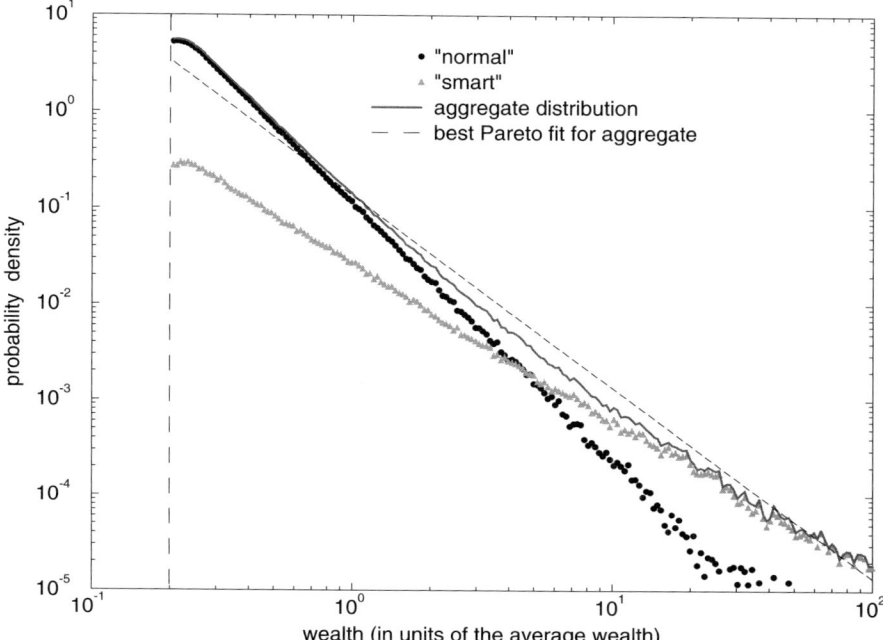

Fig. 8. Wealth distribution for two investor species with different return distributions. Model 3 was used with a lower wealth bound of $c = 20\%$ and λ was randomly drawn. For the first species λ is 1.10 or 0.95 with equal probability. For the second "more talented" species λ is 1.11 or 0.96 with equal probability. The two species were each composed of 10,000 traders with initially equal wealth (1000 dollars each). The measurement of the wealth distribution was performed after a "thermalization period" of 100,000 wealth updatings

- In the process of approaching the efficient market one also approaches the Pareto wealth distribution.

And, finally:

- By having additional knowledge on a thermodynamic system state one can extract energy (e.g. Maxwell demon Gedanken experiment).
- By having additional knowledge on a financial system one can extract abnormal returns.

Thus there is a double analogy:

$$\text{thermodynamic equilibrium} \sim \text{efficient market}$$
$$\text{Boltzmann law} \sim \text{Pareto law.}$$

This analogy also holds in the details of the microscopic interaction:

- The convergence to statistical mechanics equilibrium depends on the balance of the probability flow entering and exiting each energy level. This is usually insured microscopically by the fact that the a priori probability for a molecule to gain or loose an energy quanta in a collision is the same for any energy level with the exception of the collisions including molecules in the ground state which can only receive (but not give) energy.
- In stochastic wealth accumulation models, the convergence of the wealth distribution to the power-law is insured by the balance of flow of investors from one level of [log (relative wealth)] to another. At the individual level, this is enforced by all the individuals having the same distribution of returns (except for the individuals possessing the lowest allowed wealth). If this condition is not fulfilled, one does not get a wealth distribution power law.

These facts should guide us in the practical runs in establishing which combinations of strategies (or mechanism of strategy selection) are producing a realistic market "in the Pareto sense". In Fig. 8 one sees the wealth distribution in a model in which there are two trader species with slightly different return distributions. One sees that even a small difference in the return distributions leads to significant departures from the Pareto law which are inconsistent with the empirical observations. The absence of such departures in real life is an indication of market efficiency.

7 Leptokurtic Market Returns in LLS

It has been long known that the distribution of stock returns is leptokurtic or "fat-tailed". Furthermore, a specific functional form has been suggested for the short-term return distribution (at least in a certain finite range) – the Lévy distribution [17]. This feature is present in the LLS model, and is directly related to the Pareto distribution of wealth.

The central limit theorem insures that in a wide range of conditions the distance reached by a random walk of t steps of average squared size s^2 is a Gaussian with standard deviation $s\sqrt{t}$. Suppose that at time $t = 0$ one has N positive numbers $w_i(0)$; $i = 1, \cdots, N$ of order 1 and sum $W(0)$. Suppose that at each time step one of the numbers varies (increases or decreases) by a fraction $s_i(t) \ll 1$ extracted from a random distribution with average squared s^2 (and 0 mean). What will be the probability distribution of the sum $W(t)$ after t steps? According to the central limit theorem this would be the Gaussian:

$$P(W,t) = 1/(\sqrt{2\pi t s^2})e^{-(W(t)-W(0))^2/2ts^2}, \tag{9}$$

since it consists of t steps of average squared size s^2.

If one interprets $w_i(t)$ as the value of the stocks owned by the trader i at time t, then $W(t) = \sum_i w_i$ is the total market value of the stock and therefore $(W(t) - W(0))/W(0)$ is the rate of return on the stock for the time interval t.

One sees that if the central limit theorem would hold, one would predict a Gaussian stock returns distribution. This is in fact the case for real stocks and time intervals longer than a few weeks. For significantly shorter times t however, the distribution of returns is very different from a Gaussian. Even though there is no clear agreement as to the exact shape of the empirical return distribution, it is generally agreed that in certain ranges (typically "in the tails" – i.e. for large w_i values) this distribution is better fitted by a power law, rather than a Gaussian.

Such a situation can in principle be explained by the following scenario. Suppose that at time $t = 0$ one has an arbitrarily large number of positive numbers $w_i(0)$. Suppose moreover that the probability distribution for the sizes of $w_i(0)$ is:

$$P(w) \sim w^{-1-\alpha}. \tag{10}$$

Suppose that at each time step one of the w_i's varies (increases or decreases) by a fraction $s_i(t) \ll 1$ of average squared size s^2. What will be the probability distribution of the variation of the w_i's sum $W(t) - W(0)$ after t steps?

One is tempted to think that the correct answer is given by

$$P(W, t) = 1/(\sqrt{2\pi t s^2}) e^{-(W(t)-W(0))^2/2ts^2}$$

for some s. However this is wrong. Indeed, assuming such an s exists would imply that the probability for the sum variation $W(t) - W(0)$ to be 10 after a time $t = 1/(2s^2)$ is:

$$P(W(t) = W(0) + 10, t = 1/(2s^2)) \sim e^{-10^2} \sim 10^{-32}, \tag{11}$$

while in reality a lower bound for the probability of getting $W(t) - W(0) = 10$ in just one step it is obviously that given by:

$$P(w) \sim w^{-1-\alpha}, \tag{12}$$

i.e. $P(W(t) = W(0) + 10, t = 1/(2s^2))$ is at least of order $10^{-1-\alpha}$ which for $\alpha < 2$ means it is larger than 10^{-3} !

This coarse estimations highlights the difference between the Gaussian distributions and the distributions generated by random walks with power distributed step sizes (called Lévy distributions [17,18]): the presence of w_i's of arbitrary size implied by a power law distribution insures that the large returns distribution is dominated by the power law of the individual step sizes rather than the combinatorics of the multiple events characterizing the Gaussian system.

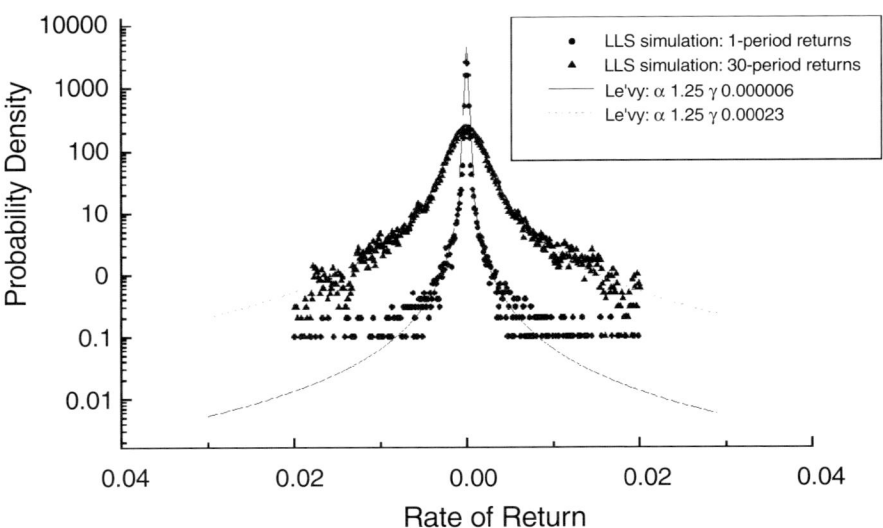

Fig. 9. The returns distribution in the LLS model in which only one trader reevaluates his/her portfolio per unit time. $c = 0.2$, $k = 3$, $U = \ln W$. The market contained 10,000 traders with initially equal wealth and portfolio composition (half in stock and half in bonds). The number of market returns in intervals of 0.001 were measured during 5,000,000 market steps (after an initial equilibration period of 1,000,000 steps)

One sees now that systems like LLS are exactly of the type one needs to explain the power tails of returns distributions:

- On the one hand according to Sect. 5, these models insure a power distribution of w_i's.
- On the other hand, the variation of the stock index $W(t)$ is the sum of the variations of the individual $w_i(t)$'s.
- These variations $w_i(t+1) - w_i(t)$ are stochastic fractions $s_i(t) = \lambda_i(t) - 1$ of w_i as above (the fact that $\lambda_i(t) - 1$ has not 0 mean is taken care by working actually with $u_i = \ln(w_i/\bar{w})$).

Therefore, according to the argument above, stochastic proportionality in LLS between individual wealth, individual investments and individual gains/losses predict that the price fluctuations in the LLS model will obey a Lévy distribution (and in particular fit a power in some range of the "tail").

There is a proviso for this argument to hold: the number of individual terms N has to be larger than the number of time steps t. Otherwise the finite size of the sample of w_i's will show up in the absence of sizes w_i larger than a certain value. In fact for $t \sim N$ one recovers (slowly) the Gaussian distribution.

In the LLS case, if the portfolio updatings are performed simultaneously by all the investors, the unit time step corresponds already to a time $t = N$. In order to verify the (truncated) Lévy distribution and the power "tail" predictions, one has to look at the dynamics at a finer time scale. We therefore performed [12] LLS runs in which at each time step only one trader i reconsiders its portfolio investment proportion x_i. In such conditions, one expects to obtain a distribution which fits in a significant range a power law (up to large w_i values where the finite N effects become important).

This is in fact confirmed by the numerical experiments. While for the global updating steps one gets a Gaussian distribution, for the trader-by-trader procedure one obtains a truncated Lévy distribution (Fig. 9).

Note that in the central region of the short time returns (before the cut-off becomes relevant) the Lévy distribution is characterized by an α equal to the exponent α of the traders' wealth distribution.

8 Summary

We have reviewed some of the main aspects of the LLS microscopic simulation stock market model. This model captures the complex interaction between different investor types, and the two-way relationship between price dynamics and wealth dynamics. The model generates different "ecological" dynamics of investor types, and also produces some of the basic phenomena observed in actual financial markets, such as booms and crashes, excess volatility, heavy volume, and the "U-shape" pattern of return autocorrelations. Finally, we discussed the deep relationship between market efficiency, the Pareto wealth distribution, and the Lévy distribution of stock returns, which are all observed in the LLS dynamics.

Acknowledgement

We thank T. Lux and D. Stauffer for very intensive and detailed correspondence on various simulation experiments using the LLS model.

References

1. M. Levy, H. Levy, S. Solomon: *Microscopic Simulation of Financial Markets* (Academic Press, New York 2000)
2. M. Levy, H. Levy, S. Solomon: Economics Letters, **45**, 103 (1994); S. Moss de Oliveira, H. de Oliveira, D. Stauffer: *Evolution, Money, War and Computers* (B.G. Teubner, Stuttgart-Leipzig 1999); S. Solomon: 'The microscopic representation of complex macroscopic phenomena'. In: *Annual Reviews of Computational Physics II*, ed. by D. Stauffer (World Scientific, Singapore 1995) pp. 243-294; S. Solomon: Comp. Phys. Comm. **121-122**, 161 (1999)

3. T. Hellthaler: Int. J. Mod. Phys. C **6**, 845 (1995); R. Kohl: Int. J. Mod. Phys. C **8**, 1309 (1997)
4. M. Levy, H. Levy, S. Solomon: J. de Phys. I **5**, 1087 (1995)
5. P.W. Anderson, J. Arrow, D. Pines: eds., *The Economy as an Evolving Complex System* (Addison-Wesley, Redwood City 1988)
6. E. Egenter, T. Lux, D. Stauffer: Physica A **268**, 250 (1999)
7. E. Fama, K. French: J. Pol. Econ. **96**, 246 (1988)
8. R.J. Shiller: American Economic Review, **71**, 421 (1981)
9. M. Levy, N. Persky, S. Solomon: Int. J. High Speed Comput. **8**, 93 (1996); see also J.D. Farmer: 'Market Force, Ecology and Evolution', e-print adap-org/9812005; and J.D. Farmer, S. Joshi: 'Market Evolution Toward Marginal Efficiency', SFI report 1999
10. M. Levy, S. Solomon: Physica A **242**, 90 (1997); M. Levy: 'Are Rich People Smarter?' UCLA, Working Paper, 1997; M. Levy: 'Wealth Inequality and the Distribution of Stock Returns' Hebrew University, Working Paper, 1999
11. S. Solomon, M. Levy: Int. J. Mod. Phys. C **7**, 745 (1996); S. Solomon: 'Stochastic Lotka-Volterra systems of competing auto-catalytic agents lead generically to truncated Pareto power wealth distribution, truncated Levy distribution of market returns, clustered volatility, booms and crashes'. In: *Computational Finance 97*, ed. by A-P.N. Refenes, A.N. Burgess, J.E. Moody (Kluwer Academic Publishers, Dordrecht 1998); O. Biham, O. Malcai, M. Levy, S. Solomon, Phys Rev E **58**, 1352 (1998)
12. M. Levy, S. Solomon: Int. J. Mod. Phys. C **7**, 595 (1996)
13. A. Blank, S. Solomon: 'Power Laws and Cities Population', cond-mat/0003240
14. V. Pareto: *Cours d'Economie Politique, Vol. 2* (Rouge, Lausanne 1897)
15. S. Solomon: 'Generalized Lotka-Volterra (GLV) Models and Generic Emergence of Scaling Laws in Stock Markets', in: *Applications of Simulation to Social Sciences*, ed. by G. Ballot and G. Weisbuch (Hermes Science Publications, Paris 2000)
16. O. Malcai, O. Biham, S. Solomon: Phys. Rev. E **60**, 1299 (1999)
17. B.B. Mandelbrot: Comptes Rendus **232**, 1638 (1951); H.A. Simon, C.P. Bonini: Amer. Econ. Rev. **48**, 607 (1958); R.N. Mantegna: Physica A **179**, 232 (1991); R.N. Mantegna, H.E. Stanley: Nature **376**, 46 (1995)
18. P. Lévy: *Théorie de l'Addition des Variables Aléatoires* (Gauthier-Villiers, Paris 1937)
19. A.J. Lotka: *Elements of Physical Biology* (Williams and Wilkins, Baltimore 1925); V. Volterra: Nature **118**, 558 (1926)
20. M.J. Gordon, G.E. Paradis, C.H. Rorke: American Economic Review **62**, 107 (1972)
21. I. Friend, M.E. Blume: American Economic Review **65**, 900 (1975)
22. H. Levy: J. Risk Uncert. **8**, 289 (1994)
23. D. Kahneman, A. Tversky: Econometrica **47**, 263 (1979)
24. W.A. Brock, C.H. Hommes: Econometrica **65**, 1059 (1997)
25. C. Chiarella, X.Z. He: Computational Economics **19**, 95 (2002)
26. E. Zschischang, T. Lux: Physica A **291**, 563 (2001)
27. W.B. Arthur, J.H. Holland, B. LeBaron, R.G. Palmer, P. Taylor: 'Asset Pricing Under Endogenous Expectations in an Artificial Stock Market', in *The Economy as an Evolving Complex System II*, ed. by W.B. Arthur, S. Durlauf, D. Lane (Addison-Wesley, Redwood City 1997)

28. J.M. Poterba, L.H. Summers: J. Fin. Econom. **22**, 27 (1988)
29. N. Jegadeesh, S. Titman: J. Fin. **48**, 65 (1993)
30. A.B. Atkinson, A.J. Harrison: *Distribution of Total Wealth in Britain* (Cambridge University Press, Cambridge 1978)
31. J. Steindl: *Random Processes and the Growth of Firms - A Study of the Pareto Law* (Charles Griffin & Company, London 1965)

Patterns, Trends and Predictions in Stock Market Indices and Foreign Currency Exchange Rates

Marcel Ausloos and Kristinka Ivanova

Specialized topics on financial data analysis from a numerical and physical point of view are discussed. They pertain to the analysis of crash prediction in stock market indices and to the persistence or not of coherent and random sequences in fluctuations of foreign exchange currency rates. A brief historical introduction to crashes is given, including recent observations on the DJIA and the S&P500. Daily data of the DAX index are specifically used for illustration. The method for visualizing the pattern thought to be the precursor signature of financial crashes is outlined. The log-periodicity of the pattern is investigated. Comparison of patterns before and after crash days is made through the power spectrum. The corresponding fractal dimension of the signal looks like that of a percolation backbone. Next the fluctuations of exchange rates (XR) of currencies forming EUR with respect to USD are analyzed. The XR power spectra are calculated before and after crashes. A detrended fluctuation analysis is performed. The characteristic exponents β and α respectively, are compared, including the time dependence of each α, found to be singular near crash dates.

1 An Introduction with Some Historical Notes as "Symptoms"

The stock market crash on Monday Oct. 19, 1987 led to the *October black Monday syndrome*. On that day, the Dow Jones Industrial Average (DJIA) lost 21.6 %. Other markets were shaken : the worst decline reached 45.8 % in Hong Kong. The downturn was spread out over two or three days in different European stock markets: the DAX lost 10 %. Nevertheless most markets had been using for a long time *breakers*, i.e. periods of trading halts and/or limitations of daily variations. This tends to suggest that the adoption of circuit breakers at the very least do delay the crash process, and not much more. In fact, Lauterbach and Ben-Zion [1] found that trading halts and price limits had no impact on the overall decline of October 1987, but merely smoothed return fluctuations in the neighborhood of the crash.

Another major characteristic of the crash of October 1987 is the phenomenon of irresistible contagion. It is well accepted that the shock arose first from Asian markets, except Japan, then propagated to the European markets before reaching the American markets after, when Asian markets were already closed [2]. A mapping of the Nikkei, DAX and DJIA daily sign fluctuations has been made onto a 1/2 Ising spin chain as if there was a continuous index calculated three times during 24 hours. This showed that the spin cluster fluctuations are rather equivalent to random fluctuations, except during pre-crash periods in which (*down-down-down*) spin clusters form with a higher probability than expected if the fluctuations are to be considered independent of each other [3]. This has allowed one to eliminate all criticism about the major responsibility of the derivative markets in the United States on the October 87 crash. Since then, the world-wide interdependence of the economy has been going on still more strongly. Thus if really a speculative bubble is occurring in some financial markets, as commonly observed in recent years, the phenomenon of propagation to other stock exchanges could be more important now. It is known that methods of negotiation have widely changed in many financial places. Markets using electronic systems of negotiation take advantage of recent improvements in their transaction capacity. It is easier today to face a substantial increase in transaction volume during a major crisis period. Moreover, the efficient use of the derivative markets could avoid useless pressures on the traditional market. These financial factual observations should be turned into quantitative measures, in order to avoid crashes if necessary.

Whence there is a need for techniques capable of rapidly following a bubble explosion or preventing it. Notice that the drop of stock market indices can not only spread out over two or three days but also over a much longer period. The example of the Tokyo stock exchange at the beginning of the 1990's is a prominent illustration. By comparison to the most famous crash of 1929, the Oct. 87 crash was spread over 2 days: the Dow Jones sank 12.8 % on October 28 and 11.7 % on the following day. (That was similar for the DAX which dropped by 8.00 % and 7.00 % on Oct. 26 and 28, 1987 respectively.) This shows that a stock market index decline does not necessarily lead to a crash in one single day. Indeed, the decline can be slow and last several days or even several months in what would be called not a crash, but a long duration bear market.

In the present econophysics research context, it is of interest to examine whether the evolutions of quotations on the main stock exchange places have similarities and whether crash symptoms can be found. Even if history generally tends to repeat itself, does it always do so in similar ways, and what are the differences? A rise in quotations can be interpreted *a posteriori* as the result of a speculative bubble but could be mere euphoria. How does this lead to a rupture of the trend? Can the duration differences be interpreted? Can we find universality classes?

Physics-like model of fracture or other phase transitions, including percolation can be turned into some economic advantage. Along the same lines of thought, the question was already touched upon in [4] within the sand pile model. This allows not only a verbal analogy of index rupture in terms of sand avalanches, but also some insight into the mechanisms. Through physical modeling and an understanding of parameters controlling the output, as in the sand pile model, symptoms can be measured, whence to suggest remedies is not impossible.

Another question raised below is the post-crash period. One might expect from a physics point of view that if a crash looks like a phase transition, and is characterized by scaling laws, as we will see it sometimes occurs [5,6], it might be expected that a relation exists between amplitudes and laws on both sides of the crash day [7]. As mentioned above, the crash might be occurring on various days, with different breaks. It might be possible that between drops some positive surge might be found. Thus some sorting of behaviors into classes should be made as well. *In fine*, some discussion on the foreign exchange market will be given in order to recall the detrended fluctuation analysis method, so often used nowadays. It is applied below to the USD exchange rate vs. the (ten) currencies forming the EUR on Jan. 01, 2000, over a time interval including the most recent years. The observation of the time variation of the power law scaling exponent of the DFA function is shown to be correlated to crash time occurrence.

1.1 Tulipomania

In 1559, the first tulip bulb (TB) was brought to Holland from China by Conrad Guenster [8]. In 1611, tulip bulbs (TBs) were stocked and sold on markets. In 1625, one tulip bulb was worth 5 dutch gulden (NLG). The flower was considered so rare that wealthy aristocrats and merchants tripped over themselves to buy the onions. Speculation ensued and the TBs became wildly overvalued. The TBs were not necessarily planted, but were just stored in the house salon. In 1635, 1 TB was worth 4 tons of wheat + 4 oxen + 8 tons rye + 8 pigs + one bed + 12 sheep + clothes + 2 wine casks + 4 tons beer + 2 tons butter + 1000 pounds cheese + 1 silver drinking cup. In 1637, 1 TB was worth 550 NLG. One average house was worth 17 800 NLG, whence about 30 TBs. However within 1637, over a 6 week time span the price of 1 TB went down 90 %.

In view of the shock, remedies had to be found and people called upon the Amsterdam Parliament for legislation. It was decided that all contracts would be void if they were dealt before Nov. 1636, and after that date the contracts kept a 10 % value. Under some protest, people appealed to the Netherlands Supreme Court which ruled that this business of selling/buying TBs was mere gambling, and no debt could be defined "by law" nor ruled upon. Nowadays one TB is worth 0.5 EUR. Too bad for long term investment strategies. The TB became the classical example for illustrating the *Extraordinary Popular Delusions and the Madness of Crowds* as described by Charles MacKay [9].

Just for the sake of physics history, let it be recalled that 1639 was the year in which Galileo Galilei (Pisa, Feb. 15, 1564; Arcetri, Jan. 8, 1642) betrayed science in saving his life.

1.2 Monopolymania

Another set of financial crises is that of the Compagnie du Mississippi [10] and that of the South Sea Company [11,12]. In 1715, John Law, a Scottish gambler, had persuaded Philippe, Regent of France, to consider a banking scheme that promised to improve the financial condition of the kingdom. In theory a private affair, the system was linked from the beginning with liquidating the national debt. When the monopoly of the Louisiana trade was surrendered in 1717, Law created a trading company known as the *Compagnie d'Occident* (or *Compagnie du Mississippi*) linked to the Royal Bank of France (first chartered in 1716 as Banque Générale) and in which government bills were accepted for the purchase of shares.

Law gained a monopoly on all French overseas trade. The result was a huge wave of speculation as the value of a share went from its initial value, i.e. 500 *livres* to 18 000 *livres*. When the paper money was presented at the bank in exchange for gold, which was unavailable, panic ensued, and shares felt by a factor of 2 in a matter of days.

In England, the Whigs represented the mercantile interests which had profited from the War of the Spanish Succession War (1703-1711), and made large profits by financing it, in doing so had created a National Debt which had to be financed by further taxation. During the wars the government handled more money than ever before in history, and they skimmed off a lot through various methods, including the invention of the Bank of England in 1694. The South Sea Company was formed in 1711 by the Tory government of Harley to trade with Spanish America, and to offset the financial support which the Bank of England had provided for previous Whig governments. They had in mind to establish a system like the Compagnie du Mississippi Monopoly, [10] using the same sort of trading privileges and monopolies, those granted to Britain after the Treaty of Utrecht. King George I of Great Britain became governor of the company in 1718, creating confidence in the enterprise, which was soon paying 100 percent interest. In 1720 a bill was passed enabling persons to whom the government owed portions of the national debt to exchange their claims for shares in company *stocks*, and to become *stock holders*. In the 1719-20 the England Public Debt went to South Sea Company stock holders, as approved by Parliament. On March 1 the stocks were valued GBP 175 and they moved quickly to GBP 200. Shortly the directors of the South Sea Company had assumed three-fifths of the national debt. The company expected to recoup itself from expanding trade, but chiefly from the foreseen rise in the value of its shares. On June 1 the shares were valued 500 and more than 1,000 in August 1720. Those unable to buy South Sea Company stocks were inveigled by overly optimistic company promoters or downright swindlers into unwise investments. Speculators took advantage of investors to obtain subscriptions for sensibly unrealistic projects. By September 1720 however, the market had collapsed, and by December 1720 South Sea Company shares were down to 124, dragging others, including government stocks with them. Many investors were ruined, and the House of Commons ordered an inquiry, which showed that ministers had accepted bribes and speculated. From a physics point of view let it be recalled that I. Newton (Woolsthorpe, Dec. 25, 1642; London, March 20, 1727) invested in such South Sea Company stocks and lost quite a bit of money [13].

1.3 WallStreetmania

In the years from 1925 to 1929 one could easily go to a broker and purchase stocks on margin, i.e. instead of buying stocks with real cash money, one could purchase them with some cash down and the rest on credit. The Coolidge administration had a *laissez-faire policy*, i.e. a government policy of non-intervention. It was almost *la façon de vivre* to play in the stock market [14]. That allowed a speculation bubble to grow unchecked. The Federal Reserve powers on economic matters were not utilized as could be done nowadays. Many successions of *mini* crashes and rallies began as early as March 1929. The summer of 1929 hearkened somewhat of the good old days of optimism. The market appeared to be stable. On Sept. 3, a bear market became firmly established, and on Thursday Oct. 24, 1929 the famous 1929 crash occurred (Fig. 1).

The 1987, 1997, and more recent 1999, 2000, 2001 crashes are reminiscent and even copies of the above ones (Fig. 1). The symptoms look similar: ar-

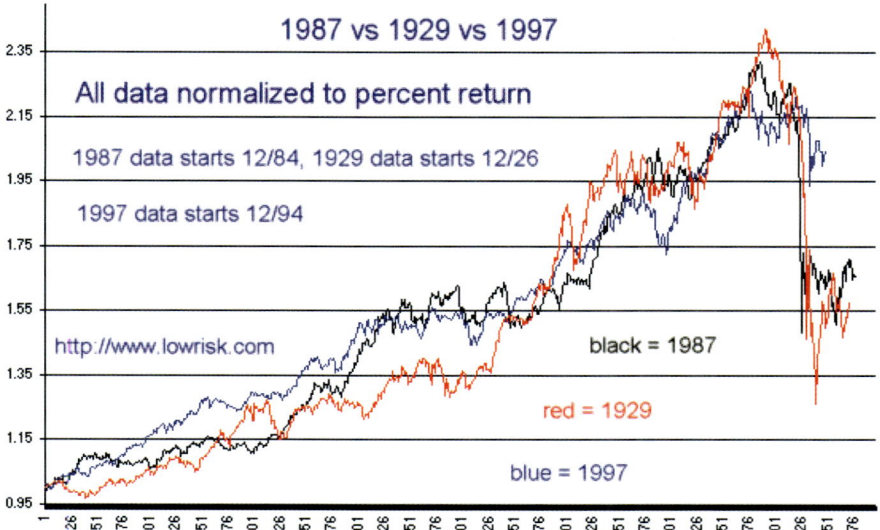

Fig. 1. Famous Wall Street Crashes : 1929, 1987, 1997; taken from color web site http://www.lowrisk.com/crash/87vs97.htm; from top to bottom on day 201 : blue (97), red (29), black (87); on day 700 : red (29), black(87), blue(97)

tificially built euphoria, malignantly established speculation, easy access to market activities, including manipulated (or rather *electronipulated*) informations ... Consider the buying frenzy on IPOs stocks at the end of the 1990's in companies for which owners do not have a coherent business plan, or are not going to make money, ... and yet see how we bought e-stocks. Nothing has changed since 1600, 1700 nor 1929.

2 Econophysics of Stock Market Indices

Econophysics [15,16] aims to fill the huge gap separating "empirical finance" and "econometric theories". Various subjects have been approached like the option pricing, stock market data analysis, market modeling and forecasting, etc. The application of statistical physics ideas to the forecasting of stock market behavior has been proposed earlier following the pioneer work of physicists interested by economic laws [47,18–23].

Even though a stock market crash is considered a highly unpredictable event, it should be reemphasized that it takes place systematically during a period of generalized anxiety spreading over the markets following a euphoria time. The crash can be seen as a natural correction bringing the market to a "normal state". Three important facts should be underlined:

(i) The series of daily fluctuations, so called *volatility*, of the stock market presents a huge clustering around the crash date, i.e. huge fluctuations are grouped around the crash date. This is well illustrated in Fig. 2 for the case of the DAX around 1987. The time span of this clustering is quite long: a few years. This clustering indicates that larger and larger fluctuations take place before crashes.

(ii) Collective effects are to be considered, be they stemming from macroeconomic information, as a set of "external fields", and leading to a bear market, or more intrinsically *self-organized*, as if microeconomic informations (or *interactions*?) were triggering the non-equilibrium state evolution.

(iii) A third remark concerns the panic-correlations appearing before crashes. This kind of collective behavior is commonly observed during a trading day. The market in Tokyo closes before London opens and thereafter New York opens. During periods of panic, financial analysts are looking for the results and evolution of the geographically preceding market. Strong correlations are found in fluctuations of different market indices before crashes.

Of course, fluctuations and correlations are both ingredients which are supposedly known to play an important role in phase transitions. Thus an analogy can be derived (Fig. 3) between phase transitions and crashes, defining the mean field (exponential-like) behavior, the time t^*, corresponding to the *Ginzburg – Levanyuk* temperature bounding the critical fluctuation region, the critical crash day t_c, etc. [6]. The character of a thermodynamic phase transition is characterized by critical exponents, following the scaling law hypothesis, exponents which are thought to depend on the symmetry of

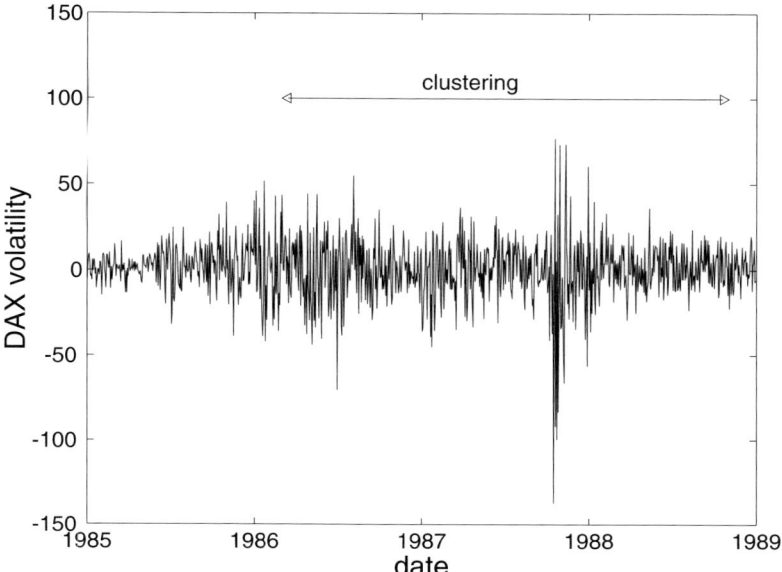

Fig. 2. DAX volatility between Jan. 01, 1985 and Dec. 31, 1988

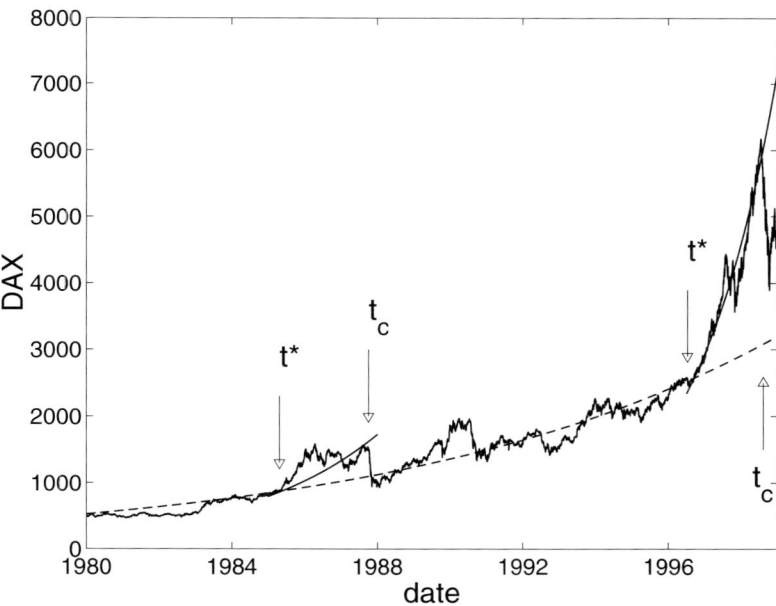

Fig. 3. DAX evolution between Jan. 01, 1980 and Dec. 31, 1998 with the mean field behavior, the time(s) t^*, corresponding to the Ginzburg-Levanyuk temperature bounding the critical fluctuation region, and the critical crash day(s) t_c

the order parameter and the underlying lattice dimensionality [14]. Similar considerations are looked for in financial crash studies.

In 1996, two independent works [25,26] have proposed that critical phenomena would be possible scenarios for describing crashes. These authors are still debating the subject [27,28]. More precisely, it has been proposed that an economic index $y(t)$ increases as a power law decorated with a log-periodic oscillation, i.e.

$$y = A + B\left(\frac{t_c - t}{t_c}\right)^{-m}\left[1 + C\sin\left(\omega\ln\left(\frac{t_c - t}{t_c}\right) + \phi\right)\right] \quad \text{for} \quad t < t_c \quad (1)$$

where t_c is the crash-time or rupture point, A, B, m, C, ω and ϕ are free parameters. This evolution $y(t)$ is in fact the real part $\Re$ of a power law behavior at $t = t_c$ with a complex exponent $m + i\omega$, i.e.

$$y \sim \Re\left\{\left(\frac{t_c - t}{t_c}\right)^{-m+i\omega}\right\}. \quad (2)$$

The law for $y(t)$ diverges at $t = t_c$ if $m > 0$. This evolution is decorated with oscillations converging at the rupture point t_c. This law is similar to that of critical points, and generalizes the situation for cases in which a hierarchical lattice structure exists, in other words a Discrete Scale Invariance (DSI) is subjacent [29].

The relationship (1) has been proposed elsewhere in order to fit experimental measurements of sound wave rate emissions prior to the rupture of heterogeneous composite stressed up to failure [30]. The same type of complex power law behavior has also been observed as a precursor of the Kobe earthquake in Japan [31]. Such log-periodic corrections have been recently reported in biased diffusion on random lattices [32].

As early as April 1997, Vandewalle and Ausloos performed a series of investigations in order to emphasize crash precursors [33–35]. The closing values of the Dow Jones Industrial Average (DJIA) and the Standard & Poor 500 (S&P500) were used for tests. A law slightly different from Eq. (1) was proposed [33,36]. A strong indication of a so-called crash event or market rupture point was numerically discovered [33–36]. Further data analysis (in Aug. 97) [37] including a risk measure [38] indicated a crash to occur in between the end of October 1997 and mid-November 1997. The crash occurred effectively on Monday October 27th, 1997 [36] !

Even though the crash of October 1997 was predicted [36,37], the scientific (physics or economy) [39,40] and media [35] community is actually divided between those who believe in such a crash prediction and those who believe that crashes are unpredictable events and such findings were mere luck or at best accidental [27,28,41]. We discuss a little bit more the predictability problem and findings in this paper going beyond a previous report [4].

2.1 Methodology and Data Analysis

In e.g. Refs. [4,42–44], the fact was underlined that there are strong physical arguments stipulating that m in Eq. (1) could be or even should be taken as "universal". The universal $m = 0$ value, i.e. a logarithmic divergence has been proposed. The logarithmic divergence of the index y for t close to t_c reads

$$y = A + B \ln\left(\frac{t_c - t}{t_c}\right)\left[1 + C\sin\left(\omega \ln\left(\frac{t_c - t}{t_c}\right) + \phi\right)\right] \quad \text{for} \quad t < t_c. \quad (3)$$

One should remark that the full period $[t_i, t_f]$ for a meaningful fit should contain the whole euphoric precursor. It has been found in [4,36,43,44] that the log-divergence is closer to the real signal than any power law divergence with $m \neq 0$.

The log-divergence in Eq. (3) contains 6 parameters. At first, it seems that non-linear fits using only the simple log-divergent function

$$y = A + B \ln(t_c - t) \quad (4)$$

with $B < 0$, thus with only 3 parameters can be performed. A good estimation of t_c can be obtained indeed following both Levenberg-Marquardt and Monte-Carlo algorithms [45]. One has observed that the estimated t_c points are close to "black" days for the first two periods [4,43,44].

Assuming that Eq. (4) is valid, one should also note that

$$\frac{dy}{dt} = \frac{-B}{(t_c - t)} \quad (5)$$

should be found in the daily fluctuation pattern (Fig. 2). This is consistent with the volatility clustering discussed here above. However, Eq. (5) fits lead to bad results with huge error bars.

The oscillating term of Eq. (3) has been quite criticized since no traditional or economical argument supports the DSI theory at this time. However, the hierarchical structure of the market has been suggested as a possible candidate for generating DSI patterns in [4,26,46,47], so is the price fixing "techniques" [48] and arbitrage methods. In order to prove that a log-periodic pattern appears before crashes, the *envelope* of the index y is constructed [42]. Two distinct curves are built: the upper envelope y_{max} and the lower one y_{min}. The former represents the maximum of y in an interval $[t_i, t]$ and the latter is the minimum of y in an interval $[t, t_f]$. One observes a remarkable pattern made of a succession of thin and huge peaks [42].

When $y_{max} - y_{min} = 0$, it means that the index y reaches some value never reached before at a time t and would never have reached if the time axis had been reversed thereafter. This corresponds to time intervals during which the value of the index y reaches new records. In fact, the pattern reflects obviously an oscillatory precursor of the crash, thus through

$$y_{max} - y_{min} = (C_1 + C_2 t)(1 - \cos(\omega \ln(t_c - t) + \phi)) \quad (6)$$

where C_1 and C_2 are parameters controlling the amplitude of the oscillations. The above relationship allows us to measure the log-frequency ω [42]. Moreover it is found that the value of ω seems to be finite and almost constant, $\omega \simeq [6, 10]$ for the major analyzed crashes. An analysis along similar lines of thought, though emphasizing the no-divergence, thus $m < 0$ in Eqs. (1)-(2), was discussed for the Nikkei [49,50] and NASDAQ April 2000 crash [51].

For illustrating the complexity of the frequency dependence of such financial signals, one can also perform a Fourier transform of a reconstructed signal. The evolution of the six strongest DAX crashes between Oct. 01, 1959 and Dec. 31, 1988 and the strongest DAX crashes in 1997, prior to the crash day, are shown in Fig. 4; y_0 denotes the index value at the closing of the crash day. Power spectra of the DAX index measured from the index value at the end of the crash day have been calculated for a time interval equal to 600 days, those prior to crashes. The 6 large DAX crash spectra for the period of interest are shown in Fig. 5. The corresponding exponents β for the best fit in the high frequency region are given in Table 1.

The roughness behavior [13,51,54] of the DAX index evolution signal before crashes can be defined trough the fractal dimension D of the

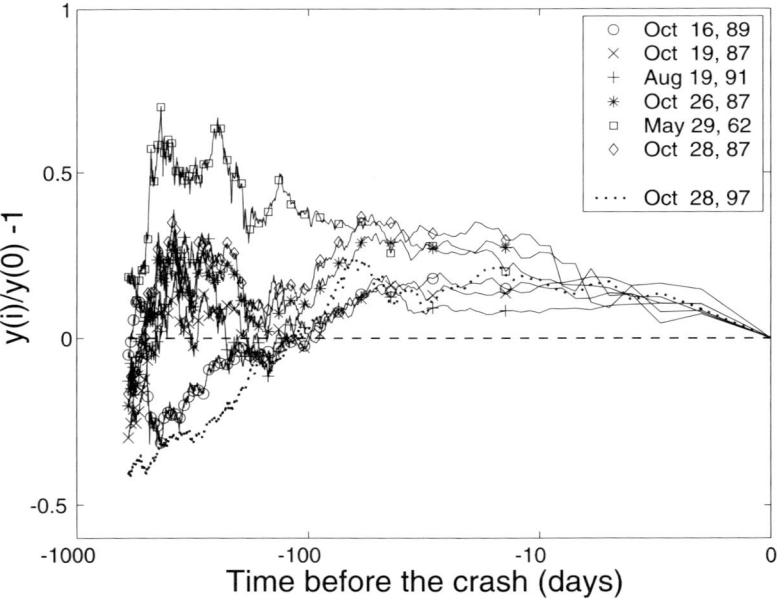

Fig. 4. The evolution of the six strongest DAX crashes between Oct. 01, 59 and Dec. 31, 88, and the strongest DAX crash in 1997, prior to the crash day ; y_0 denotes the index value at the closing of the crash day

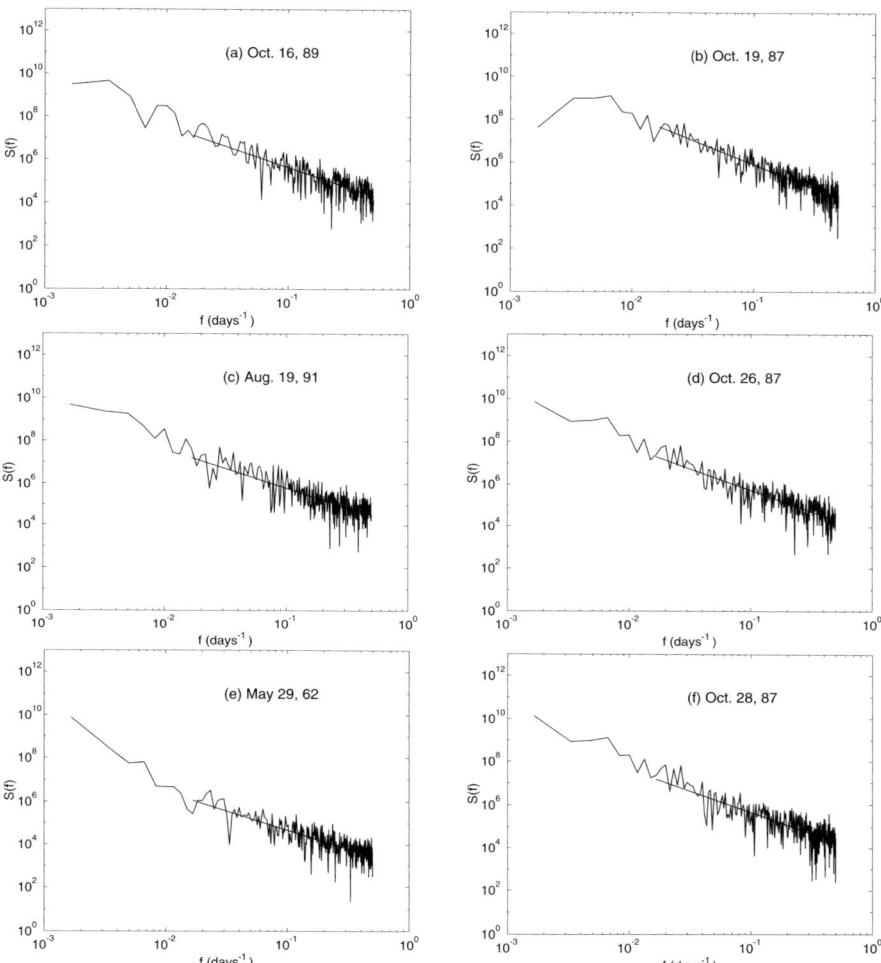

Fig. 5. Power spectrum of the 600 day DAX index evolution signal corresponding to the six major crashes between Oct. 01, 59 and Dec. 31, 96

signal, i.e. [13]

$$D = E + \frac{3-\beta}{2}, \tag{7}$$

where E is the Euclidean dimension. The values of β and D are reported in Table 1 with the crash dates and relative amplitude of the 6 major DAX crashes which occurred between Oct. 01, 1959 and Dec. 30, 1996. The same type of data is reported in Table 2 for the 3 major DAX crashes in October 1997. The power spectrum of the large Oct. 28, 97 crash is shown in Fig. 6.

Table 1. Crash dates and relative amplitude of the 6 major DAX crashes having occurred between Oct. 01, 1959 and Dec. 30, 1998; power law exponent β_- of the signal power spectrum (600 data points) and corresponding fractal dimension D_- of the signal prior to, and β_+ and D_+, after these 6 major DAX crashes.

crash dates	relative amplitude	β_-	D_-	β_+	D_+
16.10.89	-0.137	1.90±0.09	1.55	1.67±0.09	1.67
19.10.87	-0.099	2.24±0.09	1.38	1.82±0.09	1.59
19.08.91	-0.099	1.82±0.10	1.59	1.62±0.06	1.69
26.10.87	-0.080	2.00±0.09	1.50	1.70±0.08	1.65
29.05.62	-0.075	1.77±0.08	1.62	1.28±0.08	1.86
28.10.87	-0.070	1.94±0.10	1.53	1.81±0.07	1.60

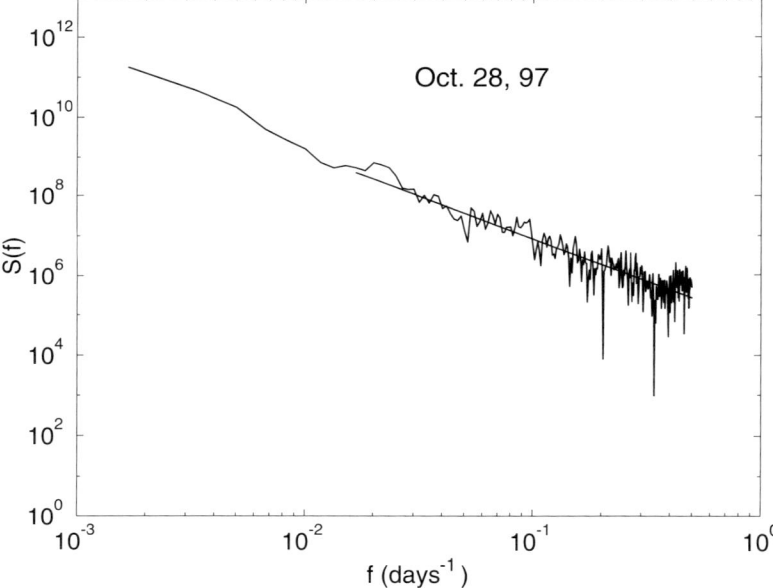

Fig. 6. Power spectrum of the 600 day DAX index evolution signal prior to the Oct. 28, 97 crash

2.2 Aftershock Patterns

The index evolution after a crash has also been analyzed through a reconstructed signal that is the difference between the DAX value signal on each day $y(i)$ and the DAX value on the crash day y_0. For the 6 largest crashes in the time interval of interest the recovery can be slow (Fig. 7). It took about

Table 2. The 3 major DAX crashes in October 1997: crash dates, relative amplitude, power law exponent β_- of the signal power spectrum (600 data points prior to these 3 major DAX crash dates) and the corresponding fractal dimension D_-.

crash dates	relative amplitude	β_-	D_-
28.10.97	-0.084	2.14±0.07	1.43
27.10.97	-0.043	1.97±0.07	1.52
23.10.97	-0.048	1.91±0.05	1.55

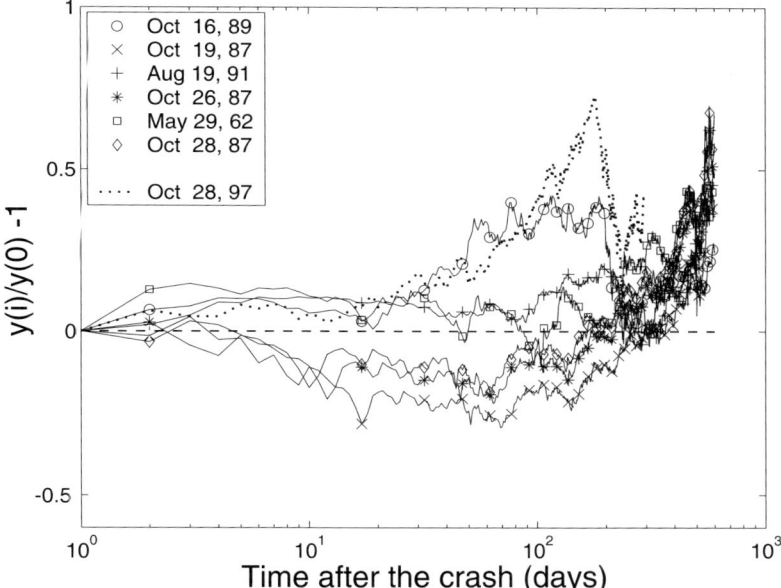

Fig. 7. The DAX recovery signal evolution after the six strongest DAX crashes having occurred between Jan. 01, 1985 and Dec. 31, 1988, and after the strongest DAX crash in 1997; y_0 denotes the index value at the closing of the crash day

one month to recover from the October 28, 1997 crash. To observe some periodic fluctuation after the crash, the power spectrum of the DAX has been computed for the 600 days following a crash day (Fig. 8 (a-f)). Note the high-frequency log-periodic oscillation regime of the power spectrum for the October 19, 1987 case on Fig. 8(b). The values of each β and corresponding fractal dimension D are reported in Table 1.

As a final point of this section, it should be noticed that the fractal dimension is close to 1.70, thus *very similar* to that of a percolation backbone. This might be a hint that hierarchical structures are present, and a cause of

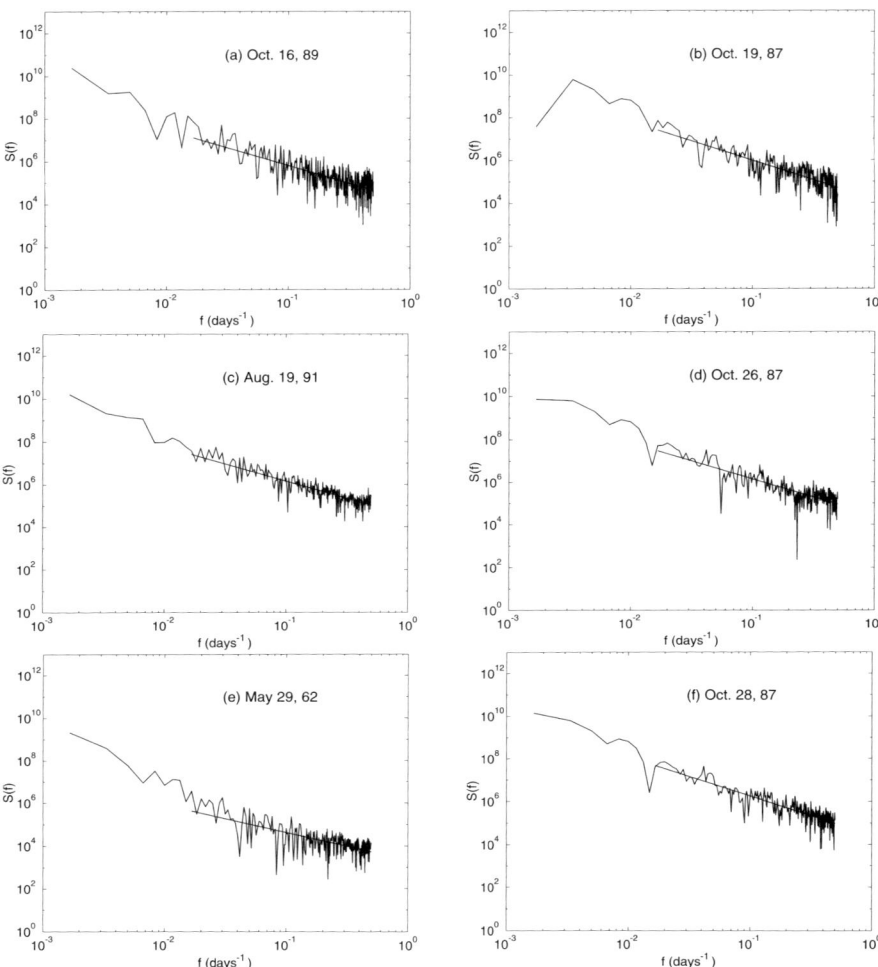

Fig. 8. Power spectrum of the 600 day DAX index evolution signal after the six major crashes having occurred between October 1, 1959 and December 31, 1996

crashes. As a consequence, the market could be viewed as a discrete fractal system, transiting at crashes like a physical system at a percolation transition. In related work, Amaral and coworkers [55] have studied the statistics of several companies as well as their respective growth. They have found that the growth of companies can be modeled using a hierarchical lattice like a Cayley tree. For simple models of hierarchically organized markets some self-regulation is found in fact [56]. On such systems the fractal dimension can be considered to have an imaginary part which is related to the log-periodic oscillations; in fact it is the signature of the branching ratio [57].

In conclusion of this section, we may conjecture that stock markets are also hierarchical objects where each level may have a different weight, connectivity, and characteristic time scales (the horizons of the investors) [4]. The hierarchical tree might be fractal at crashes and its geometry might control the type of criticality. This gives some argument in favor of the sand pile model on a fractal basis [58] as a microscopic model actually able to simulate a crash [4].

3 Foreign Currency Exchange Rates

Beside the crash cases discussed here above numerous examples of scale invariance seem to be widespread in natural and social systems [51,59]. A fundamental problem is the existence and width of the scaling range for long-range power-law correlations (LRPLC) in economic systems, as well as the presence of economic cycles. Indeed, traditional methods (like spectral methods) have corroborated the evidence that the Brownian motion idea or ordinary random walk is quite away from reality and LRPLC quite frequent [18,20,22,23]. Different approaches [54] have been envisaged to measure the LRPLC or analyze them in financial data: tails of partial distribution functions of the volatility, wavelet analysis, Detrended Fluctuation Analysis (DFA) [31], etc.

3.1 DFA Analysis

The DFA method [31] consists in dividing the whole data sequence $y(n)$ of length N into N/τ non overlapping boxes, each containing τ points. Then, the *local trend*

$$z(n) = an + b \tag{8}$$

in each box is defined to be the ordinate of a linear least-square fit of the data points in that box. One should remark that a trend $z(n)$ different from a first-degree polynomial can also be used like the cubic trend [70]. Other detrending functions may improve the accuracy of the DFA technique, sort out the reason for crossovers between scaling regimes, and pinpoint noise and intrinsic trends [61].

The so-defined detrended fluctuation function $F(\tau)$ is then calculated following

$$F(\tau)^2 = \frac{1}{\tau} \sum_{n=kt+1}^{(k+1)\tau} |y(n) - z(n)|^2, \qquad k = 0, 1, 2, \cdots, \left(\frac{N}{\tau} - 1\right). \tag{9}$$

Averaging $F(\tau)$ over the N/τ intervals gives a function depending on the box size τ. The above calculation is repeated for different box sizes τ. If the $y(n)$

data are randomly uncorrelated variables or short range correlated variables, the behavior is expected to be a power law

$$\langle F(\tau)^2 \rangle^{1/2} \sim \tau^\alpha \tag{10}$$

with an exponent $1/2$ [31] if the excursion is governed by a mere random walk. An exponent $\alpha \neq 1/2$ in a certain range of τ values implies the existence of LRPLC in that time interval. Mathematically, the correlation of a future increment $y(n) - y(0)$ with a past increment $y(0) - y(-n)$ is given by

$$\Gamma(n) = \frac{\langle (y(0) - y(-n))(y(n) - y(0)) \rangle}{\langle (y(n) - y(0))^2 \rangle} = 2^{2\alpha - 1} - 1, \tag{11}$$

where the correlations are normalized by the variance of $y(n)$. For $\alpha > 1/2$, there is *persistence*, i.e. $\Gamma > 0$. In this case, if in the immediate past the signal has a positive increment, then on the average an increase of the signal in the immediate future is expected. In other words, persistent stochastic processes exhibit rather clear trends with relatively little noise. An exponent $\alpha < 1/2$ means *antipersistence*, i.e. $\Gamma < 0$. In this case, an increasing value in the immediate past implies a decreasing signal in the immediate future, while a decreasing signal in the immediate past makes an increasing signal in the future probable. In so doing, data records with $\alpha < 1/2$ appear very *noisy* (rough). The $\alpha = 0$ situation corresponds to the so-called *white noise*. Finally, one should note that α is nothing else than Ha, the so-called Hausdorff exponent for fractional Brownian motion [51,54]. It can be useful to recall [54] that the power spectrum of such random signals is characterized by a power law with an exponent $\beta = 2\alpha - 1$.

3.2 Data and Analysis

We have considered the daily evolution of several currency exchange rates with respect to the USD from January 1990 till December 1999 including only all open banking days. This represents about $N = 3000$ data points. The data are those obtained from [63], at the closing time of the foreign exchange market in London for the ten currencies C_i, (i=1,10) forming the EUR on Jan. 01, 1999.

The evolution of such C_i/USD exchange rate from January 1, 1993 to June 30, 2000 is drawn in Fig. 9. In Fig. 10, a log-log plot of the 10 functions $\langle F(t)^2 \rangle^{1/2}$ is shown for the whole data of Fig. 9. Moreover we plot the result for a false EUR, i.e. a linear combination of the ten currencies forming the EUR [64–66,60] Except for IEP, the functions are very close to a power law with an exponent $\alpha = 0.51 \pm 0.02$ holding over two decades in time, i.e. from about one week to two years. This finding clearly shows the non-existence of LRPLC in the foreign exchange market with respect to the USD. Other cases showing marked deviations from Brownian motion have been discussed elsewhere [65,60,58,69]. It can then be observed that a wide variety

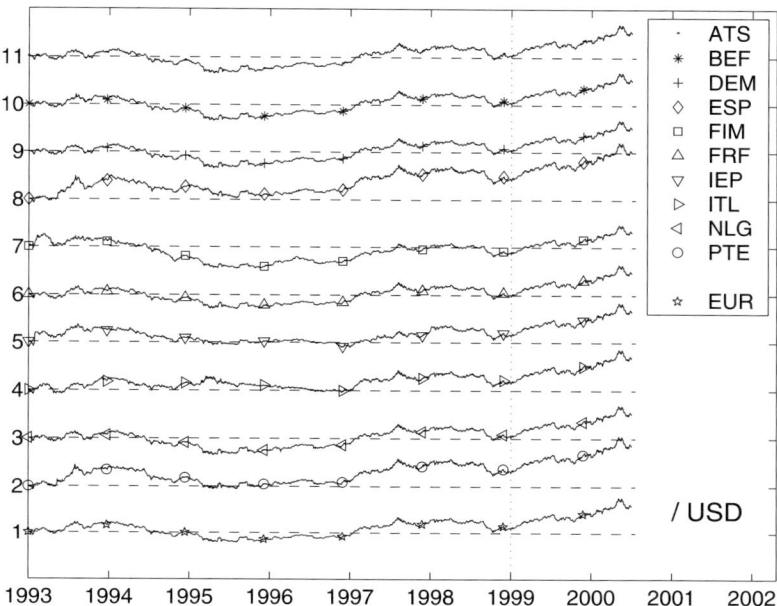

Fig. 9. Normalized EUR and C_i (i=1,10) currency forming the EUR exchange rates with respect to the USD between Jan. 01, 1993 and June 30, 2000. The data are artificially multiplied by two and then displaced along the vertical axis in order to make the fluctuations noticeable. The vertical dash line marks the date for the EUR official introduction

of behaviors is found in the foreign currency exchange market. Exponent values and the range over which a power law holds drastically vary from a currency exchange rate to another. It appears that the currency exchange rates can be classified into three different categories from the LRPLC point of view.

First, consider the rates which exhibit an exponent α larger than $1/2$ (persistent behavior). This case corresponds to currency exchange rates between leading currencies (e.g., USD, JPY, EUR) and so called *weaker* ones [69,70].

A second category concerns the rates exhibiting strict randomness ($\alpha = 1/2$) within error bars. This is the case for example of the USD/C_i rates as shown above.

A third category represents the currency exchange rates with antipersistent behavior ($\alpha < 1/2$) as e.g. DEM/BEF [58]. These currencies most often concern currency exchange rates between (European) countries which are submitted to strict monetary rules and to strict regulatory corrections by central banks due to international multilateral conventions. It should be pointed out that in general the range, over which the antipersistency signa-

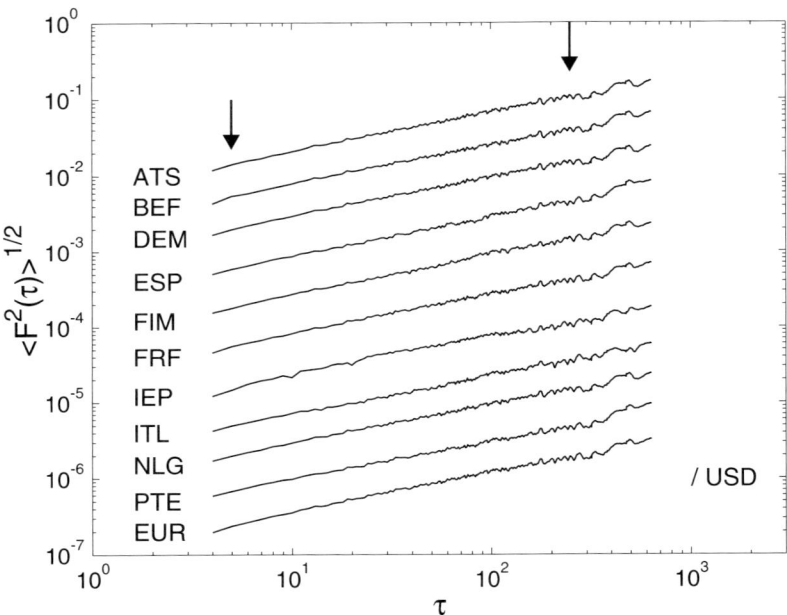

Fig. 10. Log-log plot of the DFA function showing how to obtain the α exponent for the 11 exchange rates of interest for EUR/USD. Since only the fitted slope is of interest, the DFA function data has been arbitrarily displaced along the vertical axis

ture, i.e. the power law is valid, occurs over a limited time span in this third category. In fact, there is a crossover around $\tau^* \approx 10$ weeks. For longer time scales ($\tau >> \tau^*$), the signal becomes again persistent or random.

3.3 Probing the Local Correlations

It is also of interest to know whether the LRPLC are stable along the data. In order to probe the local strength of the correlations, one constructs a so-called observation box of width T placed at the beginning of the data, and calculates α for the data contained in that box. Then, the box is moved along the data by some step toward the right along the financial sequence and α is again calculated. Iterating this procedure one obtains a "local measurement" of the degree of "long-range correlations" over T. It is crucial to choose the most adequate box size T, i.e. to choose T of the same order of magnitude as the maximum range τ over which the above power law is valid.

The evolution of the EUR/USD and C_i/USD α's for the 1995-2000 period is illustrated in Fig. 11. In order to probe the local values of α, we have used a window of size $T = 2$ years. The exponent α varies around $1/2$, i.e. the horizontal dashed line in Fig. 10. The local value of α seems to

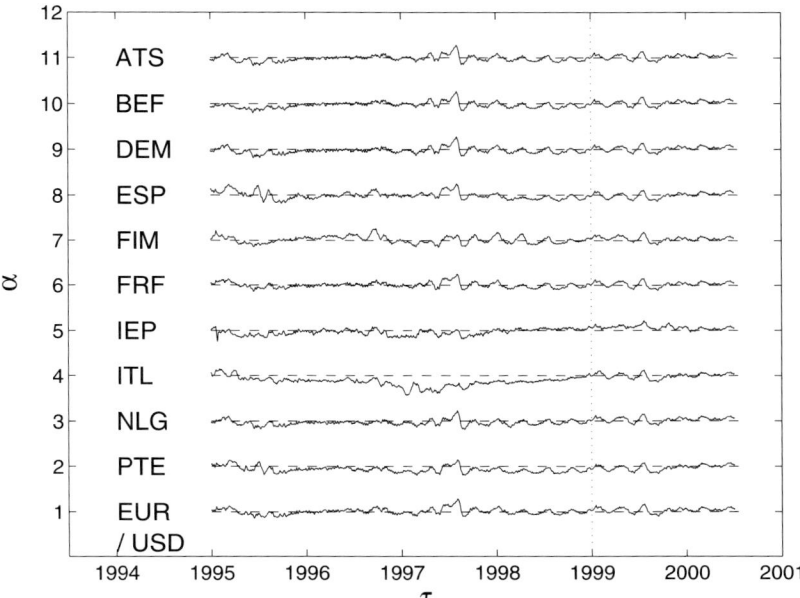

Fig. 11. Time dependence of the DFA *local* α-exponent for EUR and each currency forming the EUR exchange rate with respect to USD. The α-values are artificially multiplied by two and then displaced along the vertical axis in order to make the fluctuations noticeable. For each time dependent α a horizontal dashed line is drawn to indicate a reference to Brownian fluctuations

decrease at first and regrows in 95, is stable in 96, has a big fluctuation in mid/fall 97 and becomes pretty stable thereafter. The ITL case evolution is slightly different. The minute differences have probably to be associated to national political or economic events having an impact on the international monetary policy. It seems interesting to notice that the large fluctuations in α occur just before the crash dates of stock market indices. See also the marked singularity in mid 1999, a signature of the XR's adjustments prior to the EUR introduction. In order to further *prove* this point, a linear DFA analysis of the Dow Jones Industrial stock index around the 1987 October crash was performed [71]. A similar pattern is found in that case.

Other XR time series have been examined in order to check the non-stationarity of α [65,60]. This does support the idea that the foreign currency exchange markets are mainly governed by random conditions [72] or is said to be *efficient* in more usual economic language. However, this unconditional randomness cannot be extrapolated to speculative times nor emerging currencies [73]. Different universality classes thereby emerge. It may be useful to recall that Hartmann [74] has examined the competition between USD and EUR in a more general (political and economic) framework.

4 Conclusions

The DAX has been analyzed from the point of view of crashes, in particular the correlations in the signal volatility, before and after the critical days. The search for the crash day is separated into two numerical problems, that of the index divergence itself and that of the index oscillation frequency acceleration on the other hand. By considering the envelope of the DAX, we have demonstrated that before crashes, a log-periodic pattern exists. Even though error bars are intrinsically large, it is surprising to see that a rupture point is easily predicted. A hierarchical structure close to a fractal percolation backbone (or tree) seems intrinsic at crashes. The stability of this result should be tested in real time for the best future of our economic system. A few foreign exchange currency rates with respect to the USD have been examined in order to illustrate the DFA technique, the intrinsic structure of the DFA exponent, and its implications with respect to crashes.

Acknowledgements

Luc T. Wille is gladly thanked for inviting us to present the above results and considerations, and enticing us into writing this report. MA thanks N. Vandewalle for numerous discussions.

References

1. B. Lauterbach, U. Ben-Zion: J. Finance **48**, 1909 (1993)
2. R. Roll, 'The International Crash of October 1987', R. W. Kamphuis et al., eds.: *Black Monday and the Future of Financial Markets*, (Dow Jones/Irwin, Homewood, IL, 1989)
3. N. Vandewalle, Ph. Boveroux, F. Brisbois: Eur. J. Phys. B **15**, 547 (2000)
4. M. Ausloos, K. Ivanova: 'Crashes : symptoms, diagnoses and remedies'. In *Empirical sciences in financial fluctuations, Tokyo, Japan, Nov. 15-17, 2000 Proceedings* (Springer Verlag, Berlin, 2001)
5. D. Sornette, A. Johansen, J.P. Bouchaud: J. Physique. I (France) **6**, 167 (1996)
6. N. Vandewalle, Ph. Boveroux, A. Minguet, M. Ausloos: Physica A **255**, 201 (1998)
7. B.M. Roehner: Eur. Phys. J. B **17**, 341 (2000)
8. http://www.historyhouse.com/stories/tulip.htm
9. http://www.litrix.com/madraven/madne004.htm
10. http://www.enlou.com/people/bio-lawj.htm
11. http://landow.stg.brown.edu/victorian/history/ssbubble.html
12. http://www.britannica.com/eb/article?eu=70665&tocid=0
13. R. Westfall, *The Life of Isaac Newton* (Cambridge Univ. Press, Cambridge 1994)
14. http://mypage.direct.ca/r/rsavill/Thecrash.html
15. M. Ausloos: Europhysics News **29**, 70 (1998)

16. R. Cont, In *Statistical Physics on the Eve of the 21st Century*, ed. by M.T. Batchelor, L.T. Wille (World Scientific, Singapore 1999), pp. 47-64
17. B.B. Mandelbrot: J. Business **36**, 349 (1963)
18. R.N. Mantegna, H.E. Stanley: Nature **376**, 46 (1995)
19. J.P. Bouchaud, D. Sornette: J. Phys. I (France) **4**, 863 (1994)
20. E.E. Peters: *Fractal Market Analysis: Applying Chaos Theory to Investment and Economics* (Wiley Finance Editions, New York 1994)
21. E.E. Peters: *Chaos and Order in the Capital Markets: A New View of Cycles, Prices, and Market Volatility* (Wiley Finance Editions, New York 1996)
22. R.N. Mantegna, H.E. Stanley: *An Introduction to Econophysics* (Cambridge University Press, Cambridge 2000)
23. J. Voit: *The Statistical Mechanics of Financial Markets*, (Springer Verlag, Berlin 2001)
24. H.E. Stanley: *Phase Transitions and Critical Phenomena* (Clarendon Press, London 1971)
25. D. Sornette, A. Johansen, J.P. Bouchaud: J. Phys. I (France) **6**, 167 (1996)
26. J.A. Feigenbaum, P.G.O. Freund: Int. J. Mod. Phys. B **10**, 3737 (1996)
27. D. Sornette, A. Johansen: Quantitative Finance **1** 452 (2001)
28. J.A. Feigenbaum: Quantitative Finance **1** 346 (2001)
29. D. Sornette: Phys. Rep. **297**, 239 (1998)
30. J.C. Anifrani, C. Le Floc'h, D. Sornette, B.Souillard: J. Phys. I (France) **5**, 631 (1995)
31. A. Johansen, D. Sornette, H. Wakita, U. Tsunogai, W.I. Newman, H. Saleur: J. Phys. I (France) **6**, 1391 (1996)
32. D. Stauffer, D. Sornette: Physica A **252**, 271 (1998)
33. H. Dupuis: Trends/Tendances **22**(38), 26 (1997)
34. G. Legrand: Cash **4**(38), 3 (1997)
35. D. Daoût: Le Vif, L'Express **xx**, 124 (1997)
36. N. Vandewalle, M. Ausloos: Eur. J. Phys. B **4**, 139 (1998)
37. H. Dupuis: Trends/Tendances **22**(44), 11 (1997)
38. G. Legrand: Cash **4**(44), 3 (1997)
39. J.P. Bouchaud, P. Cizeau, L. Laloux, M. Potters: Physics World **12**, 25 (1999)
40. L. Laloux, M. Potters, R. Cont, J.P. Aguilar, J.P. Bouchaud: Europhys. Lett. **45**, 1 (1999)
41. F. Brisbois, P. Boveroux, M. Ausloos, N. Vandewalle: Int. J. Theor. Appl. Finance **3**, 423 (2000)
42. N. Vandewalle, M. Ausloos, P. Boveroux, A. Minguet: Eur. J. Phys. B **9**, 355 (1999)
43. M. Ausloos: Physica A **285**, 48 (2000)
44. M. Ausloos, N. Vandewalle, K. Ivanova: 'Time is Money'. In: *Noise, Oscillators and Algebraic Randomness* ed. by M. Planat, (Springer, Berlin, 2000) pp. 156–171
45. W.H. Press, B.P. Flannery, S.A. Teukolsky, W.T. Vetterling, *Numerical Recipes – the Art of Scientific Computing* 2nd edition, (Cambridge University Press, Cambridge 1992)
46. R.N. Mantegna: Eur. J. Phys. B **11**, 193 (1999)
47. J.A. Feigenbaum, P.G.O. Freund: Mod. Phys. Lett. B **12**, 57 (1998)
48. B.M. Roehner: Eur. J. Phys. B **14**, 395 (2000)
49. A. Johansen, D. Sornette: Int. J. Mod. Phys. C **10**, 563 (1999)

50. D. Stauffer, R.B. Pandey: Int. J. Theor. Appl. Finance **3**, 479 (2000)
51. A. Johansen, D. Sornette: Eur. J. Phys. B **17**, 319 (2000)
52. M. Schroeder: *Fractals, Chaos, Power Laws* (Freeman, New York 1991)
53. B.J. West, B. Deering: *The Lure of Modern Science: Fractal Thinking* (World Scientific, Singapore 1995)
54. M. Ausloos: 'Financial Time Series and Statistical Mechanics'. In *Vom Billardtisch bis Monte Carlo - Spielfelder der Statistischen Physik*, ed. by K. H. Hoffmann and M. Schreiber, (Springer, Berlin 2001)
55. L.A.N. Amaral, S.V. Buldyrev, S. Havlin, H. Leschhorn, P. Maass, M.A. Salinger, H.E. Stanley, M.H.R. Stanley: J. Phys. I France **7**, 621 (1997)
56. V.V. Gafiychuk, I.A. Lubashevsky, Y.L. Klimontovich: Complex Systems **12**, 103 (2000)
57. N. Vandewalle, M.Ausloos: Phys. Rev. E **55**, 94 (1997)
58. N. Vandewalle, R. D'hulst, M. Ausloos: Phys. Rev. E **59**, 631 (1999)
59. P. Bak: *How Nature Works*, (Copernicus, New York 1996)
60. C.K. Peng, S.V. Buldyrev, S. Havlin, M. Simmons, H.E. Stanley, A.L. Goldberger: Phys. Rev. E **49**, 1685 (1994)
61. N. Vandewalle, M. Ausloos: Int. J. Comput. Anticipat. Syst. **1**, 342 (1998)
62. K. Hu, P.C. Ivanov, Z. Chen, P. Carpena, H.E. Stanley: *private communication* http://arXiv.:physics/0103018
63. http://pacific.commerce.ubc.ca/xr/data.html
64. M. Ausloos, K. Ivanova: Physica A **286**, 353 (2000)
65. M. Ausloos, K. Ivanova: Eur. Phys. J. B **20**, 537 (2001)
66. K. Ivanova, M. Ausloos: '*False* EUR Exchange Rates vs. DKK, CHF, JPY and USD. What is a strong currency?'. In: *Empirical sciences in financial fluctuations, Tokyo, Japan, Nov. 15-17, 2000 Proceedings* (Springer Verlag, Berlin, 2001)
67. M. Ausloos, K. Ivanova: Int. J. Mod. Phys. C **12,** 169 (2001)
68. N. Vandewalle, M. Ausloos: Physica A **246**, 454 (1997)
69. N. Vandewalle, M. Ausloos: Int. J. Mod. Phys. C **9**, 711 (1998)
70. K. Ivanova, M. Ausloos: unpublished
71. M. Ausloos, N. Vandewalle: unpublished
72. R. Friedrich, J. Peincke, Ch. Renner: Phys. Rev. Lett. **84**, 5224 (2000)
73. K. Ivanova, M. Ausloos: Eur. Phys. J. B **8**, 665 (1999); Err. **12**, 613 (1999)
74. P. Hartmann: *Currency Competition and Foreign Exchange Markets. The Dollar, the Yen and the Euro* (Cambridge Univ. Press, Cambridge 1998)
75. http://lowrisk.com/crash/crashcharts.htm; http://lowrisk.com/crash/87vs97.htm; http://lowrisk.com/crash/1929crash.htm

Toward an Understanding of Financial Markets using Multi-agent Games

Neil F. Johnson, David Lamper, Paul Jefferies, and Michael L. Hart

We report on our use of multi-agent games to understand financial market behavior. In addition to discussing the background to the multi-agent games themselves, we report a technique which may prove useful for forecasting future movements of financial time-series. A third-party game is trained on a black-box time-series, and is then run into the future to extract next-step and multi-step predictions. Such predictions have potential use as the basis for improved risk management and portfolio optimization strategies.

1 Introduction

Our approach to understanding financial market dynamics involves agent-based market simulations. In these simulations, traders equipped with simple buy/sell strategies and limited information compete in speculative trading. This work builds on the theoretical discussion given in Ref. [1]. Reference[1] describes a multi-agent game with an out-of-equilibrium clearing process, in which the resulting dynamics for the simulated 'price' $P(t)$ closely resemble real financial movements; in particular there are fat-tailed price increments, clustered volatility and high volume autocorrelation.

Our main focus here concerns the forecasting aspect of this research agenda [2]. Section 2 introduces the basic Minority Game (MG) while Sec. 3 introduces important generalizations which are needed to provide a minimal market model. This model is called the Grand Canonical MG and allows the number of agents playing the game to vary in time. Section 4 considers next-step prediction while Sect. 5 considers prediction for a general number of timesteps into the future. Section 6 discusses briefly the aspect of risk in light of the resulting non-Gaussian distributions for price returns - in particular, a modified 'optimal' hedging strategy can be derived in order to minimize the inherent risk. Section 7 provides a conclusion.

2 The Basic MG

Agent-based models of financial markets are attracting significant attention in both the economics and physics communities - a recent example of this cross-disciplinary interest is provided by the joint-paper of Lux (economics) and

Marchesi (physics) [3]. The rationale for the interest in agent-based models of financial markets is that the fluctuations observed in financial time-series should, at some level, reflect the interactions, feedback, frustration and adaptation of the markets' many (and diverse) participants.

The Minority Game (MG) introduced by Challet and Zhang [4] offers possibly the simplest paradigm for a system containing these key features. Unlike the sophisticated model of Lux [3] there is no external noise process simulating information arrival. Nor is there any element of agents sharing local information. The MG simply comprises an odd number of agents N choosing repeatedly between the options of buying (1) and selling (0) a quantity of a risky asset. The resource level of this asset is finite and therefore the agents will compete to buy low and sell high. This gives the game its 'minority' nature; an excess of buyers will force the price of the asset up, consequently the minority of agents who have placed sell orders receive a good price at the penalty of the majority who end up buying at an over-inflated price. The MG agents act with inductive reasoning, using strategies that map the series of recent (binary) asset price fluctuations to an investment decision for the next time-step. In an attempt to learn from their past mistakes the agents constantly update the 'score' of their strategies and use only the most successful one to make their prediction. The memory m of an agent is the number of bits of the most recent past global history that are used by a strategy in order to form a prediction. The agents are assigned their s strategies randomly at the start of the game and are not allowed to replace them at any point. Each agent uses the historically most successful of her strategies to form a prediction. The predictions of all agents are then pooled and the global history is updated with the prediction of the minority group.

We now give more details concerning the binary structure of the basic MG. A single strategy maps each of the 2^m possible histories to a prediction. Thus there are 2^{2^m} different possible binary strategies. Consider $m = 2$; there are $2^{2^m} = 16$ possible strategies, each of which can be represented by a string of 4 bits $[ijk\ell]$ with $i, j, k, \ell = 0$ or 1 corresponding to the decisions based on the $2^m = 4$ possible histories 00, 01, 10, 11, respectively. For example, strategy [0000] ([1111]) corresponds to deciding to pick 0 (1) irrespective of the $m = 2$ history bit-string. [1010] corresponds to deciding to pick 1 given the histories 00 or 10, and pick 0 given the histories 01 or 11. However, many of the strategies in this space of 2^{2^m} strategies are largely similar to one another (i.e. are separated by a small Hamming distance). It has been shown [5] that the principal features of the MG are reproduced in a smaller *Reduced Strategy Space* of 2^{m+1} strategies wherein any two strategies are separated by a Hamming distance of either 2^m or 2^{m-1} (i.e. are *anti-correlated* or *uncorrelated*). If the number of strategies in play $N.s$ is greater than 2^{m+1} then the game is said to be in the *crowded* phase, in contrast $N.s \ll 2^{m+1}$ represents the *dilute* phase.

The properties of the crowded and dilute phases of the game are quite different and could be thought of as representing different regimes of a market. In the crowded phase there will at any one time be a large number of agents who are using the same (best) strategy and so will flood into the market as large groups, producing large swings in supply and demand and a consequently high volatility. If the memory of the agents is larger such as to render $N.s \sim 2^{m+1}$ then the groups of agents using the same (best) strategy (*crowds*) will be smaller. There will also be groups of agents who are forced to use the anti-correlated (worst) strategy, these can be thought of as *anti-crowds* as they cancel the market action of the *crowds*. This cancellation effect causes a reduction in the market volatility. In the dilute phase it is very unlikely that any agents will hold the same strategies and so the market behaves more randomly and can be modelled well as a group of independent coin-tossers.

The main result which emerged from early numerical simulations [6] concerns the volatility (i.e. standard deviation) of the time-series $x(t)$ corresponding to the number of agents making a given choice, say 0, as the game proceeds. When s is small, the volatility σ exhibits a pronounced minimum as a function of memory m. Around this minimum, the volatility σ is substantially *smaller* than the value obtained for the case where each agent makes his decision by tossing a coin [6]. We now give a simple order-of-magnitude description of the crowd-anticrowd effect, which captures the essential physics underlying the observed volatility variation in the basic MG [6]. Consider the oversimplified case of N independent agents each deciding by tossing a coin. Using standard random-walk results, the total variance σ^2 of agents choosing 0, say, is the sum of the variances produced by the N agents: $\sigma^2 = \sum_{i=1}^{N} \sigma_i^2 = \frac{N}{4}$, where $\sigma_i^2 = \frac{1}{2}(1 - \frac{1}{2}) = \frac{1}{4}$. However in reality on any given turn of the minority game, there are a number of agents using the same, or similar, strategies. Consider the subset of agents n_i using a particular strategy i. Although there is no information available to a given agent about other individual agents nor is any direct communication allowed between agents, this subset of agents n_i using a particular strategy i will all act in the same way and constitute a *crowd*. Since the corresponding random-walk 'step-size' has become n_i, one might think that σ_i^2 should be given by $\frac{1}{4}n_i^2$. Given that there is no *a priori* best strategy, however, it is important to realize that there may also be a subset of agents $n_{\bar{i}}$ who are using the opposite, or at least very dissimilar, strategies to the subset n_i. We call this second subgroup the *anticrowd*, and the strategy $\bar{i}$ that they use is anti-correlated to strategy i (e.g. if i is [0000] then $\bar{i}$ is [1111]). The *anticrowd* chooses the opposite room to the *crowd* and hence behaves as a crowd itself. Over the timescale during which the two opposing strategies are being played, the fluctuations are determined only by the *net* crowd-size $N_i = n_i - n_{\bar{i}}$. Hence σ_i^2 should instead be given by $\frac{1}{4}N_i^2$.

Suppose strategy i^* is the highest scoring at a particular moment: the anticorrelated strategy $\bar{i}^*$ is therefore the lowest scoring at that same moment. In the limit of small m, the size of the strategy space is small. For most values of s, each agent carries a considerable fraction of all possible strategies. Therefore, even if an agent picks $\bar{i}^*$ among his s strategies, he will most likely have a high scoring strategy in his toolbag. Therefore, many agents will choose to use either i^* itself (if they hold it) or a similar one. Very few agents will have such a poor set of s strategies that they are forced to use a strategy similar to $\bar{i}^*$. In this regime there are practically no anticrowds, and the crowds dominate. Therefore $N_i \sim N\delta_{ii^*}$ yielding $\sigma_i^2 \sim \frac{N^2}{4}\delta_{ii^*}$ and $\sigma^2 \sim \frac{N^2}{4}$, which is larger than the independent agent limit of $\frac{N}{4}$ as observed numerically [6]. In the limit of large m, the strategy space is very large and agents will have a low chance of holding the same strategy. In addition, even if an agent has s low-scoring strategies, the probability of his best strategy being strictly anticorrelated to another agent's best strategy (hence forming a crowd-anticrowd pair) is small. All the crowds and anticrowds are of size 0 or 1 implying that the crowds and anticrowds have effectively disappeared and the agents act independently. We thus have $N_i = 0$ or 1 with $\sum_i N_i \sim N$ giving $\sigma^2 \sim \frac{N}{4}$ which is again consistent with the numerical simulations. In the intermediate m region where the numerical minimum exists for small s, the size of the strategy space is relatively large so that some agents may get stuck with s strategies which are all low scoring. They can hence form anticrowds. The presence of finite-size anticrowds implies that $\sum_i N_i < N$. Considering the extreme case where the crowd and anticrowd are of similar size, we have $N_i \sim 0$ and hence $\sigma_i^2 \sim 0$. The volatility is therefore small which is again consistent with the numerical results. For a fixed value of m, this regime of small volatility will arise for small s since in this case the number of strategies available to each agent is small, hence some of the agents may indeed be forced to use a strategy which is little better than the poorly-performing $\bar{i}^*$. In other words, the cancellation effect of the crowd and anticrowd becomes most effective in this intermediate region. It is this interplay between the size of the strategy space and the probabilities of agents forming crowds and anticrowds which gives rise to the rich and non-trivial behaviour of the volatility observed in Ref. [6].

A theory based on these crowding effects reproduces quantitative results for the market volatility in the basic and so called 'thermal' MG across the full range of parameters N, m, s. For more details of this theory the reader is referred to [7]. This crowd-anticrowd theory may also be put to use in the formulation of an entirely analytical set of dynamical mapping equations that reproduce the MG. These equations can be analyzed in several interesting limiting cases to unveil the dynamics underlying microscopic behavior in different regimes of the game. They may also be used in the analysis of approaches to unstable behavior in these types of games (and possibly the real market itself).

3 Grand Canonical Minority Game

The development and study of market models from a physicist's standpoint is motivated by the desire to learn what key interactions are responsible for phenomena observed in the real-world system, the financial marketplace. The basic MG formulation captures some of the behavioral phenomena that are thought to be of importance in financial markets; those of competition, frustration, adaptability and evolution. It is also a 'minimal' system of only few parameters. However, the MG as a realistic market model has many shortcomings:

- All agents trade at each time-step
- All agents trade equal quantities
- The system resource level is fixed
- Agent diversity is typically limited

In the basic MG agents must either buy or sell at every time-step. In a real market however, traders are likely to wait on the sidelines until they are reasonably confident that they are able to make a profit with their next trade. They will observe the market passively, mentally updating their various strategies, until their confidence overcomes some threshold value - then they will jump in and make a trade. Suppose we have a total of N_{tot} agents: at each timestep only a subset $N = N_0 + N_1$ of the population, who are sufficiently confident of winning, actually play. If $N_0 = N_1$ the tie is decided by a coin-toss. Hence N and the excess demands $N_{0-1} = N_0 - N_1$ provide, to a first approximation, a 'volume' $N(t)$ and 'price-change' $\Delta P(t)$ at time t: the price-forming process is discussed in more detail in Ref. [1]. Here we concentrate on the resulting price-series $P(t)$: we will not exploit any possible additional information contained in $N(t)$. Agents have a time horizon T over which virtual points (i.e. reward for their strategies) are collected and have a threshold probability ('confidence') level r for trading. Active strategies are those with a historic probability of winning $\geq r$ [1]. We focus on the regime where the number of strategies in play is comparable to the total number available, and where $r \sim 0.5$. In addition to producing realistic dynamical features (see Fig. 1) this regime yields many of the statistical 'stylized facts' of real markets: fat-tailed price increments, clustered volatility and high volume autocorrelation[1]. As an illustration, Fig. 1 shows the resulting price-series $P(t)$ for a particular run of the Grand Canonical MG, in the region of a large change. Such large movements in the price do of course arise in real markets (they are so-called market corrections or 'crashes') and are obviously of great practical interest.

In major markets, exogenous events such as external news arrival are relatively infrequent compared to the typical transaction rate - in addition, most news is neither uniformly 'good' or 'bad' for all agents. This suggests that the majority of movements in high-frequency market data might be self-generated, i.e. produced by the internal activity of the market itself. The

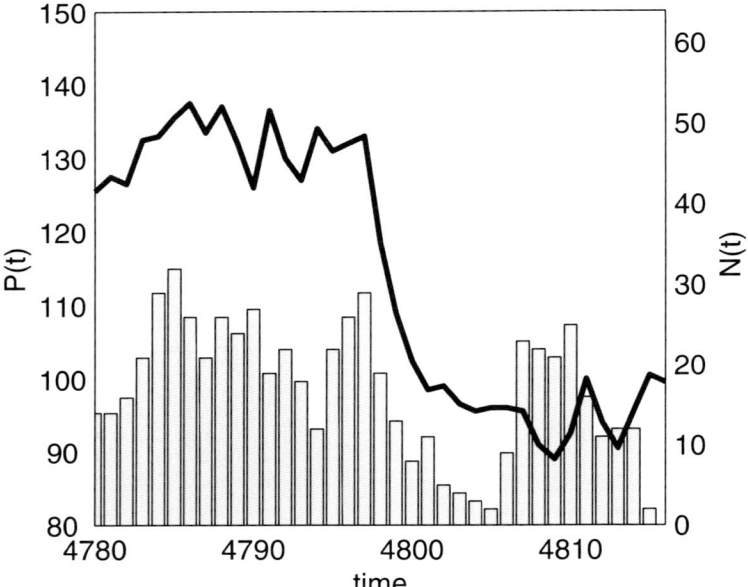

Fig. 1. Simulated price $P(t)$ (solid line) and volume $N(t)$ (bars). Here $N_{\text{tot}} = 101$, $s = 2$, $T = 100$, $r = 0.53$; memory $m = 3$

price-series $P(t)$ can therefore be thought of as being produced by a 'black-box' multi-agent game whose parameters, starting conditions ('quenched disorder'), and evolution are unknown. Using 'third-party' games trained on historic data, we aim to generate future probability distribution functions

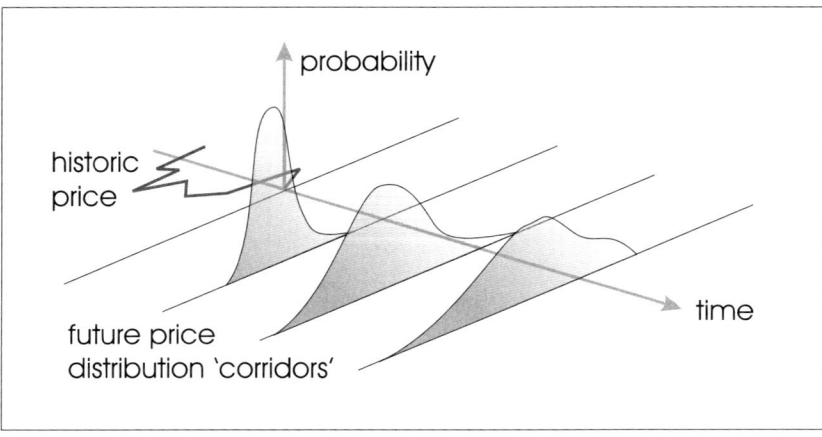

Fig. 2. Predicted distributions for future price movements

(pdfs) by driving these games forward (see Fig. 2). Typically the resulting distributions will be fat-tailed and have considerable time-dependent skewness, in contrast to more standard economic models.

4 Next Timestep Prediction

The approach is to replace the artificial price history used in the game, by real market data. The agents may then be able to collectively 'learn' to identify moments in the market where profit is attainable. As an illustration of next timestep prediction, we examine the sign of movements and hence convert $\Delta P(t)$ into a binary sequence corresponding to up/down movements. For simplicity, we also consider a confidence threshold level $r = 0$ such that all agents play all the time.

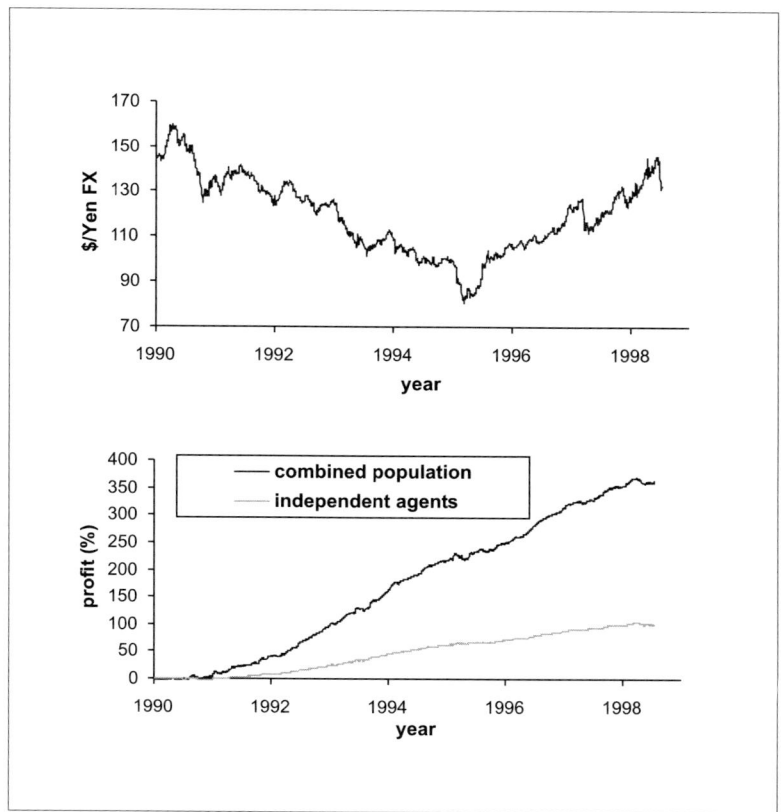

Fig. 3. Top: $/Yen FX-rate 1990-9. Bottom: cumulative profit for multi-agent game (black line) and for independent agents (shaded line)

The first step in this process is to generate binary information from the given financial time-series. This can be done in many ways in order to investigate the predictability of different aspects of the movement. We choose here to examine the sign of movements and hence our information history $h[t]$ becomes:

$$h[t] = H[p_{\text{real}}[t] - p_{\text{real}}[t-1]] \tag{1}$$

where $H[x]$ is the Heaviside function. If $p_{\text{real}}[t] = p_{\text{real}}[t-1]$ then we assign $h[t]$ a 0 or 1 randomly.

Figure 3 shows hourly Dollar \$/Yen exchange-rates for 1990-9, together with the profit attained from using the game's predictions to trade hourly. A simple trading strategy is employed each hour: buy Yen if the game predicts

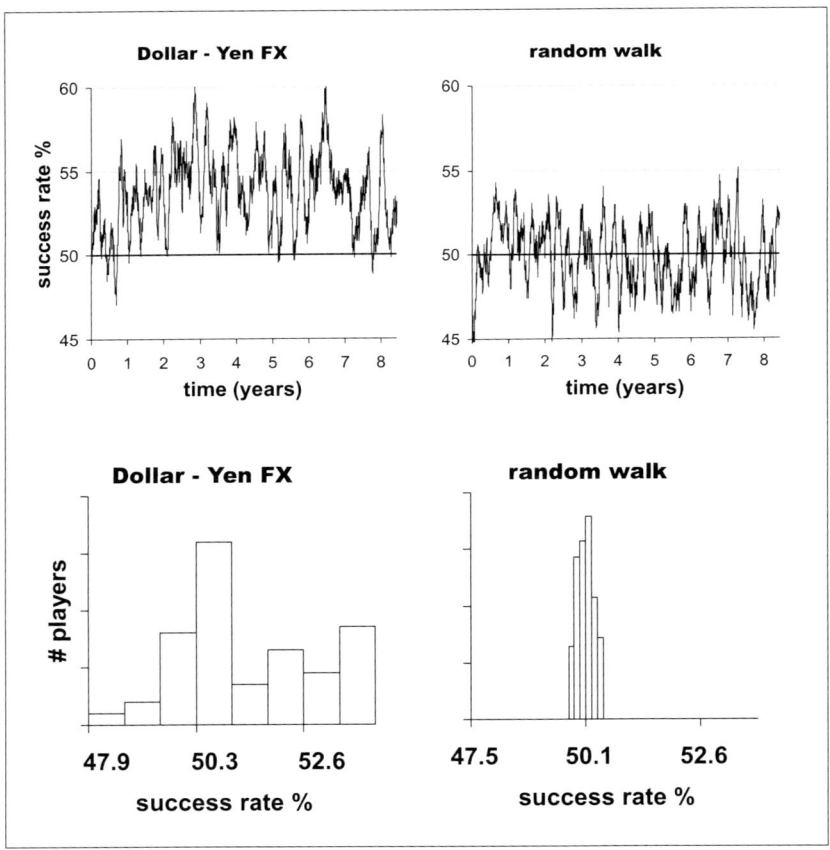

Fig. 4. Moving average of the multi-agent game's success rate for the real price-series of Fig. 3 (top left) and a random walk price-series (top right). Bottom: histogram of individual agents' time-averaged success rate

the rate to be favourable and sell at the end of each hour, banking any profit. This trading strategy is unrealistic since transaction costs would be prohibitive, however it demonstrates that the multi-agent game performs better than random ($\sim 54\%$ prediction success rate). Also shown is the profit in the case when the investment is split equally between all agents who then act independently. Acting collectively, the N-agent population shows superior predictive power and acts as a 'more intelligent' investor. As a check, Fig. 4 shows that the game's success returns to 50% for a random walk price-series.

5 Corridors for Future Price Movements

We now consider prediction over several (e.g. ten) future timesteps. As an example, we will try to predict the large movement ('crash') in Fig. 1 starting around $t = 4796$. As in the case of real prices [8], it seems impossible that this drop could have been foreseen given the prior history $P(t)$ for $t < 4796$. Even if complete knowledge of the game were available, it still seems impossible that subsequent outcomes should be predictable with significant accuracy since the coin-toss used to resolve ties in decisions (i.e. $N_0 = N_1$) and active-strategy scores, continually injects stochasticity. We start by running $P(t)$ through a trial third-party game to generate an estimate of S_0 and S_1 at each timestep, the number of active strategies predicting a 0 or 1 respectively. Provided the black-box game's strategy space is reasonably well covered by the agents' random choice of initial strategies, any bias towards a particular outcome in the active strategy set will propagate itself as a bias in the value of N_{0-1} away from zero. Thus N_{0-1} should be approximately proportional to $S_0 - S_1 = S_{0-1}$. In addition, the number of agents taking part in the game at each timestep will be related to the total number of active strategies $S_0 + S_1 = S_{0+1}$, hence the error (i.e. variance) in the prediction of N_{0-1} using S_{0-1} will be approximately proportional to S_{0+1}. We have confirmed this to be true based on extensive simulations. We then identify a third-party game that achieves the maximum correlation between the price-change $\Delta P(t)$ and our explanatory variable S_{0-1}, with the unexplained variance being characterized by a linear function of S_{0+1}. The predicted pdf for an arbitrary number j of timesteps into the future, is then generated by calculating the net value of S_{0-1} along all possible future routes of the third-party game.

Figure 5 shows the 'predicted corridors' for $P(t)$, generated at $t = 4796$ for $j = 10$ timesteps into the future. Remarkably $P(t)$ subsequently moves within these corridors. About 50% of the large movements observed in $P(t)$ occur in periods with tight predictable corridors, i.e. narrow pdfs with a large mean. Both the magnitude and sign of these extreme events are therefore predictable. The remainder correspond to periods with very wide corridors, in which the present method still predicts with high probability the sign of the change. We checked that the predictions generated from the third-party game were consistent with all such extreme changes in the actual (black-box)

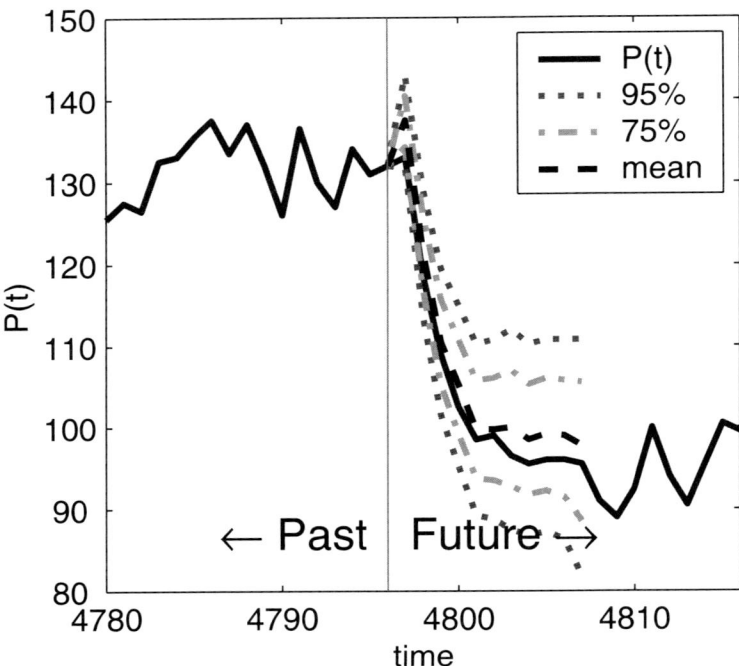

Fig. 5. Predicted corridors for 10 future timesteps, and actual $P(t)$ from Fig. 1. The confidence intervals and mean of the future distributions are shown

time series $P(t)$, likewise no predictions were made that were inconsistent with $P(t)$.

6 Real-World Risk

We have demonstrated some success of the agent-based models in direct prediction of the sign of the next price increment. However, we can also implement the models in a different way by 'training' them on historical data of a particular asset movement and then using the artificial market-making process to run the models forward into the future. If this is done with an ensemble of such models, each having a different initial allocation of strategies, we can form a distribution of likely future asset price levels. Typically the resulting distributions are fat tailed and can have considerable skewness quite in contrary to more standard economic models. This information can not only be of use in speculation but also in risk control and portfolio management.

The control of risk in financial investment should be of equal importance to the realization of profit. Most current theories of risk control rely on the implicit assumption that future behavior of the market will be like its past behavior. This assumption is continually being brought into question when banks and investors seem to be 'caught out' by events that past distributions seemed to imply were impossible. There thus may be room here for risk-control models that rely more on possible emergent future behavior than on historic data.

Using agent-based models in the way mentioned above gives us distributions for likely future price levels based on what microscopically might happen. This may be just the type of forward-casting model that could be of use here. Several years ago Bouchaud and Sornette developed a framework for examining and controlling the risk inherent in writing derivative contracts [9]. This formalism explicitly deals with future asset movements in a probabilistic, path-dependent fashion i.e. does not rely on any random-walk model etc. This makes the formalism ideal for combining with the forward-casting agent-models. The formalism examines the variation in future wealth ΔW_T from holding a certain portfolio, for example short one euro-call contract of price C_0 maturity T and strike X and long $\phi_t[S_t]$ hedging assets in the underlying which is at price S_t at time t:

$$\Delta W_T = C_0 - \max[S_T - X, 0] + \sum_{t=0}^{T} \phi_t[S_t](S_{t+1} - S_t) . \qquad (2)$$

The variance of this wealth process (which is used as a measure of risk) is then found analytically for a general underlying movement. For our models, this can be done in a Monte-Carlo fashion using each member of the model ensemble to generate a ΔW_T. Doing this we could also look at other measures of risk such as VAR etc. This process generates a more insightful measure of risk based on likely future microscopic behavior.

Bouchaud and Sornette's variance of the wealth process can be minimized with respect to the hedging strategy $\phi_t[S_t]$. The full details are given in [9]; the result is a risk-minimizing 'optimal strategy' given by:

$$\phi_t[S_t] = \int_X^\infty \frac{(S_T - S_t)\langle \delta S_{S_t,t \to S_T,T}\rangle}{\langle \delta S_t^2\rangle} P[S_T|S_t] \, dS_T . \qquad (3)$$

Using the forward-casting agent-models we obtain $P[S_T|S_t]$ (the probability of the underlying moving from value S_t to S_T) by counting the number of members of the (large) model ensemble that cast paths passing near both these two values (price space S is discretized for this purpose). Similarly $\langle \delta S_{S_t,t\to S_T,T}\rangle$ is found as the mean increment at time t of paths passing near S_t and S_T, $\langle \delta S_t^2\rangle$ is simply the mean squared increment at time t of all paths. The resulting reduction in risk when using this 'optimal strategy' with historical distributions is well documented [9]; similar effects are obtained when using the agent-models' future-cast distributions. The important difference to note is that the risk being minimized is now the microscopically derived future risk rather than a measure assuming the continuity of past behavior.

We now discuss the effect of transaction costs on the risk control process discussed in the previous paragraphs. Bouchaud and Sornette's formalism is easily couched in discrete time, accounting for the fact that continuous trading is un-physical due to transaction cost and brokerage inefficiencies. However, transaction costs themselves have not explicitly been accounted for in the wealth process, therefore their effect on risk-control cannot be gauged. We address this point here by adding a term to 2 in order to include a general transaction cost structure.

$$\Delta W_T \rightarrow \Delta W_T + \sum_{t=0}^{T} k_1 + (k_2 + k_3 S_t) \left| \phi_t \left[S_t \right] - \phi_{t-1} \left[S_{t-1} \right] \right| . \tag{4}$$

We again proceed to find the variance of this wealth process as a gauge of risk. We find that the approximation of $\left| \phi_t \left[S_t \right] - \phi_{t-1} \left[S_{t-1} \right] \right| \approx \frac{\partial \phi_t}{\partial S_t} \left| \delta S_t \right|$ holds reasonably well as the time dependence of $\phi_t \left[S_t \right]$ is weak. This allows us to formulate an analytical correction term to Bouchaud and Sornette's expression for risk comprising a sum of independent transaction cost variances plus the covariance between transaction costs. The covariance terms become very large as we execute more transactions. This non-local behavior leads to a divergence of the risk as we go toward continuous time. Clearly if we are to minimize risk now the answer is not to simply re-hedge more often.

The minimization of risk with respect to a choice of hedging strategy $\phi_t \left[S_t \right]$ is now highly complex and in general path-dependent. However, we may use pertubation theory to obtain approximate solutions. We find that the risk and transaction costs are reduced greatly using a volatility correction to 3 of the form:

$$\left\langle \delta S_t^2 \right\rangle \rightarrow \gamma \left[t \right] \left\langle \delta S_t^2 \right\rangle . \tag{5}$$

The form of $\gamma \left[t \right]$ as a function of time is amusingly that of a smile, much like the volatility correction in strike price to the Black-Scholes delta that is implied by 3 itself. The origins of these two 'volatility smiles' are of course very different. Using this correction, for portfolios where transaction costs are likely to be high, we see a dramatic reduction in the risk and also in the absolute transaction costs.

7 Conclusion

The focus of our work concerned the possible use of multi-agent models to address the important practical question of how profit opportunities may be identified and large price movements predicted. In the case of multi-step prediction into the future, we showed that it is possible to train a third-party multi-agent system in such a way that large movements in an (artificial) market price-series $P(t)$ can be predicted with surprising accuracy. Without knowing the details of the system which produced $P(t)$, the third-party multi-agent system yields predictable corridors for future movements of $P(t)$.

Remarkably, these predictable corridors become tighter at moments preceding large price movements.

We conclude that ensembles of multi-agent games may be useful as forecasting tools to predict future price movements in a probabilistic manner. Such predictions have potential use not only for speculative gain, but also as the basis for improved risk management and portfolio optimization strategies.

Acknowledgements

We thank P.M. Hui, D. Challet and D. Sornette for discussions. We are also extremely grateful to Dr. Jessica James of Bank One (London) for sharing the Dollar-Yen dataset with us and for detailed discussions.

References

1. P. Jefferies, M. Hart, P.M. Hui, N.F. Johnson: Eur. Phys. J. B **20**, 493 (2001); N.F. Johnson, M. Hart, P.M. Hui, D. Zheng: Int. J. of Theor. and Appl. Fin. **3**, 443 (2000)
2. Further details concerning the prediction method were presented elsewhere, see D. Lamper, S. Howison, N.F. Johnson: Phys. Rev. Lett. **88**, 017902 (2002), and are the subject of a patent application
3. T. Lux, M. Marchesi: Nature **397**, 498 (1999)
4. D. Challet, Y.C. Zhang: Physica A **246**, 407 (1997). See also the Minority Game web-page at http://www.unifr.ch/econophysics
5. D. Challet, Y.C. Zhang: Physica A **256**, 514 (1998)
6. R. Savit, R. Manuca, R. Riolo: Phys. Rev. Lett. **82**, 2203 (1999)
7. M. Hart, P. Jefferies, N.F. Johnson, P.M. Hui: Physica A **298**, 537 (2001); N.F. Johnson, M. Hart, P.M. Hui: Physica A **269**, 1 (1999); M. Hart, P. Jefferies, N.F. Johnson, P.M. Hui: Phys. Rev. E **63**, 017102 (2000); M. Hart, P. Jefferies, P.M. Hui, N.F. Johnson: Eur. Phys. J. B **20**, 547 (2000); N.F. Johnson, P.M. Hui, R. Jonson, T.S. Lo: Phys. Rev. Lett. **82**, 3360 (1999)
8. P. Ormerod: 'Surprised by depression', Financial Times, February 19, 2001
9. J.P. Bouchaud, D. Sornette: J. de Phys. I **4**, 863 (1994). See also J.P. Bouchaud, M. Potters, *Theory of Financial Risks* (Cambridge University Press, Cambridge 2000)

Towards Understanding the Predictability of Stock Markets from the Perspective of Computational Complexity

James Aspnes, David F. Fischer, Michael J. Fischer,
Ming-Yang Kao, and Alok Kumar

This paper initiates a study into the century-old issue of market predictability from the perspective of computational complexity. We develop a simple agent-based model for a stock market where the agents are traders equipped with simple trading strategies, and their trades together determine the stock prices. Computer simulations show that a basic case of this model is already capable of generating price graphs which are visually similar to the recent price movements of high tech stocks. In the general model, we prove that if there are a large number of traders but they employ a relatively small number of strategies, then there is a polynomial-time algorithm for predicting future price movements with high accuracy. On the other hand, if the number of strategies is large, market prediction becomes complete for two new computational complexity classes CPP and promise-BCPP, where $P^{NP[O(\log n)]} \subseteq BPP_{path} \subseteq promise\text{-}BCPP \subseteq CPP = PP$. These computational hardness results open up a novel possibility that the price graph of an actual stock could be sufficiently deterministic for various prediction goals but appear random to all polynomial-time prediction algorithms.

1 Introduction

The issue of market predictability has been debated for more than a century (see [8] for earlier papers and [22,15,6,20] for more recent viewpoints). In 1900, the pioneering work "Theory of Speculation" of Louis Bachelier used Brownian motion to analyze the stochastic properties of security prices [8]. Since then, Brownian motion and its variants have become textbook tools for modeling financial assets. Relatively recently, the radically different methodology of Mandelbrot used fractals to approximate price graphs deterministically [23]. In this paper, we initiate a study into this long-running issue from the perspective of computational complexity.

We develop a simple agent-based model for a stock market [9,21]. The agents are traders equipped with simple trading strategies, and their trades together determine the stock prices. We first consider a basic case of this model where there are only two strategies, namely, momentum and contrarian strategies. The choice of this base model and thus our general model is justified at two levels: (1) Experimental and empirical studies in the finance literature [1,7,10,11,19,7,4,12] show that a large number of traders primarily follow these two strategies. (2) Our own simulation results show that despite its simplicity, the base model is capable of generating price graphs which are visually similar to the recent price movements of high tech stocks (Figs. 1 and 2).

With these justifications, we then consider the issue of market predictability in the general model. We prove that if there are a large number of traders but they employ a relatively small number of strategies, then there is a polynomial-time algorithm to predict future price movements with high accuracy (Theorem 1). On the other hand, if there are also a large number of strategies, then the problem of predicting future prices becomes computationally very hard. To describe this hardness, we define two new computational complexity classes called CPP and promise-BCPP (Definitions 1 and 4). We show that some market prediction problems are hard for these two classes (Theorems 5 and 6) and that $P^{NP[O(\log n)]} \subseteq BPP_{path} \subseteq$ promise-BCPP $\subseteq$ CPP = PP.

These computational completeness results open up the possibility that the price graph of an actual stock could be sufficiently deterministic for various prediction purposes but appear random to all polynomial-time prediction algorithms. This is in contrast to the most popular academic belief that the future price of a stock cannot be predicted from its historical prices because the latter are statistically random and contain no information. This new possibility also differs from the fractal-based methodology in that the price graph of a stock could be a fractal but the fractal might not be computable in polynomial time. The findings in this paper can by no means settle the debate about market predictability. Our goal is only that our alternative approach could provide new insights to the predictability issue in a systematic manner. In particular, it could provide a general framework to investigate the many documented technical trading rules [25] and to generate novel and significant interdisciplinary research problems for computer science and finance.

The rest of the paper is organized as follows. Section 2 discusses the basic market model. Section 3 formulates the general model. Section 4 proves the complexity results for market prediction in the general model. We conclude the paper with some directions for future research in Sect. 5.

2 A Basic Market Model

In this section, we present a very simple market model, called the *deterministic-switching MC* (DSMC) model. The letter M stands for a *momentum* strategy, and the letter C for a *contrarian* strategy. These two strategies and the model itself are defined in Sect. 2.1. Some computer simulations for this model are reported in Sect. 2.2.

Intuitively, these strategies are heuristics ("rules of thumb") used by traders in the absence of reliable asset valuation models. As discussed in [12], a momentum trader may observe a sequence of "up" trades (price increments) and execute a buy trade in the anticipation that she will not be one of the last buyers, knowing very well that the asset is overpriced. Similarly, she may see some "down" trades (price decrements) and then make a sell trade in the hope that there will be more sellers after her. In contrast, after detecting a number of "up" (respectively, down) trades, a contrarian trader may submit a sell (respectively, buy) trade, anticipating a price reversal.

Both experimental and empirical studies have shown that traders look at past price dynamics to form their expectations of future prices, and a large number of them primarily follow momentum or contrarian strategies [1,7,10,11]. In addition, the traders may switch between these two diametrically opposite strategies. Momentum and contrarian strategies are dominant in the behavior of professional market timers as well [19]. The use of momentum and contrarian strategies sometimes signifies gambling tendencies among traders [7]. In fact, a market model with momentum and contrarian traders can also be interpreted as a market with noise traders and rational traders, where the noise traders essentially follow a momentum strategy while the rational traders attempt to exploit the noise traders by following a contrarian strategy [4,12].

2.1 Defining the DSMC Model

In the DSMC model, there is only one stock traded in the market. The model is completely specified by three integer parameters $m, L, k > 0$, and a real parameter $\alpha > 0$ as follows.

There are m traders in the market, and each trader's strategy set consists of momentum ($\mathcal{M}$) and contrarian ($\mathcal{C}$) strategies. At the beginning of day 1 of the investment period, each trader randomly chooses her initial strategy from $\{\mathcal{M}, \mathcal{C}\}$ and an integer $\ell_i \in [2, L]$ with equal probability, where L is the *maximum strategy switching period*. This is the only source of randomness in the DSMC model; from this point onwards, there is no random choice.

Rule 1 (Deterministic Strategy Switching Rule). For days $1, \ldots, k+1$, there is no trading. Each trader starts trading from day $k+2$ using her initial strategy. Trader i uses the same strategy for ℓ_i days and switches it at the beginning of every ℓ_i days.

The next rule defines the two strategies with respect to a given memory size k, which is the same for all traders.

Rule 2 (Trading Rule). At the beginning of day t, observe the stock prices P_f of days $f \in [t-(k+1), t-1]$. For $g \in [t-k, t-1]$, count the number k_u of days g when $P_g > P_{g-1}$; and the number k_d of days when $P_g < P_{g-1}$. The k-day trend is defied as $\text{Tr}(k,t) = k_u - k_d$. Then, if $\text{Tr}(k,t) \geq 0$ (respectively, < 0), the momentum strategy $\mathcal{M}$ buys (respectively, sells) one share of the stock at the market price determined by Rule 3 below. In contrast, the contrarian strategy $\mathcal{C}$ sells (respectively, buys) one share of the stock.

For instance, suppose that $k = 2$, and investor i picks her initial strategy $\mathcal{M}$ and $\ell_i = 2$ at the beginning of day 1. She then observes the prices of days 1, 2, 3, which are, say, $80, $82, $90. At the beginning of day 4, she issues a market order to buy one share of the stock. The orders issued by the traders on day 4 together determine the price of day 4 as specified by Rule 3. Suppose that the price of day 4 is $91, then investor i issues another market buy order at the beginning of day 5. Since her ℓ_i is 2, at the beginning of day 6, she switches her strategy from $\mathcal{M}$ to $\mathcal{C}$.

Rule 3 (Price Adjustment Rule). The prices for days $1, \ldots, k+1$ are given. On day $t \geq k+2$, let m_b and m_s be the total numbers of buys and sells, respectively. Then, the price P_t on day t is determined by the following equation:

$$P_t - P_{t-1} = \alpha \cdot (m_b - m_s),$$

where α is the unit of price change.

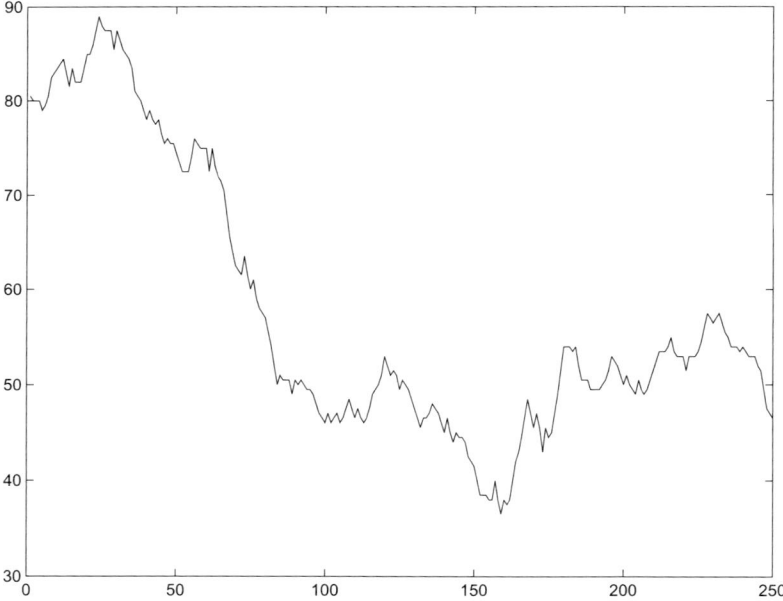

Fig. 1. A one-year price sequence generated using the DSMC model. Parameters: number of traders $m = 20$, memory size $k = 2$, maximum strategy switching period $L = 8$, unit of price change $\alpha = 0.25$, number of trading days $= 250$. The price graph appears strikingly similar to the recent price movements of high tech stocks

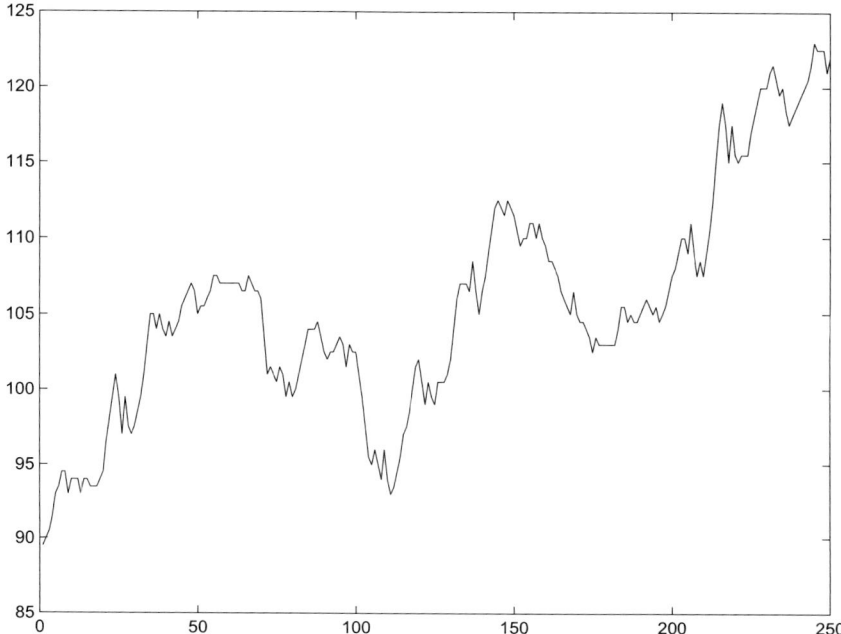

Fig. 2. A one-year price sequence generated using the DSMC model. The parameters are the same as those for Fig. 1

2.2 Computer Simulation on the DSMC Model

We have conducted some computer simulations of the DSMC model to test whether it can generate realistic price graphs. Because we had to examine the graphs visually, our time constraints limited the number of these simulations to only about six hundred. For a large fraction of them, we set $m = 20$, $L = 8$, and the initial k prices in the range of $70 to $90. We then focused on testing the effect of memory size k [24]. Two main findings are as follows:

- For $k = 1$, the price graphs were not visually real.
- For $k = 2$, about one out of four graphs were strikingly similar to those of recent high tech stocks, which was a major positive surprise to us. Two representatives of such graphs are shown in Figs. 1 and 2.

These two statements are based on our subjective impressions and limited simulations. To further understand the DSMC model, it would be useful to automate statistical analysis on the price graphs generated by this model and compare them with real stock prices.

3 A General Market Model

In this section, we define a market model, called the AS model, where the word AS stands for arbitrary strategies. It can be verified in a straightforward manner that the DSMC model is a special case of the AS model.

In the AS model, there is only one stock traded in the market. The model is completely specified as follows with five parameters: (1) the number m of traders, (2) a unit $\alpha > 0$ of price change, (3) a set $\Pi = \{\mathcal{S}^1, \ldots, \mathcal{S}^h\}$ of strategies, (4) a price adjustment rule (Eq. 1 or 2 below), and (5) a joint distribution of the population variables $X_1, \ldots, X_h$.

Rule 4 (Market Initialization). There are m traders in the market. At the beginning of day 1 of the investment period, each trader randomly chooses her initial strategy from Π. Let X_i be the number of traders who choose $\mathcal{S}^i$. Then, each X_i is a random variable, which is the only source of randomness in the model. (Unlike the DSMC model, because the allowable generality of Π, the AS model does not need strategy switching.)

Different joint distributions of the variables X_i lead to different specific models and prediction problems. In Sect. 4.2, we consider joint distributions that tend to Gaussian in the limit as the number m of traders becomes large. In Sect. 4.3, we consider the case where the variables X_i are independent, and each is 0 or 1 with equal probability.

Rule 5 (Trading Strategies). There is no trading on day 0. At the beginning of day $t \geq 1$, a trader observes the historical prices $P_0, \ldots, P_{t-1}$ and reacts by issuing a market order to buy one share of the stock, hold (i.e., do nothing), or sell one share according her strategy. Formally, a *strategy* is a collection of functions $\mathcal{S} = \{\mathcal{S}_1, \mathcal{S}_2, \ldots, \mathcal{S}_t, \ldots\}$, where each $\mathcal{S}_t$ maps $P_0, \ldots, P_{t-1}$ to +1 (buy), 0 (hold), or −1 (sell).

The price P_t of day t is determined at the end of the day by the day's m market orders using Rule 6. Since the traders choose their strategies randomly, the sequence $P_0, P_1, \ldots, P_t, \ldots$ is a stochastic process. We write $\mathcal{F}_t$ for the probability space induced by all possible sequences $\langle P_0, \ldots, P_t \rangle$ [18]. Then, we think of each function $\mathcal{S}_t$ as a random variable on $\mathcal{F}_{t-1}$.

We distinguish between strategies that react to price movements and those that ignore them.

- $\mathcal{S}$ is an *active* strategy if the functions $\mathcal{S}_t$ may or may not be constant functions. An *active* trader is one with an active strategy. Examples of active strategies include many used by day traders, who try to capture extremely short-term price trends.
- $\mathcal{S}$ is a *passive* strategy if the functions $\mathcal{S}_t$ all are constant functions. A *passive* trader is one with a passive strategy. Examples of passive strategies include two very popular ones: (1) dollar averaging, which invests an

equal amount every day over a chosen period, and (2) monthly retirement contributions by educational institutions, which are made on the same day every month.

Rule 6 (Price Adjustment). The price P_0 is given. At the end of day $t \geq 1$, the price P_t is determined by the day's market orders to buy or sell from the traders. We consider two simple rules:

With the *proportional increment* (PI) rule,

$$P_t = P_{t-1} + \alpha \cdot \sum_{i=1}^{h} X_i \cdot S_t^i, \tag{1}$$

where α is the unit of price change. Thus we can observe directly the net difference between the number of buyers and sellers on day t.

With the *fixed increment* (FI) rule,

$$P_t = P_{t-1} + \alpha \cdot \text{sign}\left(\sum_{i=1}^{h} X_i \cdot S_t^i\right). \tag{2}$$

In this case, the market moves up or down depending on whether the majority of traders are buying or selling, but the amount by which it moves is fixed at α.

For notational brevity, an *AS+FI model* refers to an AS model with the fixed increment rule, and an *AS+PI model* refers to an AS model with the proportional increment rule.

In reality, the price tends to move up if there are more buy orders than sell orders; similarly, the price tends to move down if there are more sell orders than buy orders. The FI rule is meant to model the sign but not the magnitude of the slope of this correlation, while the PI rule attempts to model both. Clearly, there can be many other increment rules, which this paper leaves for future research.

4 Predicting the Market

Informally, the *market prediction problem* at the beginning of day t is defined as follows:

- The data consists of (1) the five parameters of an AS-model, i.e., m, α, Π, X_i, and a price adjustment rule, and (2) a price history $P_0, \ldots, P_{t-1}$.
- The goal is to predict the price P_t by estimating the conditional probabilities $\Pr[P_t > P_{t-1} \mid P_0, \ldots, P_{t-1}]$, $\Pr[P_t < P_{t-1} \mid P_0, \ldots, P_{t-1}]$, and $\Pr[P_t = P_{t-1} \mid P_0, \ldots, P_{t-1}]$.

Note that $\Pr[P_t > P_{t-1} \mid P_0, \ldots, P_{t-1}]$ is symmetric to $\Pr[P_t < P_{t-1} \mid P_0, \ldots, P_{t-1}]$ and $\Pr[P_t = P_{t-1} \mid P_0, \ldots, P_{t-1}] = 1 - \Pr[P_t > P_{t-1} \mid P_0, \ldots, P_{t-1}] - \Pr[P_t < P_{t-1} \mid P_0, \ldots, P_{t-1}]$. Thus, from this point onwards, our discussion focuses on estimating $\Pr[P_t > P_{t-1} \mid P_0, \ldots, P_{t-1}]$.

From an algorithmic perspective, we sometimes assume that the price adjustment rule and the joint distribution of the variables X_i are fixed, and that the input to the algorithm is m, α, a description of Π, and the price history. This allows different algorithms for different model families as well as side-steps the issue of how to represent the possibly very complicated joint distribution of the variables X_i as part of the input. As for the description of Π, we only need $\mathcal{S}_1^i, \ldots, \mathcal{S}_t^i$ for each $\mathcal{S}^i \in \Pi$ instead of the whole Π, and the description of these functions can simplified by restricting their domains to consist of the price sequences consistent with the given price history.

4.1 Markets as Systems of Linear Constraints

In the AS+FI model with parameters m and α, a price sequence $P_0, \ldots, P_t$ and Π can yield a set of linear inequalities in the population variables X_i as follows. If the price changes on day t, we have

$$\text{sign}(P_t - P_{t-1}) \sum_{i=1}^{h} \mathcal{S}_t^i X_i > 0. \tag{3}$$

If the price does not change, we have instead the equation

$$\sum_{i=1}^{h} \mathcal{S}_t^i X_i = 0. \tag{4}$$

Furthermore, any assignment of the variables X_i that satisfies either inequality is feasible with respect to the corresponding price movement on day t. In both cases, $\mathcal{S}_t^i$ is computable from the price sequence $P_0, \ldots, P_{t-1}$. The same statements hold for days $1, \ldots, t-1$. Therefore, given m and α, we can extract from Π and $P_0, \ldots, P_t$ a set of linear constraints on the variables X_i. The converse holds similarly. We formalize these two observations in Lemmas 1 and 2 below.

Lemma 1. *In the AS+FI model with parameters m and α, given Π and a price sequence $P_0, \ldots, P_\beta$, there are matrices A and B with coefficients in $\{-1, 0, +1\}$, h columns each, and β rows in total. The rows of A (respectively, B) correspond to the days when $P_j \neq P_{j-1}$ (respectively, $P_j = P_{j-1}$). Furthermore, the column vectors $x = (X_1, \ldots, X_h)^\top$ consistent with Π and $P_0, \ldots, P_\beta$ are exactly those that satisfy $Ax > 0$ and $Bx = 0$. The matrices A and B can be computed in time $O(h\beta T)$, where T is an upper bound on the time to compute a single $\mathcal{S}_j^i$ from $P_0, \ldots, P_\beta$ over all $j \in [1, \beta]$ and $\mathcal{S}^i$.*

Proof. Follows immediately from Eqs. (3) and 4.

Lemma 2. *In the AS+FI model with parameters m and α, given a system of linear inequalities $Ax > 0, Bx = 0$, where A and B have coefficients in $\{-1, 0, +1\}$ with h columns each, and β rows in total, there exist (1) a set Π of h strategies corresponding to the h columns of A and B, and (2) a $(\beta+1)$-day price sequence $P_0, \ldots, P_\beta$ with the latter β days corresponding to the β rows of A and B. Furthermore, the values of the population variables $X_1, \ldots, X_n$ are feasible with respect to the price movement on day j if and only if column vector $x = (X_1, \ldots, X_n)^\top$ satisfies the j-th constraint in A and B. Also, $P_0, \ldots, P_\beta$ and a description of Π can be computed in $O(h\beta)$ time.*

Proof. Follows immediately from Eqs. (3) and (4).

In the AS+PI model we obtain only equations, of the form:

$$\sum_{i=1}^{h} S_t^i X_i = \frac{1}{\alpha}(P_t - P_{t-1}). \tag{5}$$

In this case there is a direct correspondence between market data and systems of linear equations. We formalize this correspondence in Lemmas 3 and 4 below.

Lemma 3. *In the AS+PI model with parameters m and α, given Π and a price sequence $P_0, \ldots, P_\beta$, there is a matrix B with coefficients in $\{-1, 0, +1\}$, h columns, and β rows, and a column vector b of length h, such that the column vectors $x = (X_1, \ldots, X_h)^\top$ consistent with Π and $P_0, \ldots, P_\beta$ are exactly those that satisfy $Bx = b$. The coefficients of B and b can be computed in time $O(h\beta T)$, where T is an upper bound on the time to compute a single S_j^i from $P_0, \ldots, P_\beta$ over all $j \in [1, \beta]$ and S^i.*

Proof. Follows immediately from Eq. (5).

Lemma 4. *In the AS+PI model with parameters m and α, given a system of linear equations $Bx = b$, where B is a $\beta \times h$ matrix with coefficients in $\{-1, 0, +1\}$, there exist (1) a set Π of h strategies corresponding to the h columns of B, and (2) a $(\beta+1)$-day price sequence $P_0, \ldots, P_\beta$ with the last β days corresponding to the β rows of B. Furthermore, the values of the population variables $X_1, \ldots, X_n$ are feasible with respect to the price movement on day j if and only if column vector $x = (X_1, \ldots, X_n)^\top$ satisfies the j-th constraint in B. Also, $P_0, \ldots, P_\beta$ and a description of Π can be computed in $O(h\beta)$ time.*

Proof. Follows immediately from Eq. (5).

4.2 An Easy Case for Market Prediction: Many Traders but Few Strategies

In Sect. 4.2, we show that if an AS+FI market has far more traders than strategies, then it takes polynomial time to estimate the probability that the next day's price will rise. In Sect. 4.2, we discuss why the same analysis technique does not work for an AS+PI market.

Predicting an AS+FI Market. For the sake of emphasizing the dependence on m, let $\Pr_m[E]$ be the probability that event E occurs when there are m traders in the market.

This section makes the following assumptions:

E1 The input to the market prediction problem is simply a price history $P_0, \ldots, P_{t-1}$. The output is $\lim_{m\to\infty} \Pr_m[P_t > P_{t-1} \mid P_0, \ldots, P_{t-1}]$.
E2 The market follows the AS+FI model.
E3 Π is fixed. The values $\mathcal{S}_j^i$ over all $i \in [1, h]$ are computable from the input in total time polynomial in j.
E4 Each of the m traders independently chooses a random strategy $\mathcal{S}^i$ from Π with fixed probability $p_i > 0$, where $p_1 + \cdots + p_h = 1$.

The parameter α is irrelevant.

Notice that the column vector $X = (X_1, \ldots, X_h)^\top$ is the sum of m independent identically-distributed vector-valued random variables with a center at $p = m \cdot (p_1, \ldots, p_h)^\top$. We recenter and rescale X to $Y = (X - m \cdot (p1, \ldots, p_h)^\top)/\sqrt{m}$. Then, by the Central Limit Theorem (see, e.g., [3, Theorem 29.5]), as $m \to +\infty$, Y converges weakly to a normal distribution centered at the h-dimensional vector $(0, \ldots, 0)^\top$. In Theorem 1 below, we rely on this fact to estimate the probability that the market rises for price histories that occur with nonzero probability.

Theorem 1. *Assume that $\lim_{m\to\infty} \Pr_m[P_0, \ldots, P_{t-1}] > 0$. Then there is a fully polynomial-time approximation scheme for estimating $\lim_{m\to\infty} \Pr_m[P_t > P_{t-1} \mid P_0, \ldots, P_{t-1}]$ from $P_0, \ldots, P_{t-1}$. The time complexity of the scheme is polynomial in (1) the length t of the price history, (2) the inverse of the relative error bound ϵ, and (3) the inverse of the failure probability η.*

Remark. We omit the explicit dependency of the running time in h and $p_1, \ldots, p_h$ in order to concentrate on the main point that market prediction is easy with this section's four assumptions. The parameters h and $p_1, \ldots, p_h$ are constant under the assumptions.

Proof. We use Lemma 1 to convert the price history $P_0, \ldots, P_{t-1}$ and the strategy set Π into a system of linear constraints $AX > 0$ and $BX = 0$, with the next day's price change $P_t - P_{t-1}$ determined by $\text{sign}(c \cdot X')$ for some c. Since the values $\mathcal{S}_j^i$ are computable in time polynomial in j, this conversion takes time polynomial in t.

Then, $\Pr_m[P_0, \ldots, P_{t-1}] = \Pr_m[AX > 0 \wedge BX = 0]$. Since $\lim_{m\to\infty} \Pr_m[AX > 0 \wedge BX = 0] > 0$, the constraints in B must be vacuous; in other words, for each $P_i = 0$ with $i \in [0, t-1]$, the corresponding constraint in B is $0 \cdot X_1 + \cdots + 0 \cdot X_h = 0$. Therefore, $\Pr_m[P_0, \ldots, P_{t-1}] = \Pr_m[AX > 0]$. Furthermore, since both A and c are constant with respect to m,

$$\lim_{m\to\infty} \Pr_m[P_t > P_{t-1} \mid P_0, \ldots, P_{t-1}] = \frac{\lim_{m\to\infty} \Pr_m[AX > 0 \wedge c \cdot X > 0]}{\lim_{m\to\infty} \Pr_m[AX > 0]}. \tag{6}$$

So to compute the desired $\lim_{m\to\infty} \Pr_m[P_t > P_{t-1} \mid P_0, \ldots, P_{t-1}]$, we compute $\lim_{m\to\infty} \Pr_m[AX > 0 \wedge c \cdot X > 0]$ and $\lim_{m\to\infty} \Pr_m[AX > 0]$ as follows.

To avoid the degeneracy caused by $\sum_{i=1}^{h} X_i = m$, we work with $X' = (X_1, \ldots, X_{h-1})^\top$ instead of X by replacing X_h with $m - \sum_{i=1}^{h-1} X_i$ and making related changes. Let $p' = (p_1, \ldots, p_{h-1})^\top$, which is the center of X'. As is true for Y, as $m \to +\infty$, the vector $Y' = (X' - m \cdot p')/\sqrt{m}$ converges weakly to a normal distribution centered at the $(h-1)$-dimensional point $(0, \ldots, 0)^\top$. Under the assumption that each p_i is nonzero, the distribution of Y' is full-dimensional (within its restricted $(h - 1)$-dimensional space), as in the limit the variance of each coordinate Y'_i is nonzero conditioned on the values of the other coordinates, which implies that the smallest subspace containing the distribution must contain all $h - 1$ axes. We can calculate the covariance matrix of Y' directly from the p_i, as it is equal to the covariance matrix for a single trader: on the diagonal, $C_{ii} = p_i - p_i^2$; and for off-diagonal elements, $C_{ij} = -p_i p_j$. Given C, Y' has density $\rho(x) = ae^{x^\top C x}$ for some constant a, and we can evaluate this density in $O(h^2)$ time given x, which is $O(1)$ time under our assumption that Π is fixed.

Let A_i be the i-th constraint of A, i.e., $A_{i,1} X_1 + \cdots + A_{i,h} X_h > 0$. Let A'_i denote the constraint $(A_{i,1} - A_{i,h}, \ldots, A_{i,h-1} - A_{i,h})$. Let $c' = (c_1 - c_h, \ldots, c_{h-1} - c_h)$.

We next convert the constraints of A on X into constraints on Y'. First of all, notice that $A_i X = \sqrt{m} \cdot (A'_i Y') + m \cdot A_i p$. So $A_i X > 0$ if and only if $A'_i Y' > -\sqrt{m} \cdot A_i p$. The term $-\sqrt{m} \cdot A_i p$ may not be constant. In such a case, as $m \to \infty$, the hyper plane bounding the half space $A'_i Y' > -\sqrt{m} \cdot A_i p$ keeps moving away from the origin, which presents some technical complication. To remove this problem, we analyze the term in three cases. If $A_i p < 0$, then since $m \cdot p$ is the center of X, as $m \to \infty$, $\Pr_m[A_i X < 0]$ converges to 1. In other words, A_i is infeasible with probability 1 in the limit. Then, since $\lim_{m\to\infty} \Pr_m[P_0, \ldots, P_{t-1}] > 0$, such A_i cannot exist in A. Similarly, if $A_i p > 0$, then $\lim_{m\to\infty} \Pr_m[A_i X > 0] = 1$ and A_i is vacuous. The interesting constraints are those for which $A_i p = 0$; in this case, by algebra, $A_i X > 0$ if and only if $A'_i Y' > 0$. Thus, let D be the matrix formed by these constraints; D can be computed in $O(ht)$ time. Then, since D is constant with respect to m, $\lim_{m\to\infty} \Pr_m[AX > 0] = \lim_{m\to\infty} \Pr_m[DY' > 0]$. Similarly, $\Pr_m[AX > 0 \wedge c \cdot X > 0]$ converges to (1) 0, (2) $\Pr_m[DY' > 0]$, or (3) $\Pr_m[DY' >$

$0 \wedge c' \cdot Y' > 0]$ for case (1) $c \cdot p < 0$, case (2) $c \cdot p > 0$, or case (3) $c \cdot p = 0$, respectively.

Therefore, by Eq. (6), $\lim_{m \to \infty} \Pr_m[P_t > P_{t-1} \mid P_0, \ldots, P_{t-1}]$ equals 0 for case (1) and equals 1 for case (2). Case (3) requires further computation.

$$\lim_{m \to \infty} \Pr_m[P_t > P_{t-1} \mid P_0, \ldots, P_{t-1}] = \frac{\lim_{m \to \infty} \Pr_m[DY' > 0 \wedge c' \cdot Y' > 0]}{\lim_{m \to \infty} \Pr_m[DY' > 0]}. \tag{7}$$

The numerator and denominator of the ratio in Eq. (7) are both integrals of the distribution of Y' in the limit over the bodies of possibly infinite convex polytopes. To deal with the possible infiniteness of the convex bodies $DY' > 0 \wedge c' \cdot Y' > 0$ and $DY' > 0$, notice that the density drops exponentially. So we can truncate the regions of integration to some finite radius around the $(h-1)$-dimensional origin $(0, \ldots, 0)^\top$ with only exponentially small loss of precision. Finally, since the distribution of Y' in the limit is normal, by applying the Applegate-Kannan integration algorithm for log-concave distributions [2] to the numerator and denominator separately, we can approximate $\lim_{m \to \infty} \Pr_m[P_t > P_{t-1} \mid P_0, \ldots, P_{t-1}]$ within the desired time complexity.

Remarks on Predicting an AS+PI Market. The probability estimation technique based on taking m to ∞ does not appear to be applicable to the AS+PI model for the following reasons.

First of all, by Lemma 3, the input price history induces a system of linear equations $BX = b$. If any equation in $BX = b$ is not equivalent to $X_1 + \cdots + X_h = m$ or $0 \cdot X_1 + \cdots + 0 \cdot X_h = 0$, then $\lim_{m \to \infty} \Pr_m[P_0, \ldots, P_{t-1}] = 0$.

A natural attempt to overcome this seemingly technical difficulty would be to (1) solve $BX = b$ to choose a maximal set U of independent variables X_i and (2) evaluate $\Pr_m[P_0, \ldots, P_{t-1}]$ in the probability space induced by this set. Still, a single constraint such as $B_{i,1} \cdot X_1 + \cdots + B_{i,h} \cdot X_h = \alpha \cdot m_0$ with $B_{i,j} \geq 0$ for all $j \in [1, h]$ and $B_{i,j'} > 0$ for some $X_{j'} \in U$ forces $\lim_{m \to \infty} \Pr_m[P_0, \ldots, P_{t-1}] = 0$ in the new probability space. This is due to the fact that m_0 is constant with respect to m.

A further attempt would be to evaluate

$$\lim_{m \to \infty} \Pr_m[P_t > P_{t-1} \mid P_0, \ldots, P_{t-1}]$$

by directly working with the probability space induced by $P_0, \ldots, P_{t-1}$. This also does not work because we show below that the market prediction problem can be reduced to the case where taking a limit in m has no effect on the distribution of the strategy counts. Suppose that we are given a market which follows the assumptions E1, E3, and 4 of Sect. 4.2 except that this market uses the PI rule and has m_0 traders. We construct a new market with any $m \geq m_0$ traders with the following modifications:

1. The price history $P_0, \ldots, P_{t-1}$ is extended with an extra day into $P'_0, \ldots, P'_{t-1}, P'_t$, where $P'_j = P_j$ for $0 \leq j \leq t-1$. Each strategy $\mathcal{S}_i$ is extended into a new strategy $\mathcal{S}'_i$ where (1) on day $j \in [1, t-1]$, $\mathcal{S}'_i(P_0, \ldots, P_{j-1}) = \mathcal{S}_i(P_0, \ldots, P_{j-1})$, (2) on day t, $\mathcal{S}'_i$ always buys, and (3) on day $t+1$, $\mathcal{S}'_i(P'_0, \ldots, P'_t) = \mathcal{S}_i(P_0, \ldots, P_{t-1})$. Thus, $P'_t = P'_{t-1} + \alpha \cdot m_0$.
2. Add a passive strategy $\mathcal{S}'_{h+1}$ that always holds.
3. Let $p'_i = \frac{1}{2} p_i$ for $1 \leq i \leq h$ and $p'_{h+1} = \frac{1}{2}$.

Note that since $P'_t - P'_{t-1} = \alpha \cdot m_0$, $m - m_0$ traders choose the passive strategy $\mathcal{S}'_{h+1}$. Also, the new market and the new price history can accommodate any $m \geq m_0$ traders. Note that because of the constraint $P'_t - P'_{t-1} = \alpha \cdot m_0$, the probability distribution of $(X_1, \cdots, X_h)^\top$ conditioned on $P'_0, \ldots, P'_t$ in the new market for each $m \geq m_0$ is identical to the probability distribution of $(X_1, \cdots, X_h)^\top$ conditioned on $P_0, \ldots, P_{t-1}$ in the original market with $m = m_0$. Furthermore, $\Pr_m[P'_{t+1} > P'_t \mid P'_0, \ldots, P'_t] = \Pr_{m_0}[P_t > P_{t-1} \mid P_0, \ldots, P_{t-1}]$. So we have obtained the desired reduction.

Consequently, we are left with a situation where the number of active strategies may be comparable to the number of traders. Such a market turns out to be very hard to predict, as shown next in Sect. 4.3.

4.3 A Hard Case for Market Prediction: Many Strategies

Section 4.2 shows that predicting an AS+FI market is easy (i.e., takes polynomial time) when the number m of traders vastly exceeds the number h of strategies. In this section, we consider the case where every trader may have a distinct strategy, and show that predicting an AS+FI or AS+PI market becomes very hard indeed.

We now define two decision-problem versions of market prediction. Both versions make the following assumption:

- Each X_i is independently either 0 or 1 with equal probability.

The *bounded* market prediction problem is:

- Input: a set of n passive strategies and a price history spanning n days such that the probability that the market rises on day $n + 1$ conditioned on the price history is either (1) greater than 2/3 or (2) less than 1/3.
- Question: Which case is it, case (1) or case (2)?

The output of bounded market prediction is not defined when the input does not yield a bounded probability of a rise or fall on the next day. Bounded market prediction is thus an example of a *promise problem* [14,13], defined as a pair of predicates (Q, R) where Q, the *promise*, specifies which inputs are permitted, and R specifies which inputs in Q are contained in the language.

The *unbounded* market prediction problem is:

- Input: a set of n passive strategies and a price history spanning n days.
- Question: Is the probability that the market rises on day $n+1$ conditioned on the price history greater than $1/2$ (without the usual ϵ term)?

The unbounded market prediction problem has less financial payoff than the bounded one due to different probability thresholds. For each of these two problems, there are in effect two versions, depending on which price increment rule is used; however, both versions turn out to be equally hard. These two problems can be analyzed by similar techniques, and our discussion below focuses on the bounded market prediction problem with a hardness theorem for the unbounded market prediction problem in Sect. 4.3.

We show in Sect. 4.3 how to construct passive strategies and price histories such that solving bounded market prediction is equivalent to estimating the probability that a Boolean circuit outputs 1 on a random input conditioned on a second circuit outputting 1. In Sect. 4.3, we show that this problem is hard for $P^{NP[O(\log n)]}$ and complete for a class that lies between $P^{NP[O(\log n)]}$ and PP. Thus bounded market prediction is not merely NP-hard, but cannot be solved in the polynomial-time hierarchy at all unless the hierarchy collapses to a finite level.

Reductions from Circuits to Markets. Lemma 5 converts a circuit into a system of linear inequalities, while Lemma 6 converts a system of linear inequalities into a system of linear equations. These systems can then be converted into AS+FI and AS+PI market models using Lemmas 2 and 4, respectively.

Note that the restriction in Lemma 5 to circuits consisting of 2-input NOR gates is not an obstacle to representing arbitrary combinatorial circuits (with constant blow-up), as 2-input NOR gates are universal.

Lemma 5. *For any n-input Boolean circuit C consisting of m 2-input NOR gates, there exists a system $Ax > 0$ of $3m+2$ linear constraints in $n+m+2$ unknowns and a length $n+m+2$ column vector c with the following properties:*

1. *Both A and c have coefficients in $\{-1, 0, +1\}$ that can be computed in time $O((n+m)^2)$.*
2. *Any 0-1 vector $(x_1, \ldots, x_n)$ has a unique 0-1 extension*

$$x = (x_1, \ldots, x_n, x_{n+1}, \ldots x_{n+m+2})$$

   satisfying $Ax > 0$.
3. *If $Ax > 0$, then $cx > 0$ if and only if $C(x_1, x_2, \ldots, x_n) = 1$.*

Proof. Let x_{n+k} represent the output of the k-th NOR gate, where $1 \leq k \leq m$. Without loss of generality we assume that gate m is the output gate.

The variables x_{n+m+1} and x_{n+m+2} are dummies to allow for a zero right-hand-side in $Ax > 0$; our first two constraints are $x_{n+m+1} > 0$ and $x_{n+m+2} > 0$.

Suppose gate k has inputs x_i and x_j. The NOR operation is implemented by the following three linear inequalities:

$$x_i \quad\;\; + x_{n+k} < 2;$$
$$x_j + x_{n+k} < 2;$$
$$x_i + x_j + x_{n+k} > 0.$$

The first two constraints ensure that the output is never 1 if an input is 1, while the last requires that the output is 1 if both inputs are 0; the constraints are thus satisfied if and only if $x_{n+k} = \neg(x_i \vee x_j)$. Using the dummy variables, the first two constraints are written as

$$-x_i \quad\;\; - x_{n+k} + x_{n+m+1} + x_{n+m+2} > 0;$$
$$-x_j - x_{n+k} + x_{n+m+1} + x_{n+m+2} > 0.$$

Let $Ax > 0$ be the system obtained by combining all of these inequalities. Then for each $(x_1, \ldots, x_n)$, $Ax > 0$ determines x_{n+k} for all $k \geq 1$. The vector c is chosen so that $cx = x_{n+m}$.

One might suspect that the fixed increment rule's ability to hide the exact values of the left-hand side of each constraint is critical to disguise the inner workings of the circuit. However, by adding slack variables we can translate the inequalities into equations, allowing the use of a proportional increment rule without revealing extra information.

Lemma 6. *Let $Ax > 0$ be a system of m linear inequalities in n variables where A has coefficients in $\{-1, 0, +1\}$. Then there is a system $By = 1$ of $mn - m + 1$ linear equations in $2mn - 3m + n + 1$ variables with the following properties:*

1. *B has coefficients in $\{-1, 0, +1\}$ that can be computed in time $O((mn)^2)$.*
2. *There is a bijection $f : x \mapsto y$ between the 0-1 solutions x to $Ax > 0$ and the 0-1 solutions y to $By = 1$, such that $x_j = y_j$ for $1 \leq j \leq n$ whenever $y = f(x)$.*

Proof. For each $1 \leq i \leq m$, let A_i be the constraint $\sum_j A_{ij} x_j > 0$. To turn these inequalities into equations, we add slack variables to soak up any excess over 1, with some additional care taken to ensure that there is a unique assignment to the slack variables for each setting of the variables x_j.

We will use the following 0-1 variables, which we think of as alternate names for y_1 through $y_{2mn-3m+n+1}$:

Variables	Purpose	Indices	Count
x_j	original variables	$1 \leq j \leq n$	n
u	constant 1	none	1
s_{ij}	slack variables for A_i	$1 \leq i \leq m, 1 \leq j \leq n-1$	$m(n-1)$
t_{ij}	slack variables for $s_{ij} \geq s_{i,j+1}$	$1 \leq i \leq m, 1 \leq j \leq n-2$	$m(n-2)$

Name	Equation	Purpose	Indices	Count
U	$u = 1$	set u	none	1
B_i	$\sum_j A_{ij} x_j - \sum_j s_{ij} = 1$	represent A_i	$1 \leq i \leq m$	m
S_{ij}	$s_{ij} - s_{i,j+1} - t_{ij} + u = 1$	require $s_{ij} \geq s_{i,j+1}$	$1 \leq i \leq m,\ 1 \leq j \leq n - 2$	$m(n-2)$

Observe that for each i, $\sum_j s_{ij}$ can take on any integer value σ_i between 0 and $n-1$, and that for any fixed value of σ_i, the S_{ij} constraints uniquely determine the values of s_{ij} and t_{ij} for all j. So each constraint B_i permits $\chi_i = \sum_j A_{ij} x_j$ to take on precisely the same values 1 to n that A_i does, and each χ_i uniquely determines σ_i and thus the assignment of all s_{ij} and t_{ij}.

Conditional Probability Complexity Classes. Suppose that we take a polynomial-time probabilistic Turing machine, fix its inputs, and use the usual Cook's Theorem construction to turn it into a circuit whose inputs are the random bits used during its computation. Then, we can feed the resulting circuit to Lemmas 5 and 2 to obtain an AS+FI market model in which there is exactly one assignment of population variables for each set of random bits, and the price rises on the last day if and only if the output of the Turing machine is 1. By applying Lemma 6 to the intermediate system of linear inequalities, we can similarly convert a circuit to an AS+PI model. It follows that bounded market prediction is BPP-hard for either model. But with some cleverness, we can exploit the conditioning on past history to show that bounded market prediction is in fact much harder than this. We do so in Sect. 4.3, after a brief detour through computational complexity in this section.

We proceed to define some new counting classes based on conditional probabilities. One of these, BCPP, has the useful feature that bounded market prediction solves all problems in BCPP, and is complete for the "promise problem" version of BCPP, which we will write as promise-BCPP and which we define in Sect. 4.3. We will use this fact to relate the complexity of bounded market prediction to more traditional complexity classes.

The usual counting classes of complexity theory (PP, BPP, R, ZPP, $C_=$, etc.) are defined in terms of counting the relative numbers of accepting and rejecting states of a nondeterministic Turing machine. We will define a new family of counting classes by adding a third decision state that does not count for the purposes of determining acceptance or rejection.

A *noncommittal* Turing machine is a nondeterministic Turing machine with three decision states: *accept, reject,* and *abstain*. We represent a noncommittal Turing machine as a deterministic Turing machine which takes a polynomial number of random bits in addition to its input; each assignment of the random bits gives a distinct computation path. A computation path is accepting/rejecting/abstaining if it ends in an accept/reject/abstain state, respectively. We often write 1, 0, or $\perp$ as shorthand for the output of an accepting, rejecting, or abstaining path.

Conditional versions of the usual counting classes are obtained by carrying over their definitions from standard nondeterministic Turing machines to noncommittal Turing machines, with some care in handling the case of no accepting or rejecting paths. We can still think of these modified classes as corresponding to probabilistic machines, but now the probabilities we are interested in are conditioned on not abstaining.

Definition 1. The *conditional probabilistic polynomial-time* class (CPP) consists of those languages L for which there exists a polynomial-time noncommittal Turing machine M such that $x \in L$ if and only if the number of accepting paths when M is run with input x exceeds the number of rejecting paths.

Definition 2. The *bounded conditional probabilistic polynomial-time* class (BCPP) consists of those languages L for which there exists a constant $\epsilon > 0$ and a polynomial-time noncommittal Turing machine M such that (1) $x \in L$ implies that a fraction of at least $\frac{1}{2} + \epsilon$ of the total number of accepting and rejecting paths are accepting and (2) $x \notin L$ implies that a fraction of at least $\frac{1}{2} + \epsilon$ of the total number of accepting and rejecting paths are rejecting.

Definition 3. The *conditional randomized polynomial-time* class (CR) consists of those languages L for which there exists a constant $\epsilon > 0$ and a polynomial-time noncommittal Turing machine M such that (1) $x \in L$ implies that a fraction of at least ϵ of the total number of accepting and rejecting paths are accepting, and (2) $x \notin L$ implies that there are no accepting paths.

As we show in Theorems 2 and 3, CPP and CR turn out to be the same as the unconditional classes PP and NP, respectively.

Theorem 2. CPP = PP.

Proof. First of all, PP $\subseteq$ CPP because a PP machine is a CPP machine that happens not to have any abstaining paths. For the inverse direction, represent each abstaining path of a CPP machine by a pair consisting of one accepting and one rejecting path, and each accepting or rejecting path by two accepting or rejecting paths. Then the resulting PP machine accepts if and only if the CPP machine does.

Theorem 3. CR = NP.

Proof. To show NP $\subseteq$ CR, replace each rejecting path of an NP machine with an abstaining path in a CR machine. For the inverse direction, replace each abstaining path of the CR machine with a rejecting path in the NP machine.

The class BCPP is more obscure; it is equivalent to the threshold version of BPP, BPP_{path} [17].[1] The class BPP_{path} is defined as the class of all languages accepted by a *threshold machine* with threshold $\frac{1}{2} + \epsilon$ for some $\epsilon > 0$, where a threshold machine accepts or rejects if at least a fixed proportion of its computation paths accept or reject, with each computation path counted as one without regard to its probability.

---

[1] We are grateful to Lance Fortnow[16] for pointing out this equivalence.

Theorem 4. $\text{BCPP} = \text{BPP}_{\text{path}}$.

Proof. To show $\text{BCPP} \subseteq \text{BPP}_{\text{path}}$, replace each abstaining path with one accepting and one rejecting path. To show $\text{BPP}_{\text{path}} \subseteq \text{BCPP}$, we must normalize the BPP_{path} computation so that all paths include the same number of branches. Suppose that in some BPP_{path} computation, the number of branches on any path is bounded by some polynomial $T(n)$. Extend each path in the BPP_{path} machine with $k < T(n)$ branches into $2^{T(n)-k}$ paths in the BCPP machine, of which all but one are abstaining and the remaining path accepts or rejects depending on the output of the corresponding BPP_{path} path.

$\text{BCPP} = \text{BPP}_{\text{path}}$ is a much stronger class than the analogous nonconditional class BPP. For example, if one takes a NP machine and replaces each accepting path with exponentially many accepting paths and each rejecting path with an equally large family of abstaining paths sprinkled with a single rejecting path, the result is a BCPP machine that accepts the same language as the NP machine. By repeating this sort of amplification of "good" paths, BCPP can in fact simulate $O(\log n)$ queries of an NP-oracle. Because of the equivalence of BCPP and BPP_{path}, we can show this formally by using similar results for BPP_{path} from [17].

Corollary 1. $\text{P}^{NP[O(\log n)]} \subseteq \text{BCPP} \subseteq \text{PP}$.

Proof. The first inclusion is immediate from Theorem 4 and the fact that $\text{P}^{NP[O(\log n)]} \subseteq \text{BPP}_{\text{path}}$, shown in Corollary 3.4 in [17]. The second inclusion follows from Theorem 4 and the observation that $\text{BPP}_{\text{path}} \subseteq \text{PP}_{\text{path}} = \text{PP}$, also from [17].

An interesting open question is where exactly $\text{BCPP} = \text{BPP}_{\text{path}}$ lies between $\text{P}^{NP[O(\log n)]}$ and PP. It is conceivable that by cleverly exploiting the power of conditioning to amplify low-probability events one could show $\text{BCPP} = \text{PP}$. However, we will content ourselves with the much easier observation that the usual amplification technique for BPP also applies to BCPP; as with other results in this section, this observation follows from the equivalence of BCPP and BPP_{path}.

Corollary 2. *If $L \in \text{BCPP}$, then there exists a noncommittal Turing machine M such that the probability that M accepts conditioned on not abstaining is at least $1 - f(n)$ if $x \in L$ and at most $f(n)$ if $x \notin L$, where $n = |x|$ and $f(n)$ is any function of the form $2^{-O(n^c)}$ for some constant $c > 0$.*

Proof. Immediate from Theorem 4 and Theorem 3.1 of [17].

Promise Problems and Promise-BCPP. Part of the motivation for defining BCPP and CPP was to identify exactly the complexity of solving bounded and unbounded market prediction. Unfortunately, while we can show that bounded market prediction is hard for BCPP, in the sense that any problem in BCPP reduces to bounded market prediction, it is not clear that bounded market prediction is actually *contained* in BCPP.

The reason is that the definition of BCPP does not allow excluding bad inputs. Though we don't care what our BCPP machine does when given an instance of market prediction in which the next day's price movement is not predictable, the definition of the class still requires that the machine produce more than $\frac{1}{2} + \epsilon$ accepting or rejecting paths. The natural solution to bounded market prediction using a noncommittal machine does not have this property, and it is not clear that we can guarantee it in general. Instead, we define a promise-problem version of BCPP, and show (in Sect. 4.3) that bounded market prediction is complete for this class.

Definition 4. The class promise-BCPP consists of all pairs of predicates (Q, R) for which there exists a constant $\epsilon > 0$ and a polynomial-time noncommittal Turing machine M such that for all $x \in Q$, (1) $x \in R$ implies that a fraction of at least $\frac{1}{2} + \epsilon$ of the total number of accepting and rejecting paths are accepting and (2) $x \notin R$ implies that a fraction of at least $\frac{1}{2} + \epsilon$ of the total number of accepting and rejecting paths are rejecting.

A pair of predicates (Q, R), in which Q specifies which inputs are valid and R specifies which valid inputs should be accepted, is called a *promise problem* [14,13]. Polynomial-time reductions, as defined for languages, have a natural analog for promise problems: (Q, R) is polynomial-time reducible to (Q', R') if and only if there is a polynomial-time function f such that (a) $f(Q) \subseteq f(Q')$, and (b) for all $x \in Q$, $f(x) \in R'$ if and only if $x \in R$.[2] Similarly, a particular promise problem is *hard* for a class of such problems if every problem in the class reduces to it in polynomial-time, and that it is *complete* for a class if it is both hard for the class and contained in the class.

There is also a natural correspondence between promise problems and standard languages. A *solution* to a promise problem (Q, R) is a language L for which L and R agree on inputs in Q; in this way promise problems can be turned into languages. In the other direction, any standard language L can be through of as a promise problem (true, L).

With this correspondence, notice that BCPP = BPP_{path} is contained in promise-BCPP, in the sense that for any L in BCPP, (true, L) is in promise-BCPP; and that promise-BCPP is in turn contained in CPP, in the sense that any problem (Q, R) in promise-BCPP has a solution in CPP(we can just run the noncommittal machine that accepts (Q, R)). We will abuse

---

[2] There are many ways to define more complicated reductions involving promise problems; a detailed discussion of this issue can be found in [5].

notation slightly by writing $\text{BPP}_{\text{path}} \subseteq \text{promise-BCPP} \subseteq \text{CPP}$, eliding the implicit conversions between languages and promise problems.

Bounded Market Prediction is Promise-BCPP-Complete. In Sect. 4.3, we have defined the complexity class BCPP and observed that it is equal to BPP_{path}, which implies that it contains the powerful class $\text{P}^{\text{NP}[O(\log n)]}$. In this section, we show that solving bounded market prediction solves all problems in BCPP.

In a sense, this result says that market prediction is a universal prediction problem: if we can predict a market, we can predict any event conditioned on past history as long as we can sample from an underlying discrete probability space whose size is at most exponential.

It also says that bounded market prediction is very hard. That is, using Corollaries 2 and 1, even if the next day's price is determined with all but an exponentially small probability, it cannot be solved in the polynomial-time hierarchy unless the hierarchy collapses to a finite level.

Theorem 5. *The bounded market prediction problem is complete for promise-BCPP, in either the AS+FI or the AS+PI model.*

Proof. First we show that bounded market prediction is a member of promise-BCPP. Given a market, construct a noncommittal Turing machine M whose input is the price history and strategies, and whose random inputs supply the settings for the population variables X_i. Let M abstain if the price history is inconsistent with the input and population variables; depending on the model, this is either a matter of checking the linear inequalities produced by Lemma 1 or the equations produced by Lemma 3. Otherwise, M accepts if the market rises and rejects if the market falls on the next day. The probability that M accepts thus equals the probability that the market rises: either more than $2/3$ or less than $1/3$. Since the problem is to distinguish between these two cases, M solves the problem within the definition of promise-BCPP.

In the other direction, we will show how to reduce from any promise-BCPP-language L to bounded market prediction. Suppose (Q, R) is accepted by some BCPP-machine M for all $x \in Q$. We will translate M and its input x into a bounded market prediction problem. First use Corollary 2 to amplify the conditional probability that M accepts to either more than $2/3$ or less than $1/3$ as bounded market prediction demands. Then convert M into two polynomial-size circuits, one computing

$$C_{\not\perp}(r) = \begin{cases} 0 \text{ if } M(x, r) = \perp; \\ 1 \text{ if } M(x, r) \neq \perp, \end{cases}$$

and the other computing

$$C_1(r) = \begin{cases} 0 \text{ if } M(x, r) \neq 1; \\ 1 \text{ if } M(x, r) = 1. \end{cases}$$

Without loss of generality we may assume that $C_{\not\!\ell}$ and C_1 are built from NOR gates. Applying Lemma 5 to each yields two sets of constraints $A_{\not\!\ell} y > 0$ and $A_1 y > 0$ and column vectors $c_{\not\!\ell}$ and c_1 such that $c_{\not\!\ell} y > 0$ if and only if $C_{\not\!\ell} y = 1$ and $c_1 x > 0$ if and only if $C_1(x) = 1$, where y satisfies the previous linear constraints and x is the initial prefix of y consisting of variables not introduced by the construction of Lemma 5. We also have from Lemma 5 that there is a one-to-one correspondence between assignments of x and assignments of y satisfying the A constraints, so probabilities are not affected by this transformation.

Now use Lemma 2 to construct a market model in which $A_{\not\!\ell} y > 0$, $A_1 y > 0$, and $c_{\not\!\ell} y > 0$ are enforced by the strategies and price history, and $\text{sign}(c_1 y)$ determines the price change on the next day of trading. Thus the consistent settings of the variables X_i are precisely those corresponding to settings of r for which $C_{\not\!\ell}(r) = 1$, or, in other words, those yielding computation paths that do not abstain. The market rises when $C_1(r) = 1$, or when M accepts. So if we can predict whether the market rises or falls with conditional probability at least $2/3$, we can predict the likely output of M. It follows that bounded market prediction for the AS+FI model is promise-BCPP-hard.

To show the similar result for the AS+PI model, use Lemma 6 to convert the constraints $A_{\not\!\ell} y > 0$, $A_1 y > 0$ into a system of linear equations $Bz = 1$, and then proceed as before, using Lemma 4 to convert this system to a price history and letting $c_1 z$ determine the price change (and thus the sign of the price change) on the next day of trading.

Unbounded Market Prediction is CPP-Complete The unbounded market prediction problem seems harder because the probability threshold in question is $\frac{1}{2}$ with no ϵ bound in contrast to the thresholds $\frac{2}{3}$ and $\frac{1}{3}$ for the bounded market prediction problem. The following theorem reflects this intuition. However, since we do not know whether BCPP is distinct from PP, we do not know whether unbounded prediction is strictly harder.

Theorem 6. *The unbounded market prediction problem is complete for* CPP = PP, *in either the AS+FI or the AS+PI model.*

Proof. Similar to the proof of Theorem 5.

5 Future Research Directions

There are many problems left open in this paper. Below we briefly discuss some general directions for further research.

We have reported a number of simulation and theoretical results for the AS model. As for empirical analysis, it would be of interest to fit actual market data to the model. We can then use the estimated parameters to (1) test whether the model has any predicative power and (2) test the effectiveness of new or known trading algorithms. This direction may require carefully choosing "realistic" strategies for Π. Besides the momentum and contrarian strategies, there are some popular ones which are worth considering, such as those based on support levels. Investment newsletters could be a useful source of such strategies.

The AS model is an idealized one. We have chosen such simplicity as a matter of research methodology. It is relatively easy to design highly complicated models which can generate very complex market behavior. A more challenging and interesting task is to design the simplest possible model which can generate the desired market characteristics. For instance, a significant research direction would be to find the simplest model in which market prediction is computationally hard. On the other hand, it would be of great interest to find the most general models in which market prediction takes only polynomial time. For this goal, we can consider injecting more realism into the model by introducing resource-bounded learning (the generality of Π is equivalent to unbounded learning), variable memory size, transaction costs, buying power, limit orders, short sell, options, etc.

Acknowledgements

We would like to thank Lance Fortnow for pointing out the equivalence of BCPP and BPP_{path}, and Lane Hemaspaandra for explaining the mechanics of promise problems.

This work originated with David Fischer's senior project in 1999, advised by Ming-Yang Kao. David would like to thank his father and role model, Professor Michael Fischer, for teaching, mentoring, and inspiring him throughout college.

JA was supported in part by NSF Grant CCR-9820888. M-YK was supported in part by NSF Grants CCR-9531028 and CCR-9988376. A preliminary version appeared in *Proceedings of the 12th Annual ACM-SIAM Symposium on Discrete Algorithms*, 2001.

References

1. P.B. Andreassen, S. Krause: J. Forecasting **9** 347 (1990)
2. D. Applegate, R. Kannan: 'Sampling and integration of near log-concave functions'. In *Proceedings of the 23rd Annual ACM Symposium on Theory of Computing, 1999*, pp 156–163
3. P. Billingsley: *Probability and Measure*, 2nd ed. (Wiley, New York 1986)
4. F. Black: J. Finance **41** 529 (1986)
5. J.-Y. Cai, L.A. Hemachandra, J. Vyskoč: 'Promises and fault-tolerant database access' In: *Complexity Theory: Current Research*, Ed. by: K. Ambos-Spies, S. Homer, U. Schöning (Cambridge University Press, Cambridge 1993) pp 101–146

6. J.Y. Campbell, A.W. Lo, A.C. MacKinlay: *The Econometrics of Financial Markets* (Princeton University Press, Princeton 1997)
7. R.G. Clarke, M. Statman: Financial Analysts Journal **54**, 63 (1998)
8. P.A. Cootner: *The Random Character of Stock Market Prices* MIT Press, Cambridge 1964
9. R.H. Day, W.H. Huang: J. Econom. Behavior Org. **14**, 299 (1990)
10. W. De Bondt: J. Portfolio Manag. **17**, 84 (1991)
11. W. DeBondt: Int. J. Forecasting **9**, 355 (1993)
12. J.B. DeLong, A. Shleifer, L.H. Summers, R.J. Waldmann: J. Pol. Econom. **98**, 703 (1990)
13. S. Even, A. Selman, Y. Yacobi: Information and Control **61**, 159 (1984)
14. S. Even, Y. Yacobi: 'Cryptocomplexity and NP-completeness' In *Lecture Notes in Computer Science 85: Proceedings of the 7th International Colloquium on Automata, Languages, and Programming*, Ed. by J.W. de Bakker, J. van Leeuwen (Springer Verlag, New York 1980) pp 195–207
15. E. Fama: J. Finance **46**, 1575 (1991)
16. L. Fortnow: Private communication, March 2001
17. Y. Han, L.A. Hemaspaandra, T. Thierauf: SIAM J. Computing **26**, 59 (1997)
18. I. Karatzas, S.E. Shreve: *Methods of Mathematical Finance* Volume 39 of *Applications of Mathematics* (Springer Verlag, New York 1998)
19. A. Kumar: 'Behavior of momentum following and contrarian market timers' Working Paper 99-01, International Center for Finance, School of Management, Yale University, January 1999
20. A.W. Lo, A.C. MacKinlay: *A Non-Random Walk Down Wall Street* (Princeton University Press, Princeton 1999)
21. T. Lux: J. Econom. Behavior Organ. **33**, 143 (1998)
22. B.G. Malkiel: *A Random Walk down Wall Street* (Norton, New York 1990)
23. B.B. Mandelbrot: *Fractals and Scaling in Finance* (Springer Verlag, New York 1997)
24. S. Mullainathan: 'A memory based model of bounded rationality' Working paper, Department of Economics, MIT, April 1998
25. R. Sullivan, A. Timmermann, H. White: J. Finance **54** 1647 (1999)

Patterns in Economic Phenomena

H. E. Stanley, P. Gopikrishnan, V. Plerou, and M. A. Salinger

This chapter discusses some of the *similarities* between work being done by economists and by physicists seeking to find "patterns" in economics. We also mention some of the *differences* in the approaches taken and seek to justify these different approaches by developing the argument that by approaching the same problem from different points of view, new results might emerge. In particular, we review two such new results. Specifically, we discuss the two newly-discovered scaling results that appear to be "universal", in the sense that they hold for widely different economies as well as for different time periods: (i) the fluctuation of price changes of any stock market is characterized by a probability density function (PDF), which is a simple power law with exponent $\alpha + 1 = 4$ extending over 10^2 standard deviations (a factor of 10^8 on the y-axis); this result is analogous to the Gutenberg-Richter power law describing the histogram of earthquakes of a given strength; (ii) for a wide range of economic organizations, the histogram that shows how size of organization is inversely correlated to fluctuations in size with an exponent ≈ 0.2. Neither of these two new empirical laws has a firm theoretical foundation. We also discuss results that are reminiscent of phase transitions in spin systems, where the divergent behavior of the response function at the critical point (zero magnetic field) leads to large fluctuations. We discuss a curious "symmetry breaking" for values of Σ above a certain threshold value Σ_c; here Σ is defined to be the local first moment of the probability distribution of demand Ω – the difference between the number of shares traded in buyer-initiated and seller-initiated trades. This feature is qualitatively identical to the behavior of the probability density of the magnetization for fixed values of the inverse temperature.

1 Introduction to Patterns in Economics

One prevalent paradigm in economics is to marry finance with mathematics, with the fruit of this marriage the development of models. In physics, we also develop and make use of models or, as they are sometimes called, "artificial worlds," but many physicists are fundamentally empirical in their approach to science – indeed, some physicists never make reference to models at all (other than in classroom teaching situations). This empirical approach has led to advances when theory has grown out of experiment. One such example is the understanding of phase transitions and critical phenomena [3]. Might this "empirics first" physics paradigm influence the way physicists approach economics? Our group's approach to economic questions has been to follow the paradigm of critical phenomena, which also studies complex systems comprised of many interacting subunits, i.e., to first examine the empirical facts as thoroughly as possible, and search for any "patterns".

That at least *some* economic phenomena are described by "scaling patterns" has been recognized for over 100 years since Pareto investigated the statistical character of the wealth of individuals by modeling them using the scale-invariant distribution

$$f(x) \sim x^{-\alpha}, \tag{1}$$

where $f(x)$ denotes the number of people having income x or greater than x, and α is an exponent that Pareto estimated to be 1.5 [1]. Pareto noticed that his result was universal in the sense that it applied to nations *"as different as those of England, of Ireland, of Germany, of the Italian cities, and even of Peru"*. A physicist would say that the universality class of the scaling law (1) includes all the aforementioned countries as well as Italian cities, since by definition two systems belong to the same universality class if they are characterized by the same exponents.

In the century following Pareto's discovery, the twin concepts of scaling and universality have proved to be important in a number of scientific fields [2,3]. A striking example was the elucidation of the puzzling behavior of systems near their critical points. Over the past few decades it has come to be appreciated that the scale-free nature of fluctuations near critical points also characterizes a huge number of diverse systems also characterized by strong fluctuations. This set of systems includes examples that at first sight are as far removed from physics as is economics. For example, consider the percolation problem, which in its simplest form consists of placing blue pixels on a fraction p of randomly-chosen plaquettes of a yellow computer screen (Fig. 1). A remarkable fact is that the largest connected component of blue pixels magically spans the screen at a threshold value p_c. This purely geometrical problem has nothing to do with the small branch of physics called critical point phenomena. Nonetheless, the fluctuations that occur near $p = p_c$ are scale free and functions describing various aspects of the incipient spanning cluster that appears at $p = p_c$ are described by power laws. Indeed, the concepts of scaling and universality provide the conceptual framework for understanding this geometry problem.

It is becoming clear that almost any system comprised of a large number of interacting units has the potential of displaying power law behavior. Since economic systems are in fact comprised of a large number of interacting units has the potential of displaying power law behavior, it is perhaps not unreasonable to examine economic phenomena within the conceptual framework of scaling and universality [2–7]. We will discuss this topic in detail below.

So having embarked on a path guided by these two theoretical concepts, what does one do? Initially, critical phenomena research – guided by the

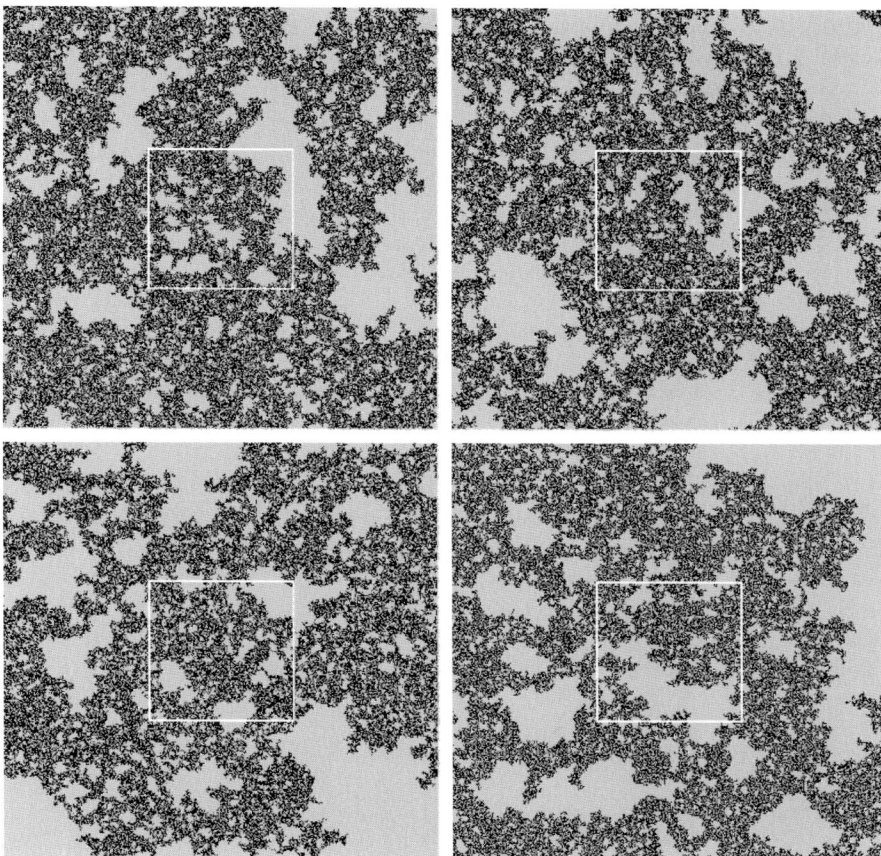

Fig. 1. We can experience the striking self-similarity of a fractal when we examine a series of pictures of a large percolation cluster created at the percolation threshold $p = p_c$. A little box is cut out of the first picture, blown up, and used as the second picture. The same little box procedure can be repeated in the second picture, creating the third picture, and in the third, creating the fourth. The untrained eye immediately recognizes that the statistical properties in all four pictures are the same, and to confirm this by a simple experiment, we can remove the labels, mix the pictures up, and then see how long it takes to put them back into sequence. It takes a remarkably long time and, significantly, can be carried out only by searching for *nonstatistical* features of the patterns, such as specific invaginations of a specific part of the cluster. An educational game is to time how long it takes each player to detect by eye which of the 24 possible panel orderings is the correct one that arranges the four panels in increasing order of magnification. This figure is courtesy of J. Kantelhardt.

Pareto principles of scaling and universality – was focused finding which systems display scaling phenomena, and on discovering the actual values of the relevant exponents. This initial empirical phase of critical phenomena research proved vital, for only by carefully obtaining empirical values of exponents such as α could scientists learn which systems have the same exponents (and hence belong to the same *universality class*). The fashion in which physical systems partition into disjoint universality classes proved essential to later theoretical developments such as the renormalization group [5] – which offered some insight into the reasons why scaling and universality seem to hold; ultimately it led to a better understanding of the critical point.

Similarly, the initial research in economics guided by the Pareto principles has largely been concerned with establishing which systems display scaling phenomena, and with measuring the numerical values of the exponents with sufficient accuracy that one can begin to identify universality classes if they exist. Economics systems differ from often-studied physical systems in that the number of subunits are considerably smaller in contrast to macroscopic samples in physical systems that contain a huge number of interacting subunits, as many as Avogadro's number 6×10^{23}. In contrast, in an economic system, one initial work was limited to analyzing time series comprising of order of magnitude 10^3 terms, and nowadays with high frequency data the standard, one may have 10^8 terms. Scaling laws of the form of (1) are found that hold over a range of a factor of $\approx 10^6$ on the x-axis [8–10].

2 Classic Approaches to Finance Patterns

As do economists, physicists view the economy as a collection of interacting units. This collection is complex; everything depends on everything else. The interesting problem is: *how* does everything depend on everything else? Physicists are looking for robust empirical laws that will describe – and theories that will help understand – this complex interaction.

To a physicist, the most interesting thing about an economic time series – e.g., the S&P 500 stock average index – is that it is dominated by fluctuations. If we make a curve of the values of the S&P 500 over a 35-year period, we see a fluctuating signal. Statistical physicists are particularly interested in fluctuating signals. The nature of this fluctuation immediately suggests to a physicist a model that was developed 100 years ago by Bachelier: the *biased random walk* [11].

A one-dimensional random walk is a drunk with a coin and a metronome. At each beat of the metronome, the coin is flipped – heads means one step to the right, tails one step to the left. If we look at our S&P 500 plot placed alongside a graph of a one-dimensional *biased* random walk – it is biased because it uses a "biased coin" that has a slight tendency to go up rather than down – we see a reasonable visual similarity. In fact, many economic pricing models – e.g., Black and Scholes – use this biased random walk.

Still there are certain points in the S&P 500 plot – such as October 19, 1987 ("Black Monday"), or the 15 percent drop over the week following the events of 11 September 2001 – that are not mirrored anywhere in the biased random walk model. Nowhere do we see a drop anywhere near the 30 percent drop of Black Monday. This could not occur in a biased random walk – the probability that a walk will move two steps in the same direction is p^2, three steps is p^3, and so on – so the probability of many steps in the same direction is exponentially rare, and virtually impossible.

Then how do we quantify these S&P 500 fluctuations? We begin by graphing the values of the fluctuations as a function of time. We place the plot of the empirical data next to the predictions of Bachelier's model. The fluctuations in the model are normalized by one standard deviation. Note that the biased random walk has a probability density function (PDF) that is a Gaussian, so the probability of having more than five standard deviations is essentially zero – you can see that a line drawn at five standard deviations is outside the range of the fluctuations.

If we normalize the empirical data we see a difference. A line drawn at five standard deviations is *not* outside the range of the fluctuations – there are many "shocks" that exceed five standard deviations. A bar placed on the positive side at five standard deviations also has 30 or 40 hits – fluctuations that equal or exceed five standard deviations in the positive direction. Some, such as Black Monday, are more than 34 standard deviations. The exponential of $(-1/2)(34)^2$ is approximately $10^{-267/2}$.

Because big economic shocks affect the economy around the world ("everything depends on everything else"), the possibility of an economic "earthquake" is one that we must take seriously. Big changes in stocks affect not only people with large amounts, but also those on the margins of society. One person's portfolio collapse is another's physical starvation; e.g., literal starvation in some areas was one result of the recent Indonesian currency collapse.

Another example is the recent Merriwether LTCM (Long Term Capital Management) collapse, caused in part by the use of models that do not take into account these catastrophic rare events. Thus there are many reasons we physicists might be interested in understanding economic fluctuations.

3 Patterns in Finance Fluctuations

One topic we physicists are interested in is symmetry. An example of traditional symmetry is sodium chloride. One can displace the lattice an amount equal to exactly two lattice constants and the configuration will remain the same. One can rotate it 90 degrees, or invert it, and the configuration will remain the same. Not only are these properties fascinating to mathematicians, they are also very relevant to solid state physics. This simple symmetry and the mathematics and physics that are built on it have led to extremely useful inventions, e.g., the transistor. The framework for our approach to systems with many interacting subunits is something that is usually called "scale invariance." These systems vary greatly from systems that do have scales.

We are all familiar with algebraic equations such as

$$x^2 = 4, \tag{2}$$

and we know the solution is a number, ± 2. Most of us are also familiar with functional equations, which are statements not about relations between numbers, but about the functional form of a function $f(x)$. Algebraic equations have solutions that are numbers, but functional equations have solutions that are functional forms. Power law functions are the solutions of certain functional equations of the form

$$f(\lambda x) = \lambda^p f(x). \tag{3}$$

In a functional equation, the converse also holds, i.e., every function that is of this power-law form also obeys this functional equation. This applies to a large number of contexts, in particular, to physical systems that have been tuned to be near critical points. An example is a binary mixture of two fluids in which the temperature has been tuned to be a special value called the critical temperature. At that temperature, there occur fluctuations in density in the binary mixture that extend over all length scales up to and including the wavelength of light. If you shine a flashlight on a tube of the binary mixture, you see an eerie glow – because the density fluctuations are so big in spatial extent, they become comparable to the wavelength of the light that is interacting with them. When that occurs, you see something that is visible – "critical opalescence." The same conceptual framework that describes this system appears to be able to describe economic systems.

Newcomers to the field of scale invariance often ask why a power law does not extend "forever" as it would for a mathematical power law of the form $f(x) = x^{-\alpha}$. This legitimate concern is put to rest by by reflecting on the fact that power laws for natural phenomena are not equalities, but rather are asymptotic relations of the form $f(x) \sim x^{-\alpha}$. Here the tilde denotes *asymptotic equality*. Thus $f(x)$ is not "approximately equal to" a power law so the notation $f(x) \approx x^{-\alpha}$ is inappropriate. Similarly, $f(x)$ is not proportional to a power law, so the notation $f(x) \propto x^{-\alpha}$ is also inappropriate. Rather, asymptotic equality means that $f(x)$ becomes increasingly like a power law as $x \to \infty$. Moreover, crossovers abound in financial data, such as the characteristic crossover from power law behavior to simple Gaussian behavior as the time horizon Δt over which fluctuations are calculated increases; such crossovers are characteristic also of other scale-free phenomena in the physical sciences [3], where the Yule distribution often proves quite useful.

For reasons of this sort, standard statistical fits to data are inappropriate, and often give distinctly erroneous values of the exponent α. Rather, one reliable way of estimating the exponent α is to form successive slopes of pairs of points on a log-log plot, since these successive slopes will be monotonic and converge to the true asymptotic exponent α.

The scale-invariance symmetry involved here is just as much a symmetry as the translational invariance symmetry in sodium chloride. We do not know how useful this scale-invariance symmetry will ultimately prove to be. Over the past 30 years physicists have used the theme of scale-invariance symmetry to understand systems near their critical points. Previous to this period of time, this class of problems was one no one could solve: there were many, many length scales, not just one. The length scales could run from one nearest-neighbor spacing out to approximately 5,000 (approximately the wavelength of light). The elements that make up this system are molecules that interact only over a short range – almost entirely with nearest neighbors. But this nearest-neighbor interaction propagates a small amount of torque through the system of nearest-neighbor interactions, so that the entire system is affected.

This is beginning to sound like economics, in which "everything affects everything else," And in economics, the first thing a physicist would do is look for the correlations. If we look at a graph of the autocorrelation function, we see a measure of the quantity G, which is a price change over some time horizon Δt. If we look at how G is now correlated with G at a time τ later, we measure that quantity as a function of τ, and as the size of τ increases, the correlation decreases. It is remarkable that this decrease happens in a regular fashion. How do we interpret this decrease? If we put the autocorrelation function in logarithmic units and the time lag in linear units, we see that the data fall on an approximate straight line This means that the function is decaying exponentially, which means it does indeed have a characteristic scale [12–14]. So the autocorrelation function is not scale invariant. This differs from systems near their critical points in which the autocorrelation functions are scale invariant.

The decay time in this finance example is short (4 minutes), so one cannot easily "make money" on these correlations [12,13]. A little less well-known is the measure of the volatility [13,14]. One way to quantify volatility is to replace G (the price change) with the absolute value of G. The data now are not at all linear on log-linear paper, but they are linear on log-log paper. And, of course, a power-law $y = x^p$ is linear on log-log paper, because $\log y = p \log x$. The slope of the log-log plot p is the value of the exponent. These exponents turn out to be fundamental quantities. In this case, $p = -0.3$. The data are straight from about 200 minutes out to about 10^5 minutes – a range of almost 1000. With the data graphed,one can see the approximate region in which the data are straight – the data are not straight over all regions. Qualitatively, we have known for a long time that there are long-range correlations in the volatility, e.g., volatility "clustering" and "persistence," but this graph helps quantify this known empirical fact.

If we cannot find an ordinary correlation near a critical point, we must try something else. For example, we might simply dump all of our data "on the floor." After we do that, the data no longer have time ordering nor do they have long-range or short-range power-law correlations in the volatility of the autocorrelation function itself. Now we pick the data points up off the floor and make a histogram. Mandelbrot did this in 1963 with 1000 data points – a tiny number by today's standards – for cotton-price fluctuations [2]. He concluded that those data were consistent with a Lévy distribution, i.e., a power-law distribution in that histogram – a so-called "fat tail."

In 1995, Mantegna and Stanley decided to test this result using data with Δt shorter than the daily data available in 1963 [12]. We used approximately 1 million data points: three orders of magnitude greater than Mandelbrot's data set. Instead of Mandelbrot's daily returns on cotton prices, we had returns as frequent as 15 seconds on the S&P 500 index. We found that on a log-linear plot (i) the histogram of the G data points for the S&P 500 clearly is not a Bachelier/Black-Scholes Gaussian, and (ii) although the *center* of the histogram agrees fairly well with Mandelbrot's Lévy distribution, it begins to disagree after a few standard deviations. This disagreement led us to develop a class of mathematical processes called *truncated* Lévy distributions – which has attracted the attention of a number of mathematicians, who have carried this concept far further than we could [15–20].

What about "universality," the notion in statistical physics that many laws seem to be remarkably independent of details? A physics example is that dramatically different materials behave exactly the same near their respective critical points. Binary mixtures, binary alloys, ferromagnets, even biological systems that involve switching, all behave the same way. An analog of this universality appears in economics. For example, Skjeltorp [21] did a study that utilized the Mantegna approach. Instead of 1,500,000 points from the S&P 500 (15-second intervals spread over six years), Skjeltorp did a parallel study for the Norwegian stock exchange and got almost exactly the same result.

We assumed that the reason we saw the truncated Lévy distribution while Mandelbrot did not was because we had more data – by three orders of magnitude. Gopikrishnan *et al.* recently acquired a data set three orders of magnitude larger still (of order 10^9) – one that records *every* transaction of every stock. They found that when their data were graphed on log-log paper, the result was linearity [22–25]. This is the log of the cumulative distribution, the same quantity Mandelbrot plotted for cotton. But where Mandelbrot's straight line had a slope of about 1.7 (well inside the Lévy regime, which stops at slope 2.0), Gopikrishnan's straight line has a slope of ≈ 3.0 (far outside the limit for a Lévy distribution). The fact that these data are approximately linear over two orders of magnitude means that fluctuations that are as much as 100 standard deviations are still conforming to the same law that describes the smaller fluctuations. This is reminiscent of the Gutenberg-Richter law that describes earthquakes [26–28]. Thus it would seem that these very rare events, which are conventionally treated as totally unexpected and unexplainable, have a precise probability describable by the same law that describes much more common events. These rare events occur with a frequency 8 orders of magnitude less than the common, everyday event.

This means that Mandelbrot's results for cotton (10^3 points) are at total odds with Gopikrishnan's results for the S&P 500 (10^9 points). Why this difference? Is it simply because Mandelbrot did not have enough data to draw reliable conclusions? Or do commodities intrinsically have fatter tails? In recent work with data from British Petroleum, it appears that commodity data may have a slightly smaller slope – consistent with the possibility that perhaps there is not one universal behavior for everything, but at least two separate universal behaviors – one for commodities and one for equities [29]. This smaller slope is still above 2, so the commodity data are not in the Lévy regime.

4 Patterns Resembling "Diffusion in a Tsunami Wave"

Over this past year, we and our collaborators have been trying to understand these exponents using procedures similar to those used in critical phenomena, e.g., we relate one exponent to another and call the relation a scaling law, or we derive some microscopic model.

In particular, there appears to be an intriguing analog not with the classic diffusion process studied in 1900 by Bachelier, [11] but rather with a generalization called anomalous diffusion. It is plausible that classical diffusion does not describe all random motion. The Brownian motion seen in the behavior of a grain of pollen in relatively calm water becomes something quite different if the grain of pollen is in a tsunami wave. The histograms would certainly be perturbed by a tsunami. A tsunami is an apt verbal metaphor for such economic "earthquakes" as the Merriwether disaster, so why not explore the stock market as an example of anomalous diffusion?

In one-dimensional classic diffusion, a particle moves at constant velocity until it collides with something. One calculates, e.g., the end position of the particle, and (of course) finds a Gaussian. Within a fixed time interval Δt, one might calculate a histogram for the number of collisions $p(N)$, and also find a Gaussian. And if one calculates a histogram of the variance W^2, one also finds a Gaussian. The fact that these are relatively narrow Gaussians means that there is a characteristic value, i.e., the width of that Gaussian, and that this is the basis for classical diffusion theory.

The corresponding quantity in the stock market to the displacement x is the price. At each transaction there is a probability that the price will change, and after a given time horizon there is a total change G. We've seen the histogram of G values – the cumulative obeyed an inverse cubic law, and therefore the pdf, by differentiation, obeys an inverse quartic law.

What about these histograms? Apparently no one had calculated these previously. Plerou et al. set about using the same data analyzed previously for G to calculate the histograms of N and W^2. They also found power laws – not Gaussians, as in classic diffusion. That means there is no characteristic scale for the anomalous diffusion case (there is a characteristic scale for the classic diffusion case) and for an obvious reason. If you are diffusing around in a medium – such as the "economic universe" we live in – in which the medium itself is changing, then the laws of diffusion change and, in particular, they adopt this scale-free form. Further, the exponents that describe $p(N)$ and $p(W^2)$ appear [30,31] to be the analogs of exponents in critical phenomena in the sense that they seem to be related to one another in interesting ways.

5 Patterns Resembling Critical Point Phenomena

Before concluding, we ask what sort of understanding could eventually develop if one takes seriously the power laws that appear to characterize finance fluctuations. It is tempting to imagine that there might be analogies between finance and known physical processes displaying similar scale-invariant fluctuations. For example, if one measures the wind velocity in turbulent air, one finds intermittent fluctuations that display some similarities with finance fluctuations [32]. However these similarities are not borne out by quantitative analysis – e.g., one finds non-Gaussian statistics, and intermittency, for both turbulence fluctuations and stock price fluctuations, but the time evolution of the second moment and the shape of the probability density functions are different for turbulence and for stock market dynamics [33,34].

More recent work pursues a rather different analogy, phase transitions in spin systems. Stock prices respond to fluctuations in demand, just as the magnetization of an interacting spin system responds to fluctuations in the magnetic field. Periods with large number of market participants buying the stock imply mainly positive changes in price, analogous to a magnetic field causing spins in a magnet to align. Recently, Plerou et al. [35] addressed the question of how stock prices respond to changes in demand. They quantified the relations between price change G over a time interval Δt and two different measures of demand fluctuations: (a) Φ, defined as the difference between the number of buyer-initiated and seller-initiated trades, and (b) Ω, defined as the difference in number of shares traded in buyer and seller initiated trades. They find that the conditional expectations $\langle G \rangle_\Phi$ and $\langle G \rangle_\Omega$ of price change for a given Φ or Ω are both concave. They find that large price fluctuations occur when demand is very small – a fact which is reminiscent of large fluctuations that occur at critical points in spin systems, where the divergent nature of the response function leads to large fluctuations. Their findings are reminiscent of phase transitions in spin systems, where the divergent behavior of the response function at the critical point (zero magnetic field) leads to large fluctuations. Further, Plerou *et al.* [36] find a curious "symmetry breaking" for values of Σ above a certain threshold value Σ_c; here Σ is defined to be the local first moment of the probability distribution of demand Ω, the difference between the number of shares traded in buyer-initiated and seller-initiated trades. This feature is qualitatively identical to the behavior of the probability density of the magnetization for fixed values of the inverse temperature.

Since the evidence for an analogy between stock price fluctuations and magnetization fluctuations near a critical point is backed up by quantitative analysis of finance data, it is legitimate to demand a theoretical reason for this analogy. To this end, we discuss briefly one possible theoretical understanding for the origin of scaling and universality in economic systems. As mentioned above, economic systems consist of interacting units just as critical point systems consist of interacting units. Two units are correlated in what might seem a hopelessly complex fashion – consider, e.g., two spins on a lattice, which are correlated regardless of how far apart they are. The correlation between two given spins on a finite lattice can be partitioned into the set of all possible topologically linear paths connecting these two spins – indeed this is the starting point of one of the solutions of the two-dimensional Ising model (see Appendix B of [3]). Since correlations decay exponentially along a one-dimensional path, the correlation between two spins would at first glance seem to decay exponentially. Now it is a mathematical fact that the total number of such paths grows exponentially with the distance between the two spins – to be very precise, the number of paths is given by a function which is a product of an exponential and a power law. The constant of the exponential *decay* depends on temperature while the constant for the exponential *growth* depends only on geometric properties of the system [3]. Hence by tuning temperature it is possible to achieve a threshold temperature where these two "warring exponentials" just balance each other, and a previously negligible power law factor that enters into the expression for the number of paths will dominate. Thus power law scale invariance emerges as a result of canceling exponentials, and universality emerges from the fact that the interaction paths depend not on the interactions but rather on the connectivity. Similarly, in economics, two units are correlated through a myriad of different correlation paths; "everything depends on everything else" is the adage expressing the intuitive fact that when one firm changes, it influences other firms. A more careful discussion of this argument is presented, not for the economy but for the critical phenomena problem, in Ref. [5].

Finally, a word of humility with respect to our esteemed economics colleagues is perhaps not inappropriate. Physicists may care passionately if there are analogies between physics systems they understand (like critical point phenomena) and economics systems they do not understand. But why should anyone else care? One reason is that scientific understanding of earthquakes moved ahead after it was recognized [26,27] that extremely rare events – previously regarded as statistical outliers regarding for their interpretation a theory quite distinct from the theories that explain everyday shocks – in fact possess the identical statistical properties as everyday events (e.g., all earthquakes fall on the same straight line on an appropriate log-log plot).

Similarly, if it were ever to be convincingly demonstrated that phenomena of interest in economics and finance possess the analogous property, then the challenge will be to develop a coherent understanding of financial fluctuations that incorporates not only everyday fluctuations but also those extremely rare "financial earthquakes."

6 Cross-Correlations in Price Fluctuations of Different Stocks

We know that a stock price does not vary in isolation from other stock prices, but that stock prices are correlated. This fact is not surprising because we know that "in economics everything depends on everything else." How do we *quantify* these cross-correlations of one stock with another? If we take the G values of four companies out of the 1000 that we have studied – corresponding to the shrinking or growing of each of these four companies in, say, a 30-minute interval. How does the behavior of these four companies during that half-hour interval affect your response to their price activity? If two of the companies were Pepsi and Coke, there would probably be some correlation in their behaviors.

In order to quantify these cross-correlations, we begin by calculating a cross-correlation matrix. If we have 1000 firms, we have a 1000×1000 matrix $\mathbf{C}$ each element C_{ij} which is the correlation of firm i with firm j. This large number of elements (1 million) does not frighten a physicist with a computer. Eugene Wigner applied random matrix theory 50 years ago to interpret the complex spectrum of energy levels in nuclear physics [37–48]. We do exactly the same thing, and apply random matrix theory to the matrix $\mathbf{C}$. We find that certain eigenvalues of that 1000×1000 matrix deviate from the predictions of random matrix theory, which has not eigenvalues greater than an upper bound of ≈ 2.0. Furthermore, the content of the eigenvectors corresponding to those eigenvalues correspond to well-defined business sectors. This allows us to define business sectors without knowing anything about the separate stocks: a Martian who cannot understand stock symbols could nonetheless identify business sectors [47,49].

7 Patterns in Firm Growth

In the economy, each firm depends on every other firm, and the interactions are not short-ranged nor are they of uniform sign. For example, Ford Motor Company is in trouble because they have been selling their Explorer vehicle with extremely unsafe tires – and the price of their stock is going down. Prospective buyers purchase General Motors cars instead. There is thus a *negative* correlation between the stock prices of the two companies. But now General Motors needs to hire more workers to make a larger number of cars, and the McDonald's near the assembly plant has many more customers at lunchtime – a *positive*. Sometime later the situation may change again. So we can say that the "spins" all interact with one another, and that these interactions change as a function of time.

Nevertheless, the general idea of a critical phenomenon seems to work. If the firms were spread out in a kind of chain, the correlations among them would decay exponentially. Instead, the firms interact with each other much the same way that subunits in critical phenomena interact with each other. This fact motivated a study carried out about five years ago by a group of physicists interacting with an economist [10,50,51]. They calculated the fluctuations in business firms from one year to the next. They found that if they broke the fluctuations into bins by size a tent-shaped distribution function was produced for each day of trading. The width of the tent was narrower for large firms than the width of the tent for small firms. This is not surprising, since a small firm has a potential to grow or shrink much more rapidly than a larger firm. When the widths of these tent-shaped distribution functions were plotted on log-log paper as a function of histogram size, the decreasing function turns out to be a straight line – corresponding to a power-law behavior in that function. The exponent in that power-law is ≈ 0.2. The linearity extends over a number of decades, indicating that the data collapse onto a single plot irrespective of scale.

Recent attempts to make models that reproduce the empirical scaling relationships suggest that significant progress on understanding firm growth may be well underway [52–55], leading to the hope of ultimately developing a clear and coherent "theory of the firm". One utility of the recent empirical work is that now any acceptable theory must respect the fact that power laws hold over typically six orders of magnitude; as Axtell put the matter rather graphically: *"the power law distribution is an unambiguous target that any empirically accurate theory of the firm must hit"* [8].

8 Universality of the Firm Growth Problem

Takayasu et al. have demonstrated that the above results are universal by moving outside the realm of US economies and studying firm behavior in other parts of the world [56].

Buldyrev et al. have shown that organizations (such as business firms) that are organized like trees will fluctuate in size [51]. The hierarchical structure is set up so that instructions from the top of the hierarchy propagate down to the branching lower levels of the structure. Within that structure is a disobedience factor – those lower down do not always obey the directives handed down from those above them. This factor is, of couse, crucial to the survival of the system. If employees always did only and exactly what they were told, any small mistake put into the system by a manager would grow and do an increasing amount of damage as it propagated through the expanding tree structure of the organization. On the other hand, the probability of an instruction being disobeyed cannot be one – or chaos would result. The propensity to disobey can be neither infinitesimal nor unity. The "obeying probability" needs to settle at a point at which the organization can maintain both its integrity and self-corrective flexibility. And the behavior of the exponent describing this probability is very similar to the behavior of critical exponents.

This result is fairly robust, not only as far as business firm fluctuations are concerned, but also in the size of countries. Lee et al. extend the same analysis used for business firms to countries – and with the same exponent [57]. Data can therefore be graphed on the same curve both for firms and for countries – where country size is measured by GDP.

We can see a similar pattern in the funding of university-based research. We researchers compete for research money the same way business firms compete for customers. Plerou et al. analyzed the funding of research groups over a 17-year period in the same way fluctuations in firm size were analyzed [58]. The results were very similar with the data collapsing onto the same curve.

As a final example, we mention the case of fluctuating bird populations in North America. In this case the exponent is 0.35 instead of 0.2. But, nevertheless, there seems to be some kind of property of organizations that we do not understand well [59].

9 "Take-Home Message"

So – what have we learned? First, that the approach we have emphasized is an empirical approach where one first seeks to uncover features of the complex economy that are challenges to understand. We find that there are two new universal scaling models in economics: (i) the fluctuation of price changes of any stock market is characterized by a PDF which is a simple power law with exponent $\alpha + 1 = 4$ that extends over 10^2 standard deviations (a factor of 10^8 on the y-axis); (ii) for a wide range of economic organizations, the histogram that shows how size of organization is inversely correlated to fluctuations in size with an exponent $\beta \approx 0.2$.

Neither of these two new laws has a firm theoretical foundation. This situation parallels the situation in the 1960s when the new field of critical phenomena also did not have a firm theoretical foundation for its new laws, but was awaiting the renormalization group. It is my hope that some of you in this room will rise to the challenge and try to find a firm theoretical foundation for the structure of the empirical laws that appear to be describing (i) finance fluctuations, and (ii) economic organizations.

Acknowledgements

We wish to thank our collaborators, L. A. Amaral, D. Canning, X. Gabaix, R. N. Mantegna, K. Matia, M. Meyer, and B. Rosenow from whom we have learned a great deal. We also thank NSF for financial support.

References

1. V. Pareto: *Cours d'Economie Politique* (Rouge, Lausanne and Paris 1897)
2. B.B. Mandelbrot: J. Business **36**, 394 (1963)
3. H.E. Stanley: *Introduction to Phase Transitions and Critical Phenomena* (Oxford University Press, Oxford 1971)
4. H. Takayasu (Ed.): *Empirical Science of Financial Fluctuations: The Advent of Econophysics* (Springer, Berlin 2002)
5. H.E. Stanley: Rev. Mod. Phys. **71**, S358 (1999)
6. R.N. Mantegna, H.E. Stanley: *An Introduction to Econophysics: Correlations and Complexity in Finance* (Cambridge University Press, Cambridge 2000)
7. J.P. Bouchaud: Quantitative Finance **1**, 105 (2001)
8. R.L. Axtell: Science **293**, 1818 (2001)
9. M.H.R. Stanley, S.V. Buldyrev, S. Havlin, R. Mantegna, M.A. Salinger, H.E. Stanley: Econ. Lett. **49**, 453 (1996)
10. M.H.R. Stanley, L.A.N. Amaral, S.V. Buldyrev, S. Havlin, H. Leschhorn, P. Maass, M.A. Salinger, H.E. Stanley: Nature **379**, 804 (1996)
11. L. Bachelier: *Théorie de la spéculation* [Ph.D. thesis in mathematics], Annales Scientifiques de l'Ecole Normale Supérieure **III-17**, 21 (1900)
12. R.N. Mantegna, H.E. Stanley: Nature **376**, 46 (1995)
13. Y. Liu, P. Gopikrishnan, P. Cizeau, M. Meyer, C.-K. Peng, H.E. Stanley: Phys. Rev. E **60**, 1390 (1999)
14. Z. Ding, C.W.J. Granger, R.F. Engle: J. Empirical Finance **1**, 83 (1993)
15. R.N. Mantegna, H.E. Stanley: Phys. Rev. Lett. **73**, 2946 (1994)
16. B. Podobnik, P.C. Ivanov, Y. Lee, A. Chessa, H.E. Stanley: Europhysics Letters **50**, 711 (2000)
17. R.N. Mantegna, H.E. Stanley: 'Ultra-Slow Convergence to a Gaussian: The Truncated Lévy Flight' in *Lévy Flights and Related Topics in Physics*, Ed. by M.F. Shlesinger, G.M. Zaslavsky, U. Frisch (Springer, Berlin 1995), pp. 300–312
18. R.N. Mantegna, H.E. Stanley: Physica A **254**, 77 (1998)
19. B. Podobnik, P.C. Ivanov, Y. Lee, H.E. Stanley: Europhysics Letters **52**, 491 (2000)
20. P.C. Ivanov, B. Podobnik, Y. Lee, H.E. Stanley: Physica A **299**, 154 (2001)
21. J.A. Skjeltorp: Physica **283**, 486 (2001)
22. T. Lux: Appl. Finan. Econ. **6**, 463 (1996)
23. P. Gopikrishnan, M. Meyer, L.A.N. Amaral, H.E. Stanley: Eur. Phys. J. B **3**, 139 (1998)
24. V. Plerou, P. Gopikrishnan, L.A.N. Amaral, M. Meyer, H.E. Stanley: Phys. Rev. E **60**, 6519 (1999)
25. P. Gopikrishnan, V. Plerou, L.A.N. Amaral, M. Meyer, H.E. Stanley: Phys. Rev. E **60**, 5305 (1999)

26. B. Gutenberg, C.F. Richter: *Seismicity of the Earth and Associated Phenomenon*, 2nd Edition (Princeton University Press, Princeton 1954)
27. D.L. Turcotte: *Fractals and Chaos in Geology and Geophysics* (Cambridge University Press, Cambridge 1992)
28. J.B. Rundle, D.L. Turcotte, W. Klein: *Reduction and Predictability of Natural Disasters* (Addison-Wesley, Reading 1996)
29. K. Matia, P. Gopikrishnan, V. Plerou, L.A.N. Amaral, S. Goodwin, H.E. Stanley: 'Scaling for the Distribution of the Commodity Market Variations', cond-mat/0202028
30. V. Plerou, P. Gopikrishnan, L.A.N. Amaral, X. Gabaix, H.E. Stanley: Phys. Rev. E **62**, 3023 (2000)
31. P. Gopikrishnan, V. Plerou, X. Gabaix, H.E. Stanley: Phys. Rev. E **62**, 4493 (2000)
32. S. Ghashgaie, W. Breymann, J. Peinke, P. Talkner, Y. Dodge: Nature **381**, 767 (1996)
33. R.N. Mantegna, H.E. Stanley: Nature **383**, 587 (1996)
34. R.N. Mantegna, H.E. Stanley: Physica A **239**, 255 (1997)
35. V. Plerou, P. Gopikrishnan, X. Gabaix, H.E. Stanley: Phys. Rev. E **66** 027104 (2002)
36. V. Plerou, P. Gopikrishnan, H.E. Stanley: 'Critical Threshold and Symmetry Breaking in Stock Demand', cond-mat/0111349
37. M.L. Mehta: *Random Matrices* (Academic Press, Boston 1991)
38. T. Guhr, A. Müller-Groeling, H.A. Weidenmüller: Phys. Reports **299**, 189 (1998)
39. E.P. Wigner: Ann. Math. **53**, 36 (1951)
40. E.P. Wigner: Proc. Cambridge Philos. Soc. **47**, 790 (1951)
41. E.P. Wigner: 'Results and theory of resonance absorption' in *Conference on Neutron Physics by Time-of-flight* (Gatlinburg, Tennessee 1956), pp. 59-70
42. M.L. Mehta, F.J. Dyson: J. Math. Phys. **4**, 713 (1963)
43. F.J. Dyson: Revista Mexicana de Física **20**, 231 (1971)
44. A.M. Sengupta, P.P. Mitra: Phys. Rev. E **60**, 3389 (1999)
45. L. Laloux, P. Cizeau, J.-P. Bouchaud, M. Potters: Phys. Rev. Lett. **83**, 1469 (1999)
46. V. Plerou, P. Gopikrishnan, B. Rosenow, L.A.N. Amaral, H.E. Stanley: Phys. Rev. Lett. **83**, 1471 (1999)
47. P. Gopikrishnan, B. Rosenow, V. Plerou, H.E. Stanley: Phys. Rev. E **64**, 035106 (2001)
48. V. Plerou, P. Gopikrishnan, B. Rosenow, L.A.N. Amaral, T. Guhr, H.E. Stanley: Phys. Rev. E **65** 066136 (2002)
49. B. Rosenow, V. Plerou, P. Gopikrishnan, H.E. Stanley: Europhys. Lett. **59**, 500 (2002)
50. L.A.N. Amaral, S.V. Buldyrev, S. Havlin, H. Leschhorn, P. Maass, M.A. Salinger, H.E. Stanley, M.H.R. Stanley: J. Phys. I France **7**, 621 (1997)
51. S.V. Buldyrev, L.A.N. Amaral, S. Havlin, H. Leschhorn, P. Maass, M.A. Salinger, H.E. Stanley, M.H.R. Stanley: J. Phys. I France **7**, 635 (1997)
52. L.A.N. Amaral, S.V. Buldyrev, S. Havlin, M.A. Salinger, H.E. Stanley: Phys. Rev. Lett. **80**, 1385 (1998)
53. J. Sutton: 'The variance of firm growth rates: the 'scaling' puzzle', working paper, London School of Economics (2001)

54. G. Bottazzi, G. Dosi, M. Riccaboni, F. Pammolli: 'The scaling of growth processes and the dynamics of business firms' Economics Department at University of Siena and Pisa San' Anna School of Advanced Studies, working paper (2001)
55. H.E. Stanley, L.A.N. Amaral, P. Gopikrishnan, V. Plerou, M.A. Salinger: 'Scale invariance and universality in economic phenomena', working paper, based on Invited Talk at the 56th ESEM Econometric Society meeting 26-29 Aug 2001 in Lausanne
56. H. Takayasu, K. Okuyama: Fractals **6**, 67 (1998)
57. Y. Lee, L.A.N. Amaral, D. Canning, M. Meyer, H.E. Stanley: Phys. Rev. Letters **81**, 3275 (1998)
58. V. Plerou, L.A.N. Amaral, P. Gopikrishnan, M. Meyer, H.E. Stanley: Nature **400**, 433 (1999)
59. T. Keitt, H.E. Stanley: Nature **393**, 257 (1998)

Part III

Bioinformatics

New Algorithms and the Physics of Protein Folding

Ulrich H.E. Hansmann

Recent years have seen an increased interest in the physics of protein folding. While most investigations focus on minimal protein models, we will show that the thermodynamics of folding can also be studied for realistic models, when modern simulation techniques such as the generalized-ensemble approach are employed. Some recent results are presented.

1 Introduction

One of the most important recent developments in life sciences is the successful enciphering of whole genomes. However, for most of the newly found sequences we do not know the function and working of the corresponding proteins. Since the biological function of proteins are closely related to their 3D shape, there is an increased interest in the prediction of the structure of proteins from their sequence of amino acids. The attempt to understand the mechanism by which a protein is driven into its unique biologically active structure is called the protein folding problem. Its importance is obvious: a detailed knowledge of the sequence-structure relation would allow us to better understand the (mal)function of enzymes, working of the imune system and could yield to more efficient ways of drug design.

One way to study the folding problem is by means of computer experiments. For instance, significant new insight was gained from simulation of minimal protein models which capture only a few, but presumably dominant interactions in proteins [1] – [2]. However, minimal models do not allow to investigate the details of structural transitions and folding in *specific* proteins. More adequate for the later purpose are detailed representations of proteins where interactions among all atoms are taken into account (for a review, see, for instance, Ref. [3]). These interactions can be divided into two groups: the conformational energy of the protein and a term describing the interaction of the protein with the surrounding solvent. Especially the later term probes to be a serious hurdle since explicit inclusions of solvent molecules are computationally very demanding. On the other hand, the intramolecular interactions of the first term are modeled in various atomic force fields such as ECEPP [4], CHARMM [5] or AMBER [6]. As example we show here the energy function of ECEPP which was used in our simulations. It is given by the sum of the electrostatic term E_{es}, the van der Waals energy E_{vdW}, and hydrogen-bond term E_{hb} for all pairs of atoms in the peptide together with the torsion term E_{tors} for all torsion angles:

$$E_{tot} = E_{es} + E_{vdW} + E_{hb} + E_{tors}, \quad (1)$$

$$E_{es} = \sum_{(i,j)} \frac{332 q_i q_j}{\epsilon r_{ij}}, \quad (2)$$

$$E_{vdW} = \sum_{(i,j)} \left(\frac{A_{ij}}{r_{ij}^{12}} - \frac{B_{ij}}{r_{ij}^{6}} \right), \quad (3)$$

$$E_{hb} = \sum_{(i,j)} \left(\frac{C_{ij}}{r_{ij}^{12}} - \frac{D_{ij}}{r_{ij}^{10}} \right), \quad (4)$$

$$E_{tors} = \sum_{l} U_l \left(1 \pm \cos(n_l \alpha_l) \right). \quad (5)$$

Here, r_{ij} is the distance between the atoms i and j, and α_l is the torsion angle for the chemical bond l. The parameters ($q_i, A_{ij}, B_{ij}, C_{ij}, D_{ij}, U_l$ and n_l) were calculated from crystal structures of amino acids using a semi-empirical molecular orbital method, etc. Since the bond lengths and bond angles are set constant here, the true degrees of freedom are rotations around these bonds characterized by dihedral angles ϕ, ψ, ω, and χ_i.

The complex form of the intramolecular forces and of the interaction with the solvent makes it extremely difficult to sample low-energy protein conformation. Containing both repulsive and attractive terms, all-atom models of proteins lead to a very rough energy landscape and a huge number of local minima. As a consequence, a typical thermal energy of the order $k_B T$ is much less than the energy barriers that the protein has to overcome, once one enters the low-temperature region. Hence, simple canonical Monte Carlo or molecular dynamics simulations will get trapped in a local minimum and not thermalize within a finite amount of available CPU time. Only small parts of the configuration space are sampled (in a finite number of simulation steps) and physical quantities cannot be calculated accurately.

Much effort was put into the development of novel simulation techniques to alleviate the above stated multiple-minima problem (for a recent review, see Ref. [7]). One successful method to enhance sampling in protein folding simulations is the so-called *generalized ensemble* approach [8]. In the following we will concentrate on this approach and demonstrate its usefulness for studies of the thermodynamics of folding and prediction of the three dimensional shapes of proteins. A number of recent applications will be presented.

2 The Generalized-Ensemble Approach

In the following section, a short introduction into the generalized-ensemble approach for protein-folding simulations will be given. A more detailed review can be found in Ref. [9].

Most computer simulations today are performed in a canonical ensemble where each configuration is weighted with the Boltzmann factor $w_B(T, E) = e^{-E/k_B T}$. Update schemes like the Metropolis algorithm realize a Markov process and generate, after thermalization, configurations which are in equilibrium with respect to the canonical distribution at temperature T:

$$P_B(T, E) = n(E)\, e^{-E/k_B T}, \tag{6}$$

where $n(E)$ is the density of states (spectral density).

On the other hand, a generalized-ensemble simulation is characterized by the condition that a Monte Carlo or molecular dynamics simulation shall lead to a uniform distribution of a pre-chosen physical quantity. Probably the earliest realization of this idea is *umbrella sampling* [10]. This idea was lately revived [11] – [13] and its usefulness for simulations of biological molecules has been increasingly recognized. The first application of these new techniques to protein simulations can be found in Ref. [14] where a Monte Carlo technique was used. Later, a formulation for the molecular dynamics method was also developed [15,16].

In the following, we will first discuss the main ideas behind generalized-ensemble methods for a prominent example, the multicanonical algorithm, [11] before turning to other variants. For simplicity we will restrict ourselves to ensembles which lead to flat distributions in only *one* variable. Extensions to higher dimensional generalized ensembles are straightforward [17,18] as are combinations with annealing techniques [19,20].

2.1 Multicanonical Sampling

In the *multicanonical algorithm* [11] configurations with energy E are assigned a weight $w(E)$ such that the distribution of energies

$$P(E) \propto n(E) w(E) = \text{const}, \tag{7}$$

where $n(E)$ is the spectral density. Since all energies appear with the equal probability, a free random walk in the energy space is enforced: the simulation can overcome *any* energy barrier and will not get trapped in one of the many local minima. In order to demonstrate the latter point the "time series" of energy is shown in Fig. 1 as a function of Monte Carlo sweeps for both a regular canonical Monte Carlo simulation at temperature $T = 50$ K (dotted curve) and a multicanonical simulation. The displayed data are from a simulation of the pentapeptide Met-enkephalin, which has the amino-acid sequence Tyr-Gly-Gly-Phe-Met. The ECEPP/2 force field [4] was used and

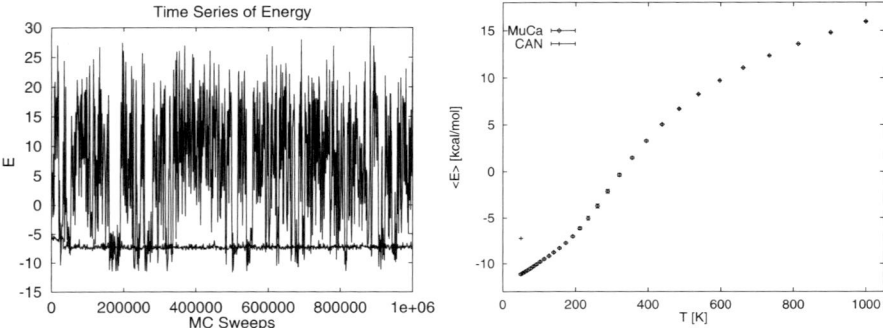

Fig. 1. *Left:* "Time series" of energy for the pentapeptide Met-enkephalin. Shown are both the results from a canonical simulation at $T = 50$ K (dotted line) and a multicanonical simulation. The results were used to calculate the average energy $<E>$ as a function of temperature T (*Right*)

the peptide bond angles fixed to their common value $\omega = 180°$. Starting from a random configuration the two simulations continued for 1,000,000 Monte Carlo sweeps. For the canonical run the curve stays around the value $E = -7$ kcal/mol with small thermal fluctuations, reflecting the low-temperature nature. The run has apparently been trapped in a local minimum, since the mean energy at this temperature is $<E> = -11.1$ kcal/mol as found in Ref. [19]. On the other hand, the multicanonical simulation covers a much wider energy range than the canonical run. It is a random walk in energy space, which keeps the simulation from getting trapped in a local minimum.

From such a multicanonical simulation one can not only locate the energy global minimum, but also calculate the expectation value of any physical quantity $\mathcal{O}$ at temperature T by re-weighting techniques [21]

$$<\mathcal{O}>_T = \frac{\int dE \, \mathcal{O}(E) P_{mu}(E) \, w_{mu}^{-1}(E) \, e^{-E/k_B T}}{\int dE \, P_{mu}(E) \, w_{mu}^{-1}(E) \, e^{-E/k_B T}} \, . \tag{8}$$

As an example, the average (ECEPP/2) energy of Met-enkephalin as a function of temperature is also shown in Fig. 1. Using the same data as displayed in the "time series", thermodynamic averages were calculated with Eq. (8) over a temperature range of 1000 K to below 100 K where canonical simulations failed (see the value for the average energy at $T = 50$ K as obtained by the canonical simulation of 1,000,000 MC sweeps).

It has to be noted that unlike in the canonical ensemble the weights $w_{mu}(E) \propto n^{-1}(E)$ are not *a priori* known (in fact, knowledge of the exact weights is equivalent to obtaining the density of states $n(E)$, i.e., solving the system) and one needs their estimates for a numerical simulation. Calculation of the multicanonical (and other generalized-ensemble weights) is usually done by a simple iterative procedure [14,19]. For the calculations in Ref. [14] about 40 % of the total CPU time was spent for this task. Several attempts were made to obtain generalized-ensemble weights in an faster way; see, for instance, Refs. [22–24].

2.2 1/k-Sampling

In multicanonical simulations the computational effort increases with the number of residues like $\approx N^4$ (when measured in Metropolis updates) [25]. In general, the computational effort in simulations increases with $\approx X^2$ where X is the variable in which one wants a flat distribution. This is because generalized-ensemble simulations realize by construction of the ensemble a $1D$ random walk in the chosen quantity X. In the multicanonical algorithm the reaction coordinate X is the potential energy $X = E$. Since $E \propto N^2$ the above scaling relation for the computational effort $\approx N^4$ is recovered. Hence, multicanonical sampling is not always the optimal generalized-ensemble algorithm in protein simulations. A better scaling of the computer time with size of the molecule may be obtained by choosing more appropriate reaction coordinate for our ensemble than the energy.

This is the motivation behind the search for other algorithms such as $1/k$ sampling [12] where the (microcanonical) *entropy* S is sampled uniformly:

$$P_{1/k}(S) = \text{const} . \tag{9}$$

This equation implies that

$$P_{1/k}(E) = P_{1/k}(S)\frac{dS}{dE} \propto \frac{dS}{dE} = \frac{d\log n(E)}{dE} . \tag{10}$$

To realize such an ensemble, Hesselbo and Stinchcombe [12] proposed that configurations are assigned a weight

$$w_{1/k}(E) = \frac{1}{k(E)} , \qquad k(E) = \int_{-\infty}^{E} dE'\, n(E') . \tag{11}$$

This implies that

$$P_{1/k}(E) = n(E) w_{1/k}(E) = \frac{n(E)}{k(E)} = \frac{d\log k(E)}{dE} . \tag{12}$$

Because the density of states $n(E)$ is a rapidly increasing function of energy, we have $\log k(E) \approx \log n(E)$ for wide range of values of E. Hence, Eqs. (10) and (12) are equivalent, and a random walk in the entropy space is realized. Since the entropy $S(E)$ is a monotonically increasing function of energy, a random walk in entropy implies a random walk in energy space (with more weight towards low-energy region.

Again, the weight $w_{1/k}(E)$ is not *a priori* known and its estimator has to be calculated. Thermodynamic quantities at any temperature can be calculated by Eq. (8), in which $P_{mu}(E)$ and $w_{mu}(E)$ are replaced by $P_{1/k}(E)$ and $w_{1/k}(E)$, respectively.

2.3 Simulated Tempering

In *simulated tempering* [13] the temperature itself becomes a dynamic variable and is sampled uniformly. Temperature and configuration are both updated with a weight:

$$w_{ST}(T, E) = e^{-E/k_B T - g(T)} , \qquad (13)$$

where the function $g(T)$ is chosen so that the probability distribution of temperature is given by

$$P_{ST}(T) = \int dE\, n(E)\, e^{-E/k_B T - g(T)} = \text{const} . \qquad (14)$$

Physical quantities have to be sampled for each temperature point separately and expectation values at intermediate temperatures are calculated by reweighting techniques.[21]

As common in generalized-ensemble simulations, the weight $w_{ST}(T, E)$ is not *a priori* known (since it requires knowledge of the parameters $g(T)$) and their estimator has to be calculated. They can be again obtained by an iterative procedure. In the simplest version the improved estimator for $g^{(i)}(T)$ for the i-th iteration is calculated from the histogram of *temperature* distribution $H_{ST}^{(i-1)}(T)$ of the preceding simulation as follows:

$$g^{(i)}(T) = g^{(i-1)}(T) + \log H_{ST}^{(i-1)}(T) . \qquad (15)$$

2.4 Other Generalized Ensembles

In protein simulations we are interested in an ensemble where not only the low-energy region can be sampled efficiently but also the high-energy states can be visited with finite probability. In this way the simulation can overcome energy barriers and escape from local minima. The probability distribution of energy should resemble that of an ideal low-temperature canonical distribution, but with a tail to higher energies. To obtain such an ensemble we proposed in Ref. [26] to update configurations according to the following probability weight:

$$w(E) = \left(1 + \frac{\beta(E - E_0)}{n_F}\right)^{-n_F} , \qquad (16)$$

where E_0 is an estimator for the ground-state energy, n_F is the number of degrees of freedom of the system, and $\beta = 1/k_B T$ is the inverse temperature with a low temperature T. Note that this weight can be understood as a special case of the weights used in Tsallis generalized mechanics formalism [27] (the Tsallis parameter q is chosen as $q = 1 + 1/n_F$). The weight reduces in the low-energy region to the canonical Boltzmann weight $\exp(-\beta E)$ for

$\beta(E - E_0)/n_F \ll 1$. On the other hand, high-energy regions are no longer exponentially suppressed but only according to a power law, which enhances excursions to high-energy regions.

Recently, another ensemble with similar properties was proposed where conformations enter with a weight $w(E) = \exp(f(E)/k_B T)$. Here, $f(E)$ is a non-linear transformation of the potential energy onto the interval $[0,1]$ and T is a low temperature. The physical idea behind such an approach is to allow the system to "*tunnel*" through energy barriers in the potential energy surface [28]. Such a transformation can be realized by

$$f(E) = e^{-(E-E_0)/n_F} , \qquad (17)$$

where E_0 is again an estimate for the ground state and n_F the number of degrees of freedom of the system. Note that the location of all minimas is preserved. Hence, at a given low temperature T, the simulation can pass through energy barriers of arbitrary height, while the low energy region is still well resolved. An exploratory study on the efficiency of this algorithm for protein-folding simulations can be found in Ref. [29], and related ideas are described in Ref. [30].

In both ensembles a broad range of energies is sampled. Hence, one can use again reweighting techniques [21] to calculate thermodynamic quantities over a large range of temperatures. In contrast to other generalized-ensemble techniques the weights are explicitly given for both new ensembles. One needs only to find an estimator for the ground-state energy E_0 which is easier than the determination of weights for other generalized ensembles.

2.5 Parallel Tempering

Even within the generalized-ensemble approach simulations of proteins can still be hampered by large correlations between the sampled conformations. In theory, this obstacle can be overcome by improved updates which allow a much faster sampling. Examples are collective updates such as the re-bridging scheme [31,32] which change local groups of variables. For the purpose of changing *whole* configurations another recently developed technique, parallel tempering [33], proved valuable in generalized-ensemble simulations.

Parallel tempering was first introduced to protein simulations in Ref. [34]. In this approach one considers an artificial system built up of N *non–interacting* copies of the molecule, each at a different temperature T_i. A state of the artificial system is specified by $\mathcal{C} = \{C_1, C_2, ..., C_N\}$, where each C_i is a set of (generalized) coordinates describing the configuration of the i-th copy. In addition to standard Monte Carlo or molecular dynamics moves which effect only one copy, parallel tempering introduces now a new *global* update [33]: the exchange of conformations between two copies i and $j = i+1$: $C_i^{new} = C_j^{old}$ and $C_j^{new} = C_i^{old}$. This replica exchange move is accepted or rejected according to the Metropolis criterion with probability:

$$w(\mathcal{C}^{old} \to \mathcal{C}^{new})$$
$$= \min(1, \exp(-\beta_i E(C_j) - \beta_j E(C_i) + \beta_i E(C_i) + \beta_j E(C_j))) \ . \tag{18}$$

It is obvious that through the exchange of conformations the Markov chain converges at low temperatures much faster towards the stationary distribution than it does in the case of a regular canonical simulation with only local moves. Note that the parallel tempering technique does not require Boltzmann-weights. The method works with any set of weights and parallel tempering can therefore easily be combined with generalized-ensemble simulations. The speed-up of generalized-ensemble simulations through use of parallel tempering moves was first demonstrated in Ref. [34].

3 The Thermodynamics of Folding

Since generalized ensembles allow to calculate thermodynamic averages over a wide range of temperatures, it follows that the approach is well suited to study the thermal behavior of proteins and peptides. This was successfully used in Ref. [20] where we studied thermodynamics of helix-coil transitions in amino-acid homopolymers. We were able to calculate estimators of the helix propagation parameter s which was also obtained by recent experiments [35]. For three characteristic amino acids, alanine, valine and glycine, we found $s(\text{Ala}) = 1.5 \sim 1.6$, $s(\text{Val}) = 0.37 \sim 0.45$, and $s(\text{Gly}) = 0.13 \sim 0.16$ around the experimentally relevant temperature ($\sim 0°$ C). These values are in remarkable agreement with the experiments,[35] where they give $s(\text{Ala}) = 1.54 \sim 2.19$, $s(\text{Val}) = 0.20 \sim 0.93$, and $s(\text{Gly}) = 0.02 \sim 0.57$.

More recently we studied the thermodynamics of folding of some proteins and peptides. Our work was motivated by the "new view" [1] of the protein folding which became increasingly popular over the last few years (for references, see, for instance, [36,37,1]). Its basic assumption is the funnel concept, namely that the energy landscape of a protein is rugged but with a sufficient overall slope towards the native structure [1]. Folding occurs by a multi-pathway kinetics and the particulars of the folding funnel determine the transitions between the different thermodynamic states [1,38]. However, these new concepts were derived from the analytical and numerical studies of minimal protein models [1,2] and need to be checked in all-atom simulations of suitable proteins. Such test motivated the research described in the next two sub-sections.

3.1 Helix-Coil Transitions in Homopolymers

A common, ordered structure in proteins is the α-helix and it is conjectured that formation of α-helices is a key factor in the early stages of protein-folding. It is long known that α-helices undergo a sharp transition towards a random coil state when the temperature is increased. The characteristics of this so-called helix-coil transition have been studied extensively [39], most recently in Refs. [25,40–43]. They are usually described in the framework of Zimm-Bragg-type theories [44] in which the homopolymers are approximated by a one-dimensional Ising model with the residues as "spins" taking values "helix" or "coil", and solely local interactions. Hence, in such theories thermodynamic phase transitions are not possible. However, in previous work [25,41–43] evidence was presented that poly-alanine exhibits a phase transition between the ordered helical state and the disordered random-coil state when interactions between all atoms in the molecule are taken into account. In the following we use generalized-ensemble simulations to demonstrate the differences in the critical behavior of our model and the Zimm-Bragg theory. In a second step we calculated a set of critical exponents which characterize the helix-coil transition. We compare then these exponents for two different protein models to test whether the fundamental mechanism of helix-coil transitions in biological molecules depends only on gross features in the energy function or also on their details, i.e. whether for this case a law of corresponding states can be applied to explain structural properties of real molecules from studies of related minimal models.

The generalized-ensemble technique, used in these investigations, was multicanonical sampling. The ECEPP/2 force field [4] was utilized in our simulations and explicit solvent molecules were neglected. The peptide-bond dihedral angles ω were fixed at the value 180° for simplicity. Note that with the electrostatic energy term E_C our model contains a long range interaction neglected in the Zimm-Bragg theory [44]. Chains of up to $N = 30$ monomers were considered. We needed between 40,000 sweeps ($N = 10$) and 500,000 sweeps ($N = 30$) for the weight factor calculations by the iterative procedure described first in Refs. [11]; and our results relied on a production run of N_{sw}=400,000, 500,000, 1,000,000, and 3,000,000 sweeps for $N = 10, 15, 20,$ and 30, respectively.

Since the multicanonical algorithm allows us to calculate estimates for the spectral density, we can construct the partition function by

$$Z(\beta) = \sum_E n(E) u^E = \sum_E P_{mu}(E) w_{mu}^{-1}(E) u^E , \qquad (19)$$

where $u = e^{-\beta}$ with the inverse temperature $\beta = 1/k_B T$. The complex solutions of the partition function determine the critical behavior of the model. They are the so-called Fisher zeros [45,46], and correspond to the complex extension of the temperature variable.

Phase transitions can also be described by means of the distribution of the zeros of the grand partition function in the complex fugacity plane [47]. In the case of the Ising model this description corresponds to a study of this model, with an external magnetic field H mathematically extended from real to complex values. In Ref. [48] Lee and Yang proved the famous circle theorem, namely that in the Ising model with ferromagnetic couplings the complex zeros lie on the unit circle in the complex y–plane, $y = e^{-2\beta H}$. For poly-alanine, the analogous of the magnetization in the Ising model is the number of helical residues M. Formally we can introduce a (non physical) external field H as the conjugate variable to this order parameter M (which plays the role of a magnetization), and study the Yang-Lee zeros by constructing the corresponding partition function:

$$Z(u,y) = \sum_{M=0}^{N-2} \sum_{E} n(E,M) u^E y^M \quad \text{with } u = e^{-\beta} \quad \text{and } y = e^{-H} \ . \quad (20)$$

If the helix-coil transition can indeed be described by the Zimm-Bragg model, then the distribution of the Yang-Lee zeros in function of the temperature should resemble that of the one dimensional Ising model (i.e. a unit circle). On the other hand, a substantially different distribution would demonstrate that the helix-coil transition in poly-alanine is not accurately described by the Zimm-Bragg theory. Our results are displayed in Fig. 2 (left) for the case of chain lengths $N = 30$. Similar distributions are obtained for smaller systems. Our results clearly exclude the possibility that the complex zeros are distributed on the unit circle. On the other hand, the Yang-Lee zeros for the 1D Ising model (which can be calculated exactly) fall on the unit circle. This is shown in Fig. 2 (right) for chain lengths $N = 5, 15$ and 30.

In the following we concentrate our analyses on the Fisher zeros of Eq. (19). From the FSS relation by Itzykson *et al.* [46] for the leading zero $u_1^0(N)$,

$$u_1^0(N) = u_c + A N^{-1/d\nu}[1 + O(N^{y/d})] \ , \quad y < 0 \quad (21)$$

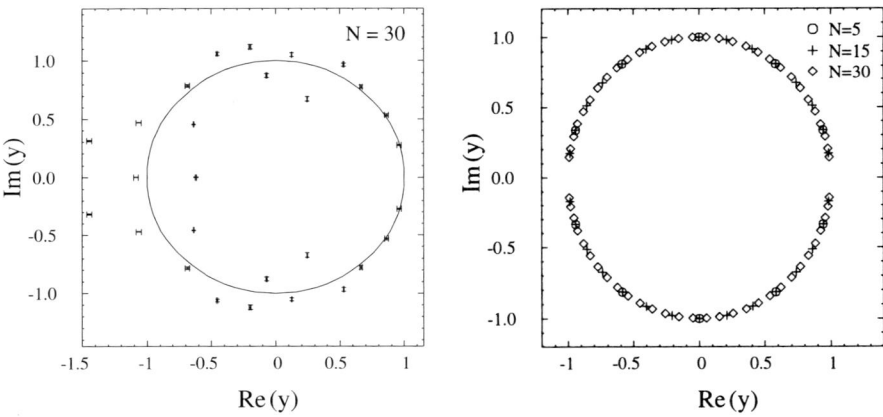

Fig. 2. *Left:* Distribution of Yang-Lee zeros at T_c for poly-alanine. We draw a unit circle to show how the zeros are distributed around it *Right:* Corresponding figure for the Ising chain with lengths $N = 5, 15$ and 30

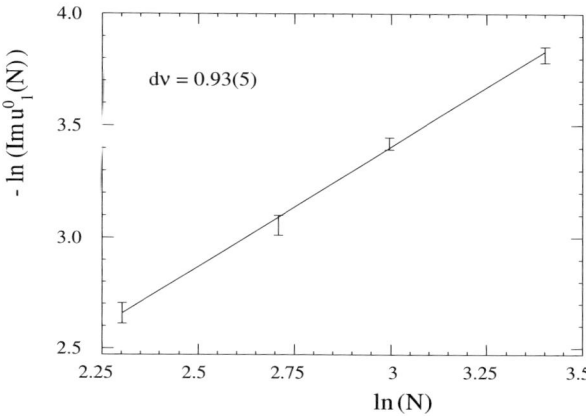

Fig. 3. Linear regression for $-\ln(\operatorname{Im} u_1^0(N))$ in the range $N = 10 - 30$

we obtain for sufficiently large N by means of the linear regression

$$-\ln|u_1^0(N) - u_c| = \frac{1}{d\nu}\ln(N) + a, \qquad (22)$$

the scaling exponent $d\nu$. Including chains of all lengths, $N = 10 - 30$, this approach leads to $d\nu = 0.93(5)$. Figure 3 displays the corresponding fit. We remark that the above scaling relation also holds where we expect effective scaling exponents $d\nu = \alpha = \gamma = 1$ [49].

Through the scaling relation for the peak of the specific heat and of the susceptibility, we can evaluate two other critical exponent, the specific heat exponent α and the susceptibility exponent γ, by:

$$C_N^{max} \propto N^{\alpha/d\nu} \quad \text{and} \quad \chi_{MAX} \propto N^{\gamma/d\nu}. \qquad (23)$$

The results for the exponent α give $\alpha = 0.89(12)$ (Q=0.9) when all chains are considered. In the same way we obtained for the exponent γ a value of $\gamma = 1.06(14)$ ($Q = 0.5$). The scaling plot for the specific heat is shown in Fig. 4: curves for all lengths of poly-alanine chains collapse on each other indicating the validity of finite size scaling of our poly-alanine data.

We extend now our previous analyses to examine the question of universality of the helix-coil transition. For this purpose, we compare our results for the detailed, all-atomic representation of a poly-alanine chain with a simple coarse-grained model describing the general features of helix-forming polymers [40]. If the two models yield the same key physical characteristics, then we at least have one concrete example of the validity of the corresponding state principle or universality hypothesis in biopolymer structures.

Our second model [40] relies on a wormlike chain to describe the backbone of the molecule, while a general directionalized interaction, in terms of a simple square well form, is used to capture the essence of hydrogen-like bonding.

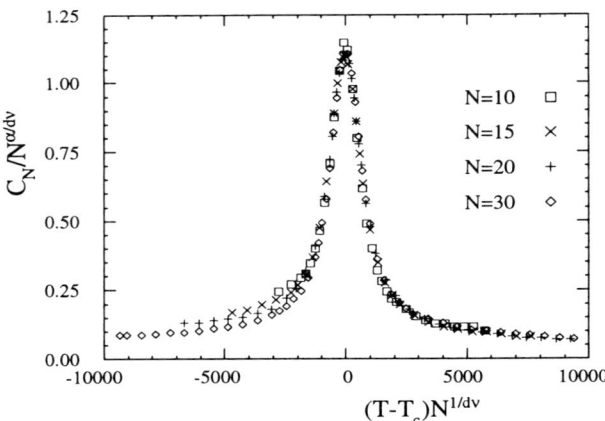

Fig. 4. Scaling plot for the specific heat $C_N(T)$ as a function of temperature T, for poly-alanine molecules of chain lengths $N = 10, 15, 20$ and 30

The interaction energy between the residue labeled i and j is modeled by,

$$V_{ij}(\mathbf{r}) = \begin{cases} \infty & r < D \\ -v & D \leq r < \sigma \\ 0 & \sigma \leq r \end{cases} \quad (24)$$

where $v = \epsilon[\hat{\mathbf{u}}_i \cdot \hat{\mathbf{r}}_{ij}]^6 + \epsilon[\hat{\mathbf{u}}_j \cdot \hat{\mathbf{r}}_{ij}]^6$, $\hat{\mathbf{u}}_i = (\hat{\mathbf{r}}_{i+1,i}) \times (\hat{\mathbf{r}}_{i,i-1})$, $\hat{\mathbf{r}}_{ij}$ is the unit vector between monomer i and j, $D = 3/2a$ is the diameter of a monomer, $\sigma = \sqrt{45/8}a$ is the bonding diameter, and a is the bond length while bond angle is fixed at $60°$.

Table 1 lists the obtained critical exponents for our minimal model, together with the corresponding values for our all-atom representation of polyalanine. Comparing the critical exponents of our two models we see that the estimates for the correlation exponent $d\nu$ agree well for the two models. Within the error bars, the estimates for the susceptibility exponent γ also agree. The estimates for the specific heat exponent α seem to disagree within the error ranges. However, in view of the fact that both analyses are based on small system size the true error ranges could be actually larger than the ones quoted here. Using these rather crude results, we can already observe a striking similarity in finite-size scalings of the two model which seems to indicate that a universality principle can be established for the coil-helix transition.

Table 1. Summary of the critical exponents obtained for the two models.

	All-atomic	Minimal
$d\nu$	0.98(11)	0.96(8)
α	0.89(12)	0.70(16)
γ	1.06(14)	1.3(2)

3.2 Energy Landscape Analysis of Peptides

Another way of testing the validity of the new theories of folding for all-atom representations of proteins is by directly investigating the free energy landscape of suitable peptides and proteins and compare it with the predictions of energy landscape theory and funnel concept. Unlike many other methods, the generalized-ensemble approach allows such a direct observation of the folding funnel.

Our system of choice was again the linear peptide Met-enkephalin. At low temperatures one finds for this peptide two major groups of well-defined compact structures which are characterized (and stabilized) by specific hydrogen bonding patterns. Structure A is the ground-state conformation in ECEEP/2 and has a Type II' β-turn between the second and last residue, stabilized by two possible hydrogen bonds. The structure B, the second-lowest energy state, is characterized by hydrogen bond between Tyr-1 and Phe-4 resulting in a Type-II β-turn between the first and fourth residue. The overlap of a given configuration with the ground state (structure A) and the second-lowest-energy state (structure B), respectively, allows to distinguish between the various low-energy conformations and defines in a natural way two order parameters for our system.

The generalized-ensemble algorithm used in this study [52] is the one described in Ref. [26] and relies on the weight in Eq. (16). We fixed the peptide bond angles ω to their common value $180°$, which left us with 19 torsion angles (ϕ, ψ, and χ) as independent degrees of freedom (i.e., $n_F = 19$). All thermodynamic quantities were then calculated from a single production run of 1,000,000 MC sweeps which followed 10,000 sweeps for thermalization. The generalized-ensemble simulation was complemented by 100 canonical Monte Carlo simulations of up to 200,000 MC sweeps for each chosen temperature.

In Fig. 5 (left) we show the free energy landscape as a function of both the overlap O_A with the ground state and the overlap O_B with structure B in the high-temperature situation (at $T = 1000$ K). The free energy has its minimum at small values of the overlap indicating that both conformers appear with only very small frequency at high temperature. We have superimposed on the free-energy landscape, as calculated from the generalized-ensemble simulation, the folding trajectory of a canonical Monte Carlo simulation (marked by dots) at the same temperature. However, we did not connect the dots, for otherwise the plot would become unreadable. It is obvious that the concentration of the dots marks the time the simulation spent in a certain region of the landscape. We see that this time is strongly correlated with the free-energy as calculated from the generalized-ensemble simulation. For instance, we have no dots for $O_A \approx 1$ (i.e. the ground state region), a region of the energy landscape suppressed by many $k_B T$.

At $T = 300K$, which is essentially the collapse temperature $T_\theta = 295 \pm 30$ K of Refs. [50,51], a large part of the space of possible configurations lies within the $2k_B T$ contour as is clear from Fig. 5 (right). Correspondingly,

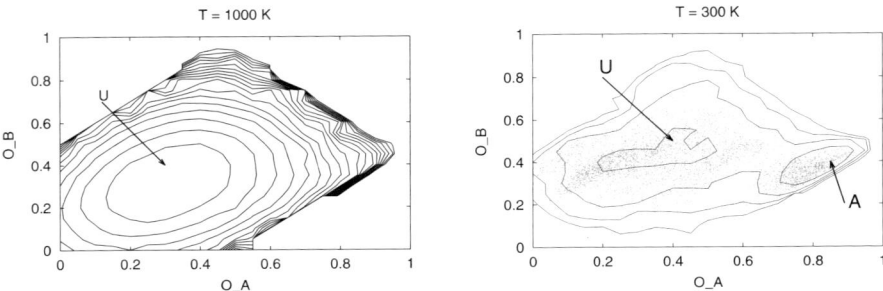

Fig. 5. Free energy landscape of Met-enkephalin for $T = 1000$ K (*left*) and for $T = 300K$ (*right*). The contour lines are spaces $1K_bT$

the dots, which mark the folding trajectory of a canonical simulation at this temperature, are equally distributed over the whole plot. We remark that at this temperature the folded conformation was found in all of the 100 canonical simulations. We found as average folding time $t_f = 19864$ MC sweeps and as the escape time out of conformer A $\tau_{es} = 2000$ MC sweeps.

At the folding temperature $T_f = 230$ K [50,51] a funnel in the energy landscape appears with a gradient towards the ground state, but Fig. 6 (left) shows that there are various other structures, the most notable of which is Conformer B (where $O_B \approx 1$), with free energies 3 $k_B T$ higher than the ground-state conformation but separated from each other and the ground state only by free energy barriers less than 1 $k_B T$. No other long-lived traps are populated. Hence, the funnel at T_f is reasonably smooth. Folding routes include direct conversion from random-coil conformations into Conformer A or some short trapping in Conformer B region before reaching Conformer A region, but at the folding temperature it is possible to reach the ground state from any configuration without getting kinetically trapped. This was indeed observed by us in the 100 canonical runs we performed at this temperature. Some of the runs went directly from the unfolded state to the folded conformation (state A), while in other runs we saw first short trapping in the region of conformer B before folding into the ground-state structure. The folding trajectory displayed in the figure is an example for the later case. We found as escape time out of conformer B roughly $\tau_{es}^B = 5000$ MC sweeps. Due to such trapping only 86 of 100 MC runs found the folded state within 200,000 sweeps, leading to an average minimal folding time of $t_f = 77,230$ MC sweeps, which is 4 times as long as for the collapse temperature $T_\theta = 300$ K. However, the escape time for the folded state also increased to $\tau_{es}^A = 19430$ MC sweeps, about 10 times as long as for $T = 300$ K. Hence, the interplay of folding time and life time of the folded state leads to the increases probability of the folded state in the free energy plot for this temperature.

Finally, Fig. 6 (right) shows the situation for $T = 150$ K where we expect onset of glassy behavior. Again one sees a funnel-like bias toward the ground

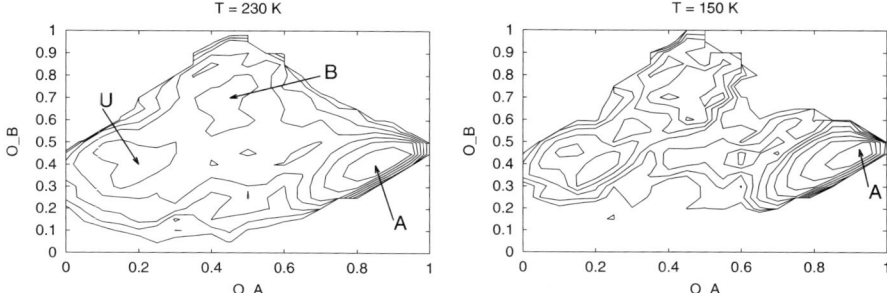

Fig. 6. Free energy landscape of Met-enkephalin for $T = 230$ K (*left*) and for $T = 150K$ (*right*). The contour lines are spaces $1K_bT$

state, however, the funnel is no longer smooth and the free energy landscape is rugged. Free energy barriers of many k_BT now separate different regions and would act as long-lived kinetic traps in a canonical simulation rendering folding at this temperature extremely difficult. This can be seen for the folding trajectory we display in that figure: the simulation got trapped in a region of the landscape far away from the folded state and never reached the folded state within the 200,000 sweeps of the simulation. Actually, only in 19 out of 100 Monte Carlo simulations of 200,000 sweeps we found the folded state and the minimal folding time is at least $t_f = 172866$ MC sweeps. This demonstrates that with increasing glassiness of the system, it becomes more and more difficult to escape the now much longer living traps.

An interesting question is at what temperature the onset of glassy behavior occurs. To measure the glass transition temperature T_g from equilibrium properties of the protein one can use the intimate connection between 'roughness' and fractality: we expect that the fractal dimension of the folding funnel will grow with increasing roughness of the *free* energy landscape, and propose to measure T_g by calculating the fractal dimension of the protein free energy landscape as a function of temperature [53].

Figure 7 displays for Met-enkephalin the temperature dependence of the fractal dimension. Various distinct regions can be observed in this graph. For the high temperature region the fractal dimension seems to be constant. With decreasing temperature the fractal dimension of the free energy landscape increases till it reaches a local maximum for $T = 280 \pm 40$ K This temperature seems to correspond to $T_\theta = 295 \pm 20$ K, the collapse temperature for Met-enkephalin. At T_θ both extended coil structures and an ensemble of collapsed structures can exist, and the free energy landscape reflects the large fluctuations at this temperature. With further decreasing temperature we observe that the fractal dimension decreases until below a temperature $T = 180 \pm 30$ K, the fractal dimension increases rapidly again, indicating the onset of glassy behavior and the appearance of long-living traps. Hence, we identify this temperature as the glass temperature and find

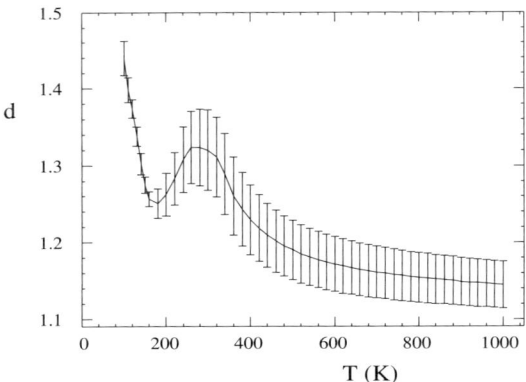

Fig. 7. Fractal dimension of the Met-enkephalin energy landscape as a function of temperature

for Met-enkephalin: $T_g = 180 \pm 30 K$.[53] It is expected that for a protein $T_f > T_g$, i.e. a good folder can be characterized by the relation $T_f/T_g > 1$ [38]. The result for $T_f = 230 \pm 30$ K [50] and our estimate $T_g = 180 \pm 30$ K lead indeed to $T_f/T_g = 1.28 > 1$. This value of the ratio demonstrates that Met-enkephalin is a good folder according to the above criterion.

4 Structure Prediction of Proteins

In the following we want to demonstrate that generalized-ensemble techniques are equally well suited for the prediction of the native state of proteins. The first example where we studied this problem was the 13-residue C-peptide of ribonuclease A [54,55]. However, any computer simulation approach to protein folding will have eventually to prove that it allows simulation of folding of stable domains in proteins. Such domains consist usually of 50-200 amino acids. As a next step in that direction we have studied the villin headpiece subdomain, 36-residue peptide (HP-36). HP-36 is one of the smallest peptides that can fold autonomously and it was chosen recently by Duan and Kollman for a 1-microsecond molecular dynamics simulation of protein folding [57]. The experimental structure was determined by NMR analyses [58]. Since it is a solvated molecule we also had to take into account the interaction between protein and solvent. Following a common practice we approximated this contribution to the overall energy by adding a solvent accessible surface term [59] to the energy function: $E = E_{Ecepp/2} + \sum_i \sigma_i A_i$. Here, the sum goes over all atoms and with A_i the solvent accessible surface areas of the atoms. The parameters σ_i were chosen from Ref. [60]. The above energy function is implemented in the program package SMMP [61] which was used in the present simulations. Our simulations rely on the generalized-ensemble technique described in Ref. [30].

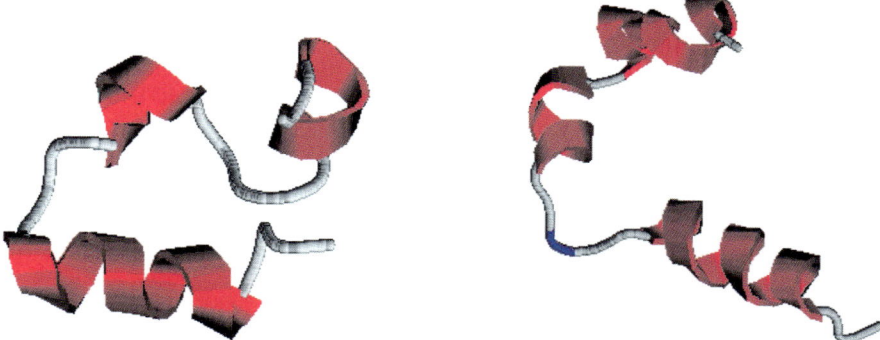

Fig. 8. *Left:* Experimental structure of HP-36 as deposited in the PDB data-bank. *Right:* Lowest energy structure of HP-36 as obtained in computer simulation

The structure of HP-36 as obtained from the Protein Data Bank (PDB code 1vii) is shown in Fig. 8 (left). The figure was created with RasMol [56]. The structure consists of three helices between residues 4-8, 15-18, and 23-32, respectively, which are connected by a loop and a turn. After regularizing this structure with the program FANTOM [62] we obtained as its energy (ECEPP/2 + solvation term) $E_{nat} = -276$ kcal/mol. Our approach led to a configuration with lowest energy $E_{min} = -277$ kcal/mol which we show in Fig. 8 (right) [30]. The above structure has a radius of gyration $R_\gamma = 10.1$ Å which indicates that the numerically obtained structure is slightly less compact than the experimental structure ($R_\gamma = 9.6$Å). It consists of three helices where the first helix stretches from residue 2 to residue 11 and is more elongated than the corresponding one in the native structure (residues 4-8). The second helix consist of residues 13-17 (compared to residue 15-18 in the native structure) and the third helix stretches from residue 23-33 (residues 23-32 in the PDB structure). The structure has 95% of the native helical content and 65% of the native contacts were formed in our structure. Both values are comparable with the results in Ref. [57] (but required orders of magnitude less computer time) where the optimal structure of a 1 μs molecular dynamic folding simulation showed 80% of native helical content and 62% of native contacts. Similarly comparable were the values of the root-mean-square deviation (RMSD) of both numerically determined conformers to the native structure: 5.8 Å versus 5.7 Å in Ref. [57] when all backbone atoms where counted.

We conclude that even for large peptides such as HP-36 our generalized-ensemble method is able to find structures that are close to the experimentally determined structures. The remaining differences between our optimal structure and the experimentally determined one point to a general problem in protein simulations: it is not clear whether the utilized cost function has indeed the biologically active structure of a given protein as its global minimum. In fact, our optimal structure has slightly lower energy than the native one. The problem becomes obvious when solvation effects are neglected. A global optimization run of 50,000 sweeps with our algorithm relying only on the ECEPP/2 force field led to a structure with an ECEPP energy of $E_{GP} = -192$ kcal/mol. That structure, build out of two helices (between residues 2-16 and 23-33) connected by a loop, differs significantly from the regularized PDB-structure with the higher potential energy $E_{nat} = -176$ kcal/mol. Hence, the native structure of the peptide HP-36 is *not* the global minimum configuration in ECEPP/2. Only the inclusion of the solvation term led to an essentially correct structure as global minimum configuration.

5 Conclusion

We gave a brief introduction into generalized-ensemble techniques and their application to the protein folding problem. We could show that our approach is superior to standard techniques for simulations of simple peptides where the interactions among all atoms were taken into account. Our results support pictures for the kinetics of protein folding which were developed from the study of simplified protein models. We could find in an unbiased simulation the correct structure of medium-sized peptides. These examples demonstrate that the generalized-ensemble algorithms are well-suited for investigations of both thermodynamics of proteins and prediction of their structure and may lead to an increased understanding of the protein folding problem.

Acknowledgements

The presented work was done in collaboration with Y. Okamoto (IMS, Okazaki, Japan), on helix-coil transition with Nelson Alves (FFCLRP, University of Sao Paulo, Brasil) and Zheng Yu "Jeff" Chen (University of Waterloo, Canada), and on energy landscape analysis with J. Onuchic (UCSD, San Diego). Financial supports from a research grant (CHE-9981874) of the National Science Foundation (USA) is gratefully acknowledged.

References

1. K.A. Dill, H.S. Chan: Nature Structural Biology **4**, 10 (1997)
2. J.N. Onuchic, Z. Luthey-Schulten, P.G. Wolynes: Annual Reviews in Physical Chemistry, **48**, 545 (1997)
3. M. Vásquez, G. Némethy, H.A. Scheraga: Chem. Rev. **94**, 2183 (1994)
4. M.J. Sippl, G. Némethy, H.A. Scheraga: J. Phys. Chem. **88**, 6231 (1984), and references therein
5. B.R. Brooks, R.E. Bruccoleri, B.D. Olafson, D.J. States, S. Swaminathan, M. Karplus: J. Comp. Chem. **4**, 187 (1983)

6. S.J. Weiner, P.A. Kollman, D.T. Nguyen, D.A. Case: J. Comp. Chem. **7**, 230 (1986)
7. U.H.E. Hansmann, Y. Okamoto: Curr. Opin. Struc. Biol. **9**, 177 (1999)
8. U.H.E. Hansmann, Y. Okamoto: J. Comp. Chem. **18**, 920 (1997)
9. U.H.E. Hansmann, Y. Okamoto: 'The Generalized-Ensemble Approach for Protein Folding Simulations', In *Annual Reviews in Computational Physics VI* Ed. by D. Stauffer (World Scientific, Singapore 1999) pp. 129–157
10. G.M. Torrie, J.P. Valleau: J. Comput. Phys. **23**, 187 (1977)
11. B.A. Berg, T. Neuhaus: Phys. Lett. B **267**, 249 (1991); Phys. Rev. Lett. **68**, 9 (1992); B.A. Berg: Int. J. Mod. Phys. **C3**, 1083 (1992)
12. B. Hesselbo, R.B. Stinchcombe: Phys. Rev. Lett. **74**, 2151 (1995)
13. A.P. Lyubartsev, A.A. Martinovski, S.V. Shevkunov, P.N. Vorontsov-Velyaminov: J. Chem. Phys. **96**, 1776 (1992); E. Marinari, G. Parisi: Europhys. Lett. **19**, 451 (1992)
14. U.H.E. Hansmann, Y. Okamoto: J. Comp. Chem. **14**, 1333 (1993)
15. U.H.E. Hansmann, Y. Okamoto, F. Eisenmenger: Chem. Phys. Lett. **259**, 321 (1996)
16. N. Nakajima, H. Nakamura, A. Kidera: J. Phys. Chem. **101**, 817 (1997)
17. J. Higo, N. Nakajima, H. Shirai, A. Kidera, H. Nakamura: J. Comp. Chem. **18**, 2086 (1997)
18. S. Kumar, P.W. Payne, M. Vásquez: J. Comp. Chem. **17**, 1269 (1996)
19. U.H.E. Hansmann, Y. Okamoto: J. Phys. Soc. (Jpn.) **63**, 3945 (1994); Physica A **212**, 415 (1994)
20. Y. Okamoto, U.H.E. Hansmann: J. Phys. Chem. **99**, 11276 (1995)
21. A.M. Ferrenberg, R.H. Swendsen: Phys. Rev. Lett. **61**, 2635 (1988); Phys. Rev. Lett. **63**, 1658(E) (1989), and references given in the erratum.
22. U.H.E. Hansmann: Phys. Rev. E **56**, 6200 (1997)
23. U.H.E. Hansmann, P. de Forcrand: Int. J. Mod. Phys. C **8**, 1085 (1997)
24. F. Wang, D.P. Landau: Phys. Rev. Let. **86**, 2050 (2001)
25. U.H.E. Hansmann, Y. Okamoto: J. Chem. Phys. **110**, 1267 (1999); **111**, 1339(E) (1999)
26. U.H.E. Hansmann, Y. Okamoto: Phys. Rev. E **56**, 2228 (1997)
27. E.M.F. Curado, C. Tsallis: J. Phys. A: Math. Gen. **27**, 3663 (1994)
28. W. Wenzel, K. Hamacher: Phys. Rev. Let. **82**, 3003 (1999)
29. U.H.E. Hansmann: Eur.Phys.J.B **12**, 607 (1999)
30. U.H.E. Hansmann, L.T. Wille: Phys. Rev. Lett. **88**, 068105 (2002)
31. N. Go, H.A. Scheraga: *Macromolecules* **3**, 170 (1970)
32. M.G. Wu, M.W. Deem: J. Chem. Phys. **111**, 6625 (1999)
33. K. Hukushima, K. Nemoto: J. Phys. Soc. (Jpn.) **65**, 1604 (1996); G.J. Geyer: Stat. Sci. **7**, 437 (1992); M.C. Tesi, E.J.J. van Rensburg, E. Orlandini, S.G. Whittington: J. Stat. Phys. **82**, 155 (1996)
34. U.H.E. Hansmann: Chem. Phys. Lett. **281**, 140 (1997)
35. A. Chakrabartty, R.L. Baldwin: In *Protein Folding: In Vivo and In Vitro* Ed. by J.L. Cleland and J. King, (ACS Press, Washington, D.C. 1993) pp. 166-177
36. J.D. Bryngelson, J.N. Onuchic, N.D. Socci, P.G. Wolynes: Proteins **21**, 167 (1995)
37. M. Karplus, M. Sali: Curr. Opin. Struc. Biol. **5**, 58 (1995)
38. J.D. Bryngelson, P.G. Wolynes: Proc. Natl. Acad. Sci. (USA) **84**, 7524 (1987)
39. D. Poland, H.A. Scheraga: *Theory of Helix-Coil Transitions in Biopolymers* (Academic Press, New York 1970)

40. J.P. Kemp, Z.Y. Chen: Phys. Rev. Lett. **81**, 3880 (1998)
41. N.A. Alves, U.H.E. Hansmann: Phys. Rev. Lett. **84** 1836 (2000)
42. J.P. Kemp, U.H.E. Hansmann, Zh.Y. Chen: Eur. Phys. J. B **15**, 371 (2000)
43. N.A. Alves, U.H.E. Hansmann: Physica A **292** 509 (2001)
44. B.H. Zimm, J.K. Bragg: J. Chem. Phys. **31**, 526 (1959)
45. M.E. Fisher: in *Lectures in Theoretical Physics*, Vol. 7c, (University of Colorado Press, Boulder 1965), p. 1
46. C. Itzykson, R.B. Pearson, J.B. Zuber: Nucl. Phys. B **220** [FS8], 415 (1983)
47. C.N. Yang, T.D. Lee: Phys. Rev. **87**, 404 (1952)
48. T.D. Lee, C.N. Yang: Phys. Rev. **87** 410 (1952)
49. M. Fukugita, H. Mino, M. Okawa, A. Ukawa: J. Stat. Phys. **59**, 1397 (1990); and references given therein
50. U.H.E. Hansmann, M. Masuya, Y. Okamoto: Proc. Natl. Acad. Sci. U.S.A. **94**, 10652 (1997)
51. U.H.E. Hansmann, Y. Okamoto, J.N. Onuchic: Proteins **34**, 472 (1999)
52. U.H.E. Hansmann, J.N. Onuchic: J. Chem. Phys. **115** 1601 (2001)
53. N.A.Alves, U.H.E. Hansmann: Int. J. Mod. Phys. C **11**, 301 (2000)
54. U.H.E. Hansmann, Y. Okamoto: J. Phys. Chem. **102**, 653 (1998)
55. U.H.E. Hansmann, Y. Okamoto: J. Phys. Chem. **103**, 1595 (1999)
56. R.A. Sayle, E.J. Milner-White: TIBS **20**, 374 (1995)
57. Y. Duan, P.A. Kollman: Science **282**, 740 (1998)
58. C.J. McKnight, D.S. Doehring, P.T. Matsudaria, P.S. Kim: J. Mol. Biol. **260**, 126 (1996)
59. T. Ooi, M. Obatake, G. Nemethy, H.A. Scheraga: Proc. Natl. Acad. Sci. USA **8**, 3086 (1987)
60. L. Wesson, D. Eisenberg: Protein Science **1**, 227 (1992)
61. F. Eisenmenger, U.H.E. Hansmann, Sh. Hayryan, C.-K. Hu: Comp. Phys. Comm. **138**, 192 (2001)
62. T. Schaumann, W. Braun, K. Wuthrich: Biopolymers **29**, 679 (1990)

Sequence Alignment in Bioinformatics

Yi-Kuo Yu

Over two billion US dollars have been budgeted for the Human Genome Project alone in the past twelve years, not to mention other similar or related projects worldwide. These investments have led to the production of enormous amount of biological data, many of which are sequence information of biomolecules – e.g. specifying proteins/DNAs by identifying each amino-acid/nucleotide in the sequential order. These sequence data, presumably containing the "digital" information of life, are hard to decipher. Extracting useful and important information out of those massive biological data has developed into a new branch of science – *bioinformatics*. One of the most important and widely used method in bioinformatics research is called "sequence alignment". The basic idea is to expedite the identification of biological functions of a newly sequenced biomolecule, say a protein, by comparing the sequence content of the new molecule to the existing ones (characterized and documented in the database).

The purpose of this chapter is to provide a comprehensive introduction to sequence alignment, some of our recent development, and some open problems. The readers should be warned, however, that this is not a standard review article which contains all the relevant papers in the reference list. On the contrary, for pedagogical purpose, the author is forced to cut out a lot of important and original references that contributed to this field significantly. Furthermore, the recent development and open problems that will be mentioned probably reflect more about the author's personal view or prejudice rather than the truth or the consensus. Nevertheless, if one is only interested in knowing what sequence alignment is and what has been done, this paper should satisfy some curiosities.

This paper is organized into two main sections. In the first section, I will introduce the mathematics, some relevant background, and algorithms involved in sequence alignment. In the second section, I will briefly mention some of the recent developments and some open problems. In appendix A, we will use a very simple evolution model to demonstrate the relationship between evolution and probabilistic alignment.

1 Introduction to Sequence Alignment

The purpose of sequence alignment is to detect *mutual similarity* between sequences of characters. Each character of a sequence is taken from a finite character set χ. For protein sequences, χ denotes the set of 20 amino acids; and for DNA sequences, χ is the set of four base nucleotides, $\{A\,T\,C\,G\}$.

Two types of alignment algorithms are found in literature: those which search for *optimal* alignment (as exemplified by the Smith-Waterman algorithm [1]), and those which identify *likely* alignments (as exemplified by the Hidden Markov Model(HMM) based "Sequence Alignment Modules" [2]). In each case, the quality of an alignment is summarized by a score $\mathsf{S_o}$, which corresponds to the *highest* alignment score in the former case and corresponds to the logarithm of relative total likelihood in the probabilistic approaches. Despite the existence of many different approaches, they do share one common important issue that we now turn to.

1.1 The Holy Grail

The "holy grail" of sequence alignment is to understand the probability distribution function (pdf) of the alignment score, i.e., pdf($\mathsf{S_o}$), (or equivalently $\Pr(\mathsf{S_o} < x) \equiv \int_{-\infty}^{x} \text{pdf}(\mathsf{S_o}) \, d\mathsf{S_o}$) for the appropriate "null models". This distribution provides us with the p–value $\equiv 1 - \Pr(\mathsf{S_o} < x)$, or equivalently the probability that a score larger than x (with $x \gg 1$) could have arisen by chance. The p-value is therefore more meaningful than the alignment score itself. Throughout this paper, we employ a frequently used "Markovian null model" where each sequence is composed of independently identically distributed characters with probabilities $p(a)$ for each character $a \in \chi$ and $\sum_{a \in \chi} p(a) = 1$. The probability of having a pair of sequences, say $\mathbf{a} = [a_1, \cdots, a_M]$ and $\mathbf{b} = [b_1, \cdots, b_N]$, under this null model is therefore given by

$$P_0[\mathbf{a}, \mathbf{b}] = P_0[\mathbf{a}] P_0[\mathbf{b}] = \prod_{1 \leq m \leq M, 1 \leq n \leq N} p(a_m) p(b_n). \tag{1}$$

Let us also introduce the notation of average over the null model

$$\langle f[\mathbf{a}, \mathbf{b}] \rangle_0 \equiv \sum_{[\mathbf{a}, \mathbf{b}]} f[\mathbf{a}, \mathbf{b}] P_0[\mathbf{a}, \mathbf{b}] \tag{2}$$

where $f[\mathbf{a}, \mathbf{b}]$ can be any function that depends on the sequences $\mathbf{a}$ and $\mathbf{b}$. Despite decades of effort, the only well understood score statistics of the "null model" is that of gapless alignment. It was rigorously shown [3,4] that the score statistics of gapless alignment follows the so-called Gumbel form [5]

$$\Pr(\mathsf{S_o} < x) = \exp\left[-KMN e^{-\lambda x}\right], \tag{3}$$

where $M \gg 1$ and $N \gg 1$ are the two sequence lengths, K and λ are called Gumbel parameters. There are explicit formulas relating the complicated alignment parameters to the two Gumbel parameter λ and K [4]. Due to the fast computation speed and well understood background statistics, gapless alignment tools, e.g. BLAST [6], have been among the top choices in large database search. Unfortunately, forbidding *insertions and deletions* (or "indels") that do occur in the natural sequence evolutions makes gapless algorithm incapable of detecting weak similarities. To detect weak similarity, gaps were introduced to alignment algorithms at the expense of losing the exact knowledge of score statistics. We will discuss in more details about the score statistics of gapped alignment in section 2. To explicitly show how sequence alignment is done, we now turn to the algorithm part of sequence alignment.

1.2 Alignment Algorithms

Let $\mathbf{a} = [a_1, a_2, ..., a_M]$ and $\mathbf{b} = [b_1, b_2, ..., b_N]$ be two sequences of lengths M and N respectively, with elements a_i and b_j taken from a finite character set χ. We start with the algorithm of optimal alignment.

Optimal Alignment
Let $\mathbf{a}_{m';m} = [a_{m'}, a_{m'+1}, ..., a_m]$ and $\mathbf{b}_{n';n} = [b_{n'}, b_{n'+1}, ..., b_n]$ denote *subsequences* of $\mathbf{a}$ and $\mathbf{b}$ respectively, with $1 \leq m' \leq m \leq M$, and $1 \leq n' \leq n \leq N$. A restricted global alignment $\widehat{\mathcal{A}}$ of the sequences $\mathbf{a}_{m';m}$ and $\mathbf{b}_{n';n}$ consists of an ordered set of pairings of their elements, with each other or with gaps, e.g., $\widehat{\mathcal{A}} = \{(a_{m'}, b_{n'}), (a_{m'+1}, -), (a_{m'+2}, b_{n'+1}), ..., (a_m, b_{n-1}), (-, b_n)\}$, for the example shown in Fig. 1(a). Here, (a_i, b_j) denotes the pairing of elements a_i with b_j, and $(a_i, -)$ and $(-, b_j)$ denote pairing of an element with a *gap*; we refer to these three types of pairings as substitutions, deletions, and insertions respectively. In typical alignment applications, pairings of gaps to each other are not allowed. It is also a common practice to restrict the order of insertions and deletions, e.g., to forbid insertions following deletions, in order to avoid overcounting of the same alignments[1]. With these restrictions, each alignment $\widehat{\mathcal{A}}$ can be uniquely represented by the set $\mathcal{R}$ of index pairs (i,j) for all paired elements (a_i, b_j), e.g., $\mathcal{R} = \{(m', n'), (m'+2, n'+1), ..., (m, n-1)\}$ for the example in Fig. 1(a). Generally, we shall use the notation

$$\mathcal{R}(m', n'; m, n) = \{(m_1, n_1), (m_2, n_2), ..., (m_l, n_l)\} \qquad (4)$$

to denote the set of l pairings in an alignment. A valid restricted global alignment $\widehat{\mathcal{A}}$ for the sequences $\mathbf{a}$ and $\mathbf{b}$ is then any set $\mathcal{R}$ of index pairs satisfying the condition $m' \leq m_1 < m_2 < ... < m_l \leq m$ and $n' \leq n_1 < n_2 < ... < n_l \leq n$. $\mathcal{R}$ can also be viewed as coordinates of a *directed path* on the alignment grid, with the "backward end" of the path fixed at the lower left corner of the cell (m', n') and the "forward end" fixed at the upper right corner of the cell (m, n); see Fig. 1(b).

---
[1] For example, the alignments $\{(a_1, -), (-, b_1), (a_2, -)\}$ and $\{(a_1, -), (a_2, -), (-, b_1)\}$ both describe the situation where the elements a_1 and a_2 are not aligned with b_1, and thus should not be multiply counted.

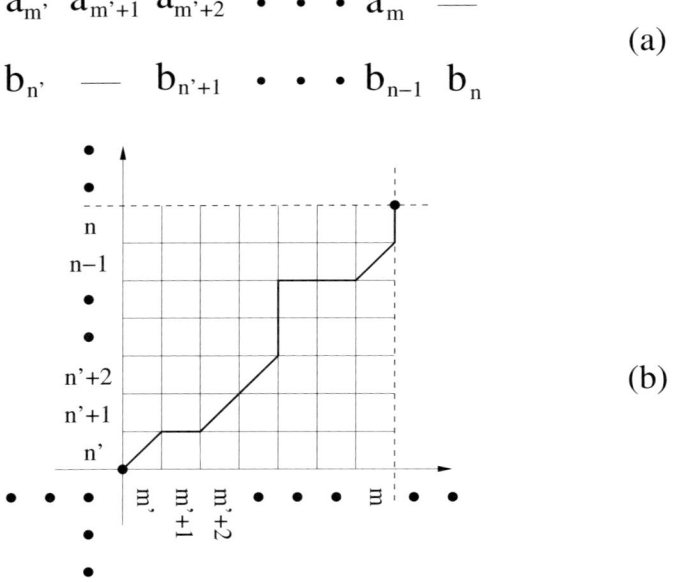

Fig. 1. (a) A possible global alignment of the sequences $\mathbf{a}_{m';m}$ and $\mathbf{b}_{n';n}$; (b) the directed path representation of the alignment shown in (a). For a restricted global alignment $\widehat{\mathcal{A}}$ of $\mathbf{a}_{m';m}$ and $\mathbf{b}_{n';n}$, the corresponding directed path must have one of its end (the "backward end") fixed at the lower left corner of the cell (m', n') and the other end (the "forward end") fixed at the upper right corner of the cell (m, n). The two dashed lines which mark the right and upper boundaries of the alignment region will be referred to as the "forward boundaries" in the text

The *score* $\mathcal{S}$ of the alignment $\widehat{\mathcal{A}}$ is obtained by summing up the individual paring scores, e.g., $s(a_i, b_j)$ for the pairing of elements a_i with b_j, and the "gap scores". For protein sequences, the frequently used pairing scores are the PAM [7] or BLOSUM [8] substitution scores – see below for more details. The frequently used affine-gap function assigns a cost of $\delta + \varepsilon \cdot (\ell - 1)$ for each consecutive run of ℓ gaps in a given sequence. An additional cost δ' can be assigned to penalize the situation where a run of gaps in one sequence is immediately followed by a run of gaps in the other sequence[2]. Let the length of the two gaps separating two pairings be ℓ_1 and ℓ_2 respectively, then the

---

[2] In the problem considered originally by Smith and Waterman [1], each run of gaps had to terminate with a pairing; this corresponds to the limit $\delta' = \infty$ in our scoring system.

gap cost function γ can be written as

$$\gamma(\ell_1, \ell_2) = \begin{cases} 0 & \ell_1 = 0, \ \ell_2 = 0 \\ \delta + \varepsilon \cdot (\ell_1 - 1) & \ell_1 \geq 1, \ \ell_2 = 0 \\ \delta + \varepsilon \cdot (\ell_2 - 1) & \ell_1 = 0, \ \ell_2 \geq 1 \\ \delta' + 2\delta + \varepsilon \cdot (\ell_1 + \ell_2 - 2) & \ell_1 \geq 1, \ \ell_2 \geq 1. \end{cases} \quad (5)$$

Given the *scoring functions* $\{\bar{s}, \gamma\}$, and the alignment path $\mathcal{R}$, the score for this alignment of **a** and **b** is uniquely determined:

$$\mathcal{S}[\mathcal{R}; \mathbf{a}, \mathbf{b}; \bar{s}, \gamma] = \sum_{k=1}^{l} s(a_{m_k}, b_{n_k})$$

$$- \sum_{k=0}^{l} \gamma(m_{k+1} - m_k - 1, n_{k+1} - n_k - 1)], \quad (6)$$

where we used $(m_0, n_0) = (m'-1, n'-1)$ and $(m_{l+1}, n_{l+1}) = (m+1, n+1)$ to compactify the notation. The alignment with the highest score (for a given sequence pair $[\mathbf{a}, \mathbf{b}]$ and given scoring functions) is the optimal restricted global alignment $\widehat{\mathcal{A}}^*$, with score

$$S_{m',n';m,n} = \max_{\mathcal{R}(m',n';m,n)} \{\mathcal{S}[\mathcal{R}; \mathbf{a}, \mathbf{b}; \bar{s}, \gamma]\}. \quad (7)$$

This score can be computed via a well-known dynamic programming algorithm provided by Needleman and Wunsch [9]. Below is the simplest example for the case of *linear* gap cost with $\delta = \varepsilon$, $\delta' = 0$, and no constraint in the order of occurrence of insertions and deletions. Readers interested in the extension to the affine gap function (5) can find detailed documentation in [10]. To compute $S_{m',n';m,n}$, one simply iterates the following recursion relation

$$S_{m',n';i,j} = \max \begin{cases} S_{m',n';i-1,j-1} + s(a_i, b_j) \\ S_{m',n';i-1,j} - \varepsilon, \ S_{m',n';i,j-1} - \varepsilon \end{cases} \quad (8)$$

for $i = m'$ to m and $j = n'$ to n, with the "boundary condition"

$$S_{m',n';i,j=n'-1} = -\varepsilon \cdot [i - (m'-1)]$$
$$S_{m',n';i=m'-1,j} = -\varepsilon \cdot [j - (n'-1)]. \quad (9)$$

This boundary condition enforces the anchoring of the "backward end" of the alignment path as shown in Fig. 1(b).

A *local* alignment $\mathcal{A}$ between the sequences **a** and **b** is *any* restricted global alignment of the subsequences $\mathbf{a}_{m';m}$ and $\mathbf{b}_{n';n}$, alignment of **a** or **b** with null, or the "null alignment" (i.e., no alignment at all). The optimal local alignment $\mathcal{A}^*$ is one whose score $\mathsf{S}_o = \mathcal{S}[\mathcal{A}^*]$ is the highest; the corresponding alignment path is denoted by $\mathcal{R}^*$. From (7), we have

$$\mathsf{S}_o[\mathbf{a}, \mathbf{b}; \bar{s}, \gamma] = \max_{\substack{1 \leq m' \leq m \leq M \\ 1 \leq n' \leq n \leq N}} \{S_{m',n';m,n}, 0\}, \quad (10)$$

where the entry '0' in (10) selects the null alignment if alignments between all possible subsequences are below a threshold, e.g., zero. The score $\mathsf{S_o}$ is called the optimal local alignment score, or simply the optimal score.

Smith and Waterman [1] developed an efficient strategy to compute the optimal score $\mathsf{S_o}$: First, define the "restricted" local alignment score $H_{m,n}$ to be

$$H_{m,n} = \max_{\substack{1 \le m' \le m \\ 1 \le n' \le n}} \{S_{m',n';m,n}, 0\}. \tag{11}$$

It records the optimal local alignment between the subsequences $\mathbf{a}_{1;m}$ and $\mathbf{b}_{1;n}$. The H's can again be computed by dynamic programming. For the simple linear gap function, it reads

$$H_{m,n} = \max \left\{ \begin{array}{l} H_{m-1,n-1} + s(a_m, b_n) \\ H_{m-1,n} - \varepsilon,\ H_{m,n-1} - \varepsilon,\ 0 \end{array} \right\}, \tag{12}$$

with the boundary condition $H_{0,n} = 0 = H_{m,0}$. The affine gap version of the algorithm can be found, for example, in Appendix A of [10]. Given $H_{m,n}$ for all $1 \le m \le M$, $1 \le n \le N$, the optimal alignment score $\mathsf{S_o}$ defined in (10) is obtained simply as

$$\mathsf{S_o}[\mathbf{a}, \mathbf{b}; \bar{s}, \gamma] = \max_{\substack{1 \le m \le M \\ 1 \le n \le N}} \{H_{m,n}\}. \tag{13}$$

The combination of (12) and (13) is the celebrated Smith-Waterman local alignment algorithm which is the representative of the optimal alignment algorithms. We now turn to the corresponding case in the probabilistic alignment.

Probabilistic Alignment

We first describe the probabilistic approach to restricted global alignment. Each restricted global alignment $\hat{\mathcal{A}}$ of the subsequences $\mathbf{a}_{m';m}$ and $\mathbf{b}_{n';n}$ is described by an alignment path $\mathcal{R}$ as in (4). Let each pairing (a_i, b_j) contribute a "weight" $w(a_i, b_j)$ towards the net weight $\mathcal{W}$ of the alignment $\hat{\mathcal{A}}$. For the gap weights, we use

$$g(\ell_1, \ell_2) = \begin{cases} 1 & \ell_1 = 0,\ \ell_2 = 0 \\ \mu \cdot \nu^{\ell_1 - 1} & \ell_1 \ge 1,\ \ell_2 = 0 \\ \mu \cdot \nu^{\ell_2 - 1} & \ell_1 = 0,\ \ell_2 \ge 1 \\ \mu' \cdot \mu^2 \cdot \nu^{\ell_1 + \ell_2 - 2} & \ell_1 \ge 1,\ \ell_2 \ge 1 \end{cases} \tag{14}$$

where μ is the weight of gap initiation and ν is the weight of gap extension, and μ' is the additional weight for the double gap configuration. In all the numerical work to be presented below, we will use $\mu' = 1$, which treats each run of gaps the same way. However, if one wishes to exclude the double-gap configuration as considered originally by Smith and Waterman [1], one can simply set μ' to zero.

The net weight $\mathcal{W}$ for a given configuration of pairings $\mathcal{R}$ is just the product of the individual weight factors w's and g's, i.e.,

$$\mathcal{W}[\mathcal{R}; \mathbf{a}, \mathbf{b}; w, g] = \prod_{k=1}^{l} w(a_{m_k}, b_{n_k}) \tag{15}$$

$$\cdot \prod_{k=0}^{l} g(m_{k+1} - m_k - 1, n_{k+1} - n_k - 1),$$

again with $(m_0, n_0) = (m'-1, n'-1)$ and $(m_{l+1}, n_{l+1}) = (m+1, n+1)$. The total weight for the global alignment is

$$W_{m',n';m,n} = \sum_{\mathcal{R}(m',n';m,n)} \mathcal{W}[\mathcal{R}; \mathbf{a}, \mathbf{b}; w, g], \tag{16}$$

where $\sum_{\mathcal{R}}$ denotes the sum over all allowed paths as defined in Sect. 2. This weight can be computed exactly by extending the dynamic programming algorithm of Needleman and Wunsch. For the simple linear gap function ($\mu = \nu$, $\mu' = 1$) and without any constraint in the order of occurrence of insertions and deletions, one can simply iterate the recursion relation

$$W_{m',n';i,j} = w(a_i, b_j) \cdot W_{m',n';i-1,j-1} + \nu \cdot [W_{m',n';i-1,j} + W_{m',n';i,j-1}] \tag{17}$$

for $i = m'$ to m and $j = n'$ to n, with the boundary conditions

$$W_{m',n';i \geq m'-1, j=n'-1} = \nu^{i-(m'-1)}$$
$$W_{m',n';i=m'-1, j \geq n'-1} = \nu^{j-(n'-1)}. \tag{18}$$

To have the probability interpretation, we need the probability conservation condition

$$\langle w(a,b) \rangle_0 + 2\nu = 1; \tag{19}$$

see Appendix for details about this condition as well as the relation between the substitution scores $\{s(a,b)\}$ and substitution weights $\{w(a,b)\}$. Generalizations of the recurrence relation (17) and the probability conservation condition (19) to the case of affine gap function (14) is given in [10]. Next, we introduce the probabilistic version of the restricted *local* alignment. The total weight of the restricted local alignment of the sequences $\mathbf{a}_{1;m}$ and $\mathbf{b}_{1;n}$ for the case of linear gap function is

$$Z_{m,n} = 1 + \sum_{m'=1}^{m} \nu^{m'} + \sum_{n'=1}^{n} \nu^{n'} + \sum_{\substack{1 \leq m' \leq m \\ 1 \leq n' \leq n}} W_{m',n';m,n}; \tag{20}$$

again the generalization to affine gap functions is given in [10]. In (20), the first term on the right-hand side is the weight of null alignment, the second and third term is the weight of aligning a subsequence of $\mathbf{a}_{m';m}$ or $\mathbf{b}_{n';n}$ with the null, and the last term gives the weight of aligning the subsequence $\mathbf{a}_{m';m}$ with $\mathbf{b}_{n';n}$, taking the weight of "skipping" the subsequences $\mathbf{a}_{1;m'-1}$ and $\mathbf{b}_{1;n'-1}$ to be 1. These skipping factors accomplish exactly the task of the Free Insertion Modules used in the HMM approach to local alignment [2]. Further using the same weighting factor of 1 for skipping the subsequences $\mathbf{a}_{m+1;M}$ and $\mathbf{b}_{n+1;N}$, the total weight of the local alignment between the sequences $\mathbf{a}$ and $\mathbf{b}$ becomes simply

$$\mathsf{W}[\mathbf{a},\mathbf{b};w,g] = 1 + \sum_{\substack{1 \leq m \leq M \\ 1 \leq n \leq N}} Z_{m,n} - M \cdot N, \qquad (21)$$

with the last term accounting for the $M \cdot N$ redundant counts of the null alignment included in the second term. Equations (20) and (21) define the algorithm for the probabilistic version of local alignment.

After stating the simplest versions of both optimal and probabilistic alignment algorithms, we next discuss their score statistics.

1.3 Score Statistics

We emphasize here again that the value of the optimal/probabilistic alignment score S_o does not in itself convey any meaning regarding the degree of homology between the sequences being aligned. One way to assess sequence homology is to compare the alignment score S_o with the typical alignment score of aligning sequences from a null model. The pdf of alignment scores for the alignment of random sequences is

$$\mathrm{pdf}(\mathsf{S}_o) = \langle \delta(\mathsf{S}_o - \mathsf{S}_o[\mathbf{a},\mathbf{b};\bar{s},\gamma]) \rangle_0, \qquad (22)$$

where $\langle \ldots \rangle_0$ denotes average over the null sequence distribution (1). Integrating the pdf (22) provides the probability, or the $p-\text{value} \equiv \int_x^\infty \mathrm{pdf}(\mathsf{S}_o) d\mathsf{S}_o$, that an alignment of two uncorrelated random sequences receives an optimal score x or higher. Before we state our current knowledge of score statistics for different type of algorithms, Let us first briefly comment on scoring functions. In general, a scoring function consists of a lot of pairwise substitution scores as well as gap functions. Since the pairwise substitution scores are constantly arranged in a matrix, hence the name "substitution matrix" or "scoring matrix", denoted by $\bar{s}$, to represent the collection of pairwise substitution scores. Commonly used substitution matrices in protein sequence comparison include the PAM matrices [7] and the BLOSUM matrices [8]. In what follows, we will first discuss some general properties of scoring matrix followed by the main subject of this section – score statistics of different alignment algorithms.

1.4 Substitution (Scoring) Matrices

A substitution matrix is designed empirically by experienced biologists to detect a certain type of mutual similarity among sequences. One way to design substitution scores is by observing mutations happened. Based on experimental data, biologists find the probability that a character mutates into another character within certain evolution distance, say 1% of the residue characters, and construct a base transition matrix T. The matrix element $T_{a,b}$ represent the transition probability for character a mutate into character b. Another matrix $T(d) \equiv T^d$, obtained by multiplying this base transition matrix d times, can be viewed as an effective transition matrix of larger evolution distance. Apparently, $\sum_b T_{a,b}(d) = 1$ for all a.

A famous such construction is the PAM matrices for proteins [7]. Other constructions, such as BLOSUM matrices for proteins [8], have also been popularly used.

The "PAM-d" substitution scores $\{s(a,b)\}$ are obtained via

$$s(a,b) = \frac{\tau}{2} \ln\left(\frac{T_{a,b}(d)}{p(b)} \cdot \frac{T_{b,a}(d)}{p(a)}\right) \qquad (23)$$

where τ being some scale factor. Note that when the transition matrix exists and detailed balance holds, $p(a)T_{a,b}(d) = p(b)T_{b,a}(d)$ is actually identical to $q(a,b)$, the joint probability of observing character pair (a,b) in homologous sequence pairs. In this case, we may write the substitution scores as

$$s(a,b) = \tau \ln(q(a,b)/p(a)p(b)). \qquad (24)$$

For the BLOSUM matrices construction, one directly works with the joint probabilities (24) without even assuming existence of the transition matrix and detailed balance at all.

Gapless Alignment

Clearly, the pdf (22) would depend generally on the sequence lengths M, N, and the scoring functions $\bar{s}$ and γ. For gapless alignment, $\delta = \infty$, the form of the distribution function is known exactly [4,11–13] in the asymptotic limit $M, N \gg 1$. For all scoring systems satisfying the condition

$$\sum_{a,b \in \chi} p(a)p(b)s(a,b) < 0 \qquad (25)$$

which includes all the PAM and BLOSUM matrices, the pdf reaches the universal form

$$\text{pdf}(\mathsf{S}_\text{o}) = KMN\lambda \exp\left[-\lambda \mathsf{S}_\text{o} - KMNe^{-\lambda \mathsf{S}_\text{o}}\right], \qquad (26)$$

known as the Gumbel distribution [5]. This distribution is specified completely by the two parameters λ and K, with a mean $\langle \mathsf{S}_\text{o} \rangle_0 \equiv \overline{\mathsf{S}_\text{o}} \approx \lambda^{-1} \ln KMN$, and an exponential tail

$$\text{pdf}(\mathsf{S}_o \gg \overline{\mathsf{S}_o}) = \lambda K M N e^{-\lambda \mathsf{S}_o}, \tag{27}$$

characterized by the parameter λ.

The theory of Karlin and Altschul [4] provides explicit formulas for these parameters in terms of the scoring function $\bar{s}$. For example, λ can be found as the unique positive root of the equation

$$\sum_{a,b \in \chi} p(a)p(b)e^{\lambda s(a,b)} = 1. \tag{28}$$

A more complicated expression exists for the calculation of K, which we will not describe here. We mention instead another important characteristics of the background statistics, which we will make use of later in the text. It is the average pairwise score of the optimal alignment of random sequences,

$$\alpha = \sum_{a,b \in \chi} s(a,b) p(a) p(b) e^{\lambda s(a,b)}. \tag{29}$$

This quantity, known as the "relative entropy", is needed in the calculation of K. It also governs the magnitude of the "finite-size" correction of λ and K from their asymptotic values; see section 2 and/or [10] for details.

Gapped Alignment

Compared to gapless alignment, the statistics of gapped alignment for the null model (1) is much more difficult to characterize. First of all, the average optimal score $\overline{\mathsf{S}_o}$ does not always have the logarithmic dependence on sequence lengths. For sufficiently small gap cost, the mean score in fact acquires a *linear* dependence on sequence length even if the condition (25) is satisfied, i.e., $\overline{\mathsf{S}_o} = v \cdot N$ (for sequences of lengths $M \approx N \gg 1$), with the proportionality factor $v \geq 0$ depending on the substitution scores and gap cost. The critical line $v = 0$ defines the loci of *phase transition* points [14–16] separating the "linear" and "logarithmic" regimes of $\overline{\mathsf{S}_o}$.

Interestingly, ample empirical evidences [17–23] suggest that the optimal score S_o of gapped alignment still obeys the Gumbel distribution (26) in the logarithmic phase. Unfortunately, the functional dependence of the Gumbel parameters λ and K on the scoring functions are not known. Figure 2 shows a histogram of alignment score obtained via numerical simulation using a linear gap function $\delta = 3$ and PAM-250 substitution matrix.

Probabilistic Alignment

We now turn our attention to the statistics of probabilistic alignment, as described by the distribution of the log-likelihood score [24], e.g., $\text{pdf}(\ln \mathsf{W})$.

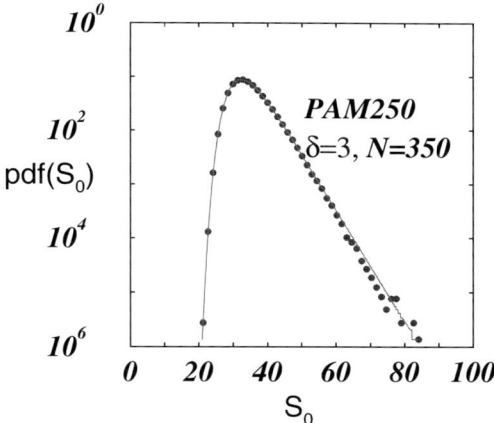

Fig. 2. The histogram of gapped alignment score and Gumbel fit. The small filled circles represent the normalized histogram (or pdf) of gapped optimal alignment score obtained from aligning sequence-pairs generated from the null model. The solid curve indicate a numerical Gumbel fit

This statistics is not well understood at all, even in comparison to the not-so-well-understood Smith-Waterman score statistics: Previously, an exponential bound on the tail of the log-odd score distribution was obtained for a simple sequence analysis problem done by Milosavljevic and Jurka [25]. Barret et. al [26] applied this bound to the HMM version of local alignment and found empirically that it did not correctly account for the false positives observed. In the mean time, there is a general expectation that the distribution of ln W might still have a Gumbel form. Here, we would like to point out that there is in fact no *a priori* reason to expect a Gumbel distribution for log-odd scores generated by probabilistic algorithms. This is because W, computed according to (21), is a *sum* of a large number of *correlated* terms, while the Gumbel distribution is typically obtained from taking the maximum of a large number of uncorrelated terms (see [10] for more details).

In the appendix, a simple prescription of converting substitution scores $\{s(a,b)\}$ into substitution weight $\{w(a,b)\}$ will be given. Figures 3(a) and (b) show respectively the distribution of ln W for the $\{PAM - 120, \mu' = 1, \mu = 2^{-5.5}, \nu = 2^{-0.5}\}$ and $\{PAM - 250, \mu' = 1, \mu = 2^{-6}, \nu = 2^{-0.5}\}$ scoring functions[3], each having 125,000 alignments of Markov sequence pairs of length 300. From the figures, it is clear that the tails of the distribution functions are neither exponential nor Gaussian. The best least-square fits to the

---

[3] On the scale of the PAM scoring system used by BLAST, where the substitution scores is defined as $s_d(a,b) = 2 \cdot \log_2[T^d_{a,b}/p(b)]$ for PAM-120, and $s_d(a,b) = 3 \cdot \log_2[T^d_{a,b}/p(b)]$ for PAM-250, our values of μ and ν translates to gap costs of $\delta = 11, \varepsilon = 1$ for the PAM-120 scoring system and $\delta = 18, \varepsilon = 1.5$ for the PAM-250 scoring system.

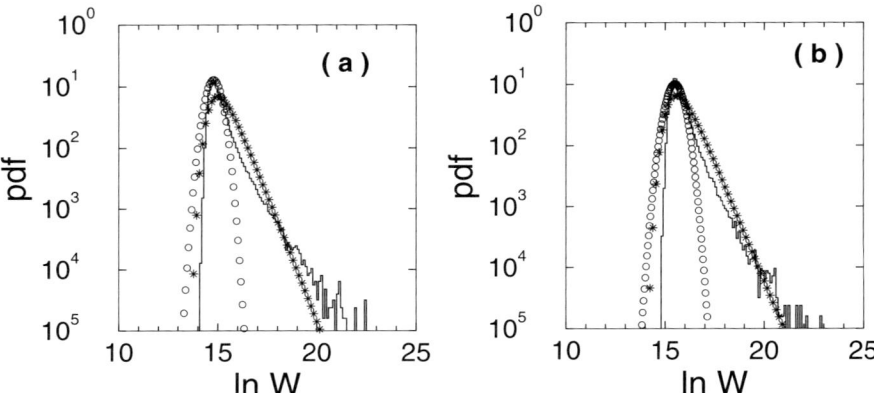

Fig. 3. The pdf of ln W for the two parameter sets (**a**) PAM-120 set and (**b**) PAM-250 set. Each pdf (shown as a staircase) is obtained by normalizing the histogram (collected for 125,000 pairwise alignments of random sequences of length 300). The Gumbel fits are shown by stars while the Gaussian fits are shown by circles

Gumbel and Gaussian distributions, shown by stars and circles respectively, are not satisfactory at all. The tails of the pdf's are in fact much broader than even the Gumbel form.

2 Some Recent Developments

2.1 Optimal Alignments

Recently, an efficient numerical method was developed by Olsen *et al.* [23] to characterize the tail of the Gumbel distribution, without doing exhaustive simulation such as shuffling. The method utilizes intermediate computational results, e.g., the restricted local alignment score $H_{m,n}$, also known as the "score landscape". The landscape consist of a collection of positive scoring "islands", e.g. clusters of positive H's, separated by a "sea" at $H = 0$. The peak scores of the islands are found to follow Poisson statistics. From this, the Gumbel distribution of the optimal score S_o can be derived. In particular, the Gumbel parameters λ and K can be obtained numerically from the island statistics in an efficient manner.

The study on island statistics indicates clearly that the key to understanding the Gumbel distribution is to characterize the probability tail of obtaining a *single* large island, the statistics of which can be more conveniently studied in the context of *global* alignment. Using the saddle point method, we give [10] a heuristic derivation of the Poisson distribution of the large island scores. The results lead to the Gumbel distribution for the optimal scores, as well as the all-important Gumbel parameter λ, in terms of the solution of the equation

$$\Omega(\lambda) \equiv \lim_{N \to \infty} \left\langle e^{\lambda h(N)} \right\rangle_0 = 1, \qquad (30)$$

where $h(N) = \max_{1 \leq j \leq N}\{S_{1,1;j,N}, S_{1,1;N,j}\}$, in the logarithmic phase where $\langle S_{1,1;N,N}\rangle_0 < 0$ for large N.

The function $\Omega(\lambda)$ contains a great deal of information and is difficult to compute in general. Only recently has it been computed [27] for a special choice of scoring functions with

$$s(a,b) = \begin{cases} 1 & \text{if } a = b \\ -2\varepsilon & \text{if } a \neq b \end{cases}$$

and linear gap cost ($\delta = \varepsilon$, $\delta' = 0$), under the approximation that the scores $s(a_i, b_j)$ are uncorrelated for different i's or j's. The result $\lambda(\varepsilon)$ obtained in this case are in excellent agreement with extensive numerical simulation [27], and demonstrates the validity of the formula (30). However, the computation of $\Omega(\lambda)$ for *arbitrary* scoring functions remains unsolved. Along the practical side, Mott and Tribe [28] produced an empirical formula for λ which works reasonably well in the large gap-cost regime. Siegmund and Yakir [29] studied a similar limit where the maximum number of gaps is finite. Despite all of these studies, the current understanding of the statistics of gapped optimal alignment remains very limited.

2.2 Hybrid Alignment

A recent advance in the score statistics of gapped alignment is achieved [10] by hybridizing sum-over-all-path alignment and optimal alignment, hence the name hybrid alignment. Note that what we called sum-over-all-path alignment is very similar to probabilistic alignment except that the probability conservation condition (19) is not required and thus include the probabilistic alignment. In the follows, we will only briefly discuss part of this progress, more details can be found in [10,32,33].

In light of a statistical physics system called "Directed Path in Random Media"(DPRM) [30,31], we have successfully implemented [10] the sum-over-all-path local alignment in dynamic programming approach (corresponding to the transfer matrix approach in DPRM) to provide an efficient computation scheme. Basically, the auxiliary variable $Z_{m,n}$ defined in (20) represents the total weight of all subsequence pairings ending at (a_m, b_n), plus the null alignments. We illustrate the recursive equation of $Z_{m,n}$ in the simplest linear gap function case:

$$Z_{m,n} = 1 + w(a_m, b_n) \cdot Z_{m-1,n-1} + \nu \cdot [Z_{m-1,n} + Z_{m,n-1}], \qquad (32)$$

with the boundary conditions $Z_{0,n} = Z_{m,0} = 1$, for $1 \leq m \leq M$ and $1 \leq n \leq N$. The corresponding affine-gap version can be found in [10]. Note that (32) is the sum-over-all-path version of (12). In the DPRM language, $Z_{m,n}$ is the restricted partition function at finite temperature of a directed path whose end point is fixed at (m, n) but allow all possible starting points.

We further introduce a maximum-log-likelihood (MLL) score

$$\mathsf{S_o} = \max_{1 \le m \le M;\ 1 \le n \le N} \ln (Z_{m,n}) \tag{33}$$

to characterize the quality of the alignment. Equations (32) and (33) define the hybrid algorithm which we proposed to use.

Our alignment method, Equations (32) and (33), is clearly a *hybrid* of sum-over-all-path and optimal algorithms. The MLL score defined in (33) allows us to develop a statistical theory for the maximum score distribution $P(\mathsf{S_o})$. Under a double-saddle point calculation, we have shown [10] that $\Pr(\mathsf{S_o} < x) = \int_{-\infty}^{x} P(\mathsf{S_o}) d\mathsf{S_o}$ of MLL scores follows the Gumbel form (3) in general. But the equation relating Gumbel parameter λ to scoring function is very difficult to solve except when the probability conservation condition holds where an analytical prediction of λ becomes possible. When the probability conservation condition holds, we have

- The Gumbel parameter $\lambda = 1$ in the limit of infinite sequence length. This conclusion even holds true in the case of *position-specific* alignment parameters (scoring functions) [32].
- The *extremal ensemble*, the collection of sequence pairs $[\mathbf{a}_e, \mathbf{b}_e]$ exhibiting similarities that a given scoring system is most sensitive to, has the pairwise joint probability $Q^*[\mathbf{a}_e, \mathbf{b}_e]$ given by [10,33]

$$Q^*[\mathbf{a}_e, \mathbf{b}_e] = P_0[\mathbf{a}_e, \mathbf{b}_e] \cdot W[\mathbf{a}_e, \mathbf{b}_e] \tag{34}$$

where $W[\mathbf{a}_e, \mathbf{b}_e]$ is the global alignment weight of the sequence pair $[\mathbf{a}_e, \mathbf{b}_e]$. The normalization of Q^* is ensured by (39) and the sequence pairs in the extremal ensemble can be *generated* by employing the mutation models similar to what described in the Appendix. *Note that comparable development for gap-allowed optimal algorithms does not exist.*
- The finite-size correction to the *more important* Gumbel parameters λ can be cast into

$$\lambda(N) = 1 + 2/\overline{\sigma}(N), \tag{35}$$

where $\overline{\sigma}(N)$ is the average score of sequence pairs of length N in extremal ensemble, see Fig. 4. For $N \gg 1$, $\overline{\sigma}(N) \approx \alpha N + c$ with constants α and c depend on scoring function used. The constant α is also called "relative entropy". Both α and c can be accurately determined by aligning about 50 sequence pairs in the extremal ensemble, which takes very little computer time.

In terms of sensitivity of homology detection, we have found [10,32] that our hybrid algorithm is comparable to or better than that of the Smith-Waterman algorithm [1]. To evaluate the performance, we hypothesize that

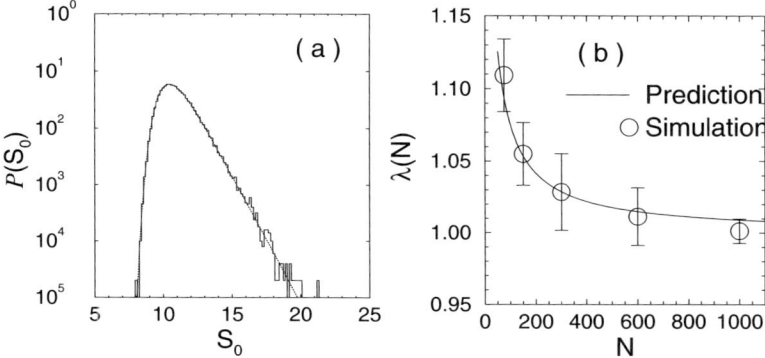

Fig. 4. (a) A MLL score histogram (steps), from using our algorithm to align 50,000 random sequence pairs of length 300 each, is fitted by a Gumbel distribution (solid line); (b) Direct comparison of the numerical values of λ and the theoretical prediction from (35). Each alignment here is performed under the scoring function set $\{\text{PAM} - 120,\ \mu' = 1,\ \mu = 2^{-5.5}, \nu = 2^{-0.5}\}$

the two sequences **a** and **b** are similar if the p-value, $\tilde{p}(\mathbf{a}, \mathbf{b})$, of their alignment score is below some threshold $\tilde{p}_0$. This is to be compared to the "superfamily" classification of the SCOP database[34] which we take as the "gold standard". The superfamily classification separates the sequence pairs into N_p pairs which are truly similar and N_n pairs which are not similar. It also separates the pairs deemed similar by the alignment algorithm (i.e., pairs with $\tilde{p}(\mathbf{a}, \mathbf{b}) < \tilde{p}_0$) into two classes: Those pairs which are also similar according to the gold standard are called "true positives", while the rest are called "false positives". Clearly, the numbers $tp(\tilde{p}_0)$ and $fp(\tilde{p}_0)$ of true and false positives respectively depend on the choice of the threshold $\tilde{p}_0$. Following Gribskov and Robinson [36], we will characterize the algorithm by its coverage rate $c(\tilde{p}_0) \equiv tp(\tilde{p}_0)/N_p$ and its false positive rate $f(\tilde{p}_0) \equiv fp(\tilde{p}_0)/N_n$. The plot of $c(\tilde{p}_0)$ against $f(\tilde{p}_0)$ is known as the Receiver Operating Characteristics (ROC) plot. The overall sensitivity is defined as the area under the ROC curve, i.e., $\int c(f)df$.

In Fig. 5(a) and (b), we show the ROC curves (solid lines) as obtained with the PDB40 and PDB90 sequences respectively, using the hybrid alignment algorithm (detailed in Appendix A of [10]) with the BLOSUM-62 substitution matrix as given in [8] and with affine gap cost $\delta = 11$, $\varepsilon = 1$. The sensitivities of the hybrid algorithm on the two databases are 0.791 for PDB-40 and 0.855 for PDB-90. A comparison of these ROC curves to the corresponding results in the study by Brenner et al. [35] reveals that the performance of hybrid alignment is comparable to the best of the algorithms tested there.

To do a more quantitative comparison, we repeat the above process using the Smith-Waterman algorithm, which is generally recognized as the best among the existing algorithms. We use the same BLOSUM-62 substitution

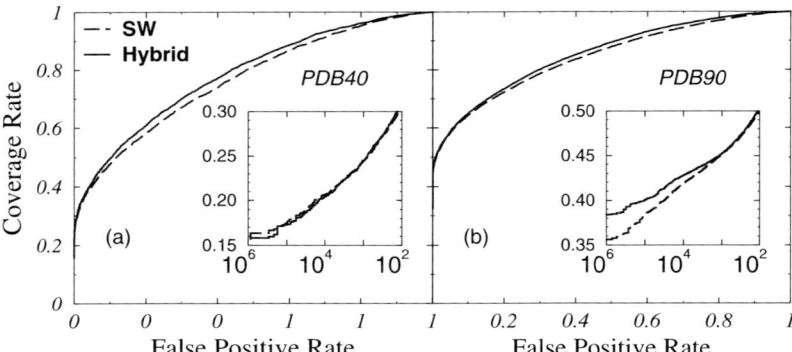

Fig. 5. Plots of ROC curves for both the hybrid algorithm (solid line) and the Smith-Waterman algorithm (dashed line) for (a) the PDB40D–B database and (b) the PDB90D–B database. We find that the hybrid algorithm in general performs comparable to the Smith-Waterman algorithm. The insets magnify the regions very close to zero false-positive rate; they correspond to the regions studied in detail by Brenner et al. [35]

matrix and the same affine gap function. Here, the conversion from alignment scores to the p-values requires the knowledge of the score distribution function which can only be obtained from large simulations: For each of the five different lengths between $N = 75$ and $N = 900$, 50,000 random sequence pairs were generated and aligned in order to obtain reliable score distributions which enable accurate Gumbel fits including length corrections.

The resulting ROC curves are plotted as the dashed lines in Fig. 5(a) and (b). It is clear that the performance of the hybrid and Smith-Waterman algorithms are comparable. Small differences in the ROC curves and in the sensitivity measure (0.767 for PDB-40 and 0.844 for PDB-90) are not deemed significant, as arbitrary removal of a subset of the sequences results in changes that are of the same order as the observed differences. The same qualitative result, that the performance of hybrid and Smith-Waterman alignment are comparable, is obtained for a number of other (uniform) scoring functions we examined, although the absolute performance may differ. For example, the use of PAM substitution matrices led to worse performance by both algorithms.

2.3 Open Problems

In the following, we list a few open problems of interest.

- Although the statistical theory of hybrid alignment has provided us knowledge of the more important Gumbel parameter λ, the other Gumbel parameter K still eludes our analysis. It will be interesting to extend the current theory to incorporate determination of the other Gumbel parameter K.

- It will be interesting if equivalent statistical theory for the gapped optimal alignment can be developed so that one can predict the Gumbel parameters. This issue, however, is very difficult.
- It will also be interesting to study the score statistics of non-Markovian null model. In this case, traditional optimal alignment algorithms do not have Gumbel distribution as the background statistics already. Both the characterization of this new type os statistics and the development new alignment methods deserve further investigations.

Appendix

In this appendix, we will first establish the connection between evolution model and probabilistic alignments and then state the conversion rule from optimal alignment scoring parameters to weight parameters used in hybrid alignment. The first observation to make is that each alignment path in a probabilistic alignment can also be viewed as an evolution path. To illustrate this idea, let us consider the following simple generative (evolution) model: Start with empty sequences **a** and **b** and go through the hidden Markov model illustrated in Fig. 6.

- Until one of the sequences reaches the desired length N, there is a probability ν_c for a "deletion step" and the same probability ν_c for an "insertion step".
    - if the "insertion mode" is selected, generate a new element a according to the background frequencies $p(a)$ and append it to sequence **a**.
    - if the "deletion mode" is selected, generate a new element b according to the background frequencies $p(b)$ and append it to sequence **b**.
    - if neither deletion nor insertion is selected, generate a *pair of elements* (a,b) according to some *joint probability distribution* $q(a,b)$[4] and append a to sequence **a** and b to sequence **b**.
- If one of the sequences reaches the desired length N generate random elements according to the background frequencies $p(a)$ and append them to the shorter of the two sequences until they both have the length N.

Apparently, each realization of such pair generation also trace out a path in Fig. 1(b). In fact, the total likelihood for two sequences **a** and **b** to be related through such generative model can be easily calculated by first computing the weight $\mathcal{W}[\mathbf{a}, \mathbf{b}]$.

The "weight" $\mathcal{W}[\mathbf{a}, \mathbf{b}]$ for two sequences **a** and **b** related through the generative model can be computed iteratively [37] by using an auxiliary variable $\mathcal{W}_{m,n}$:

$$\mathcal{W}_{m,n} = w_c(a_m, b_n) \cdot \mathcal{W}_{m-1,n-1} + \nu \cdot [\mathcal{W}_{m-1,n} + \mathcal{W}_{m,n-1}], \tag{36}$$

---
[4] The joint probability distribution $q(a,b)$ is often chosen as $\mathcal{T}_{a,b}\, p(a)$ where $\mathcal{T}_{a,b}$ is the *transition probability* for a mutation from element a into element b.

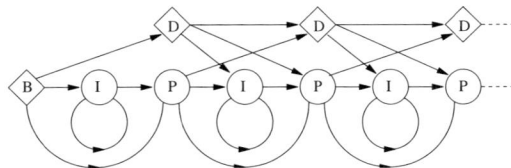

Fig. 6. Schematics of the hidden Markov model $\mathcal{M}$ for sequence evolution. The different states are B for the "begin" state, I for the "insertion" states, D for the "deletion" states, and P for the "pair emission" state. The arrows indicate the allowed transitions between the states, with transition probabilities as given in the text. Sequence elements are "emitted" according to the following rules: An element a is emitted into sequence $\mathbf{a}$ with probability $p(a)$ every time the state I is visited, and an element b is emitted into sequence $\mathbf{b}$ with probability $p(b)$ every time the state D is visited. In state P a pair (a,b) is emitted according to some joint distribution $q(a,b)$ and elements a and b are appended to sequences $\mathbf{a}$ and $\mathbf{b}$ respectively

with the net substitution probability $w_c(a,b)$ being

$$w_c(a,b) = (1 - 2\nu_c)\, q(a,b) \,/\, [p(a)\, p(b)] \tag{37}$$

where the factor $(1-2\nu_c)$ is the probability of getting into emission state. Note that the substitution weight satisfy

$$\langle w_c(a,b) \rangle_0 + 2\nu_c = 1 \tag{38}$$

due to the fact that $\sum_{a,b} q(a,b) = 1$. This demonstrate the reason of our choosing the probability conservation condition (19) for the substitution weight used in probabilistic alignment.

The auxiliary quantity $\mathcal{W}_{m,n}$ can be regarded as the partial weight flow into (m,n). $\mathcal{W}[\mathbf{a},\mathbf{b}]$ is then obtained as the sum of all partial weight flowing into the boundary specified by $(0 \leq m \leq M, n = N)$ and $(m = M, 0 \leq n \leq N)$ [38].

Note that the condition of the type (37) assure us that

$$\sum_{[\mathbf{a},\mathbf{b}]} \mathcal{W}[\mathbf{a},\mathbf{b}] \cdot P_0[\mathbf{a},\mathbf{b}] = 1, \tag{39}$$

which is a manifestation of probability conservation for the different ways sequences can be mutated into each other. In fact, $\mathcal{W}[\mathbf{a},\mathbf{b}] \cdot P_0[\mathbf{a},\mathbf{b}]$ is the likelihood of generating the sequence pair by the mutation model [39,40].

From the similarity between (36) and (17) and the similarity between (37) and (19), we see that the probabilistic alignment is actually looking for sequence pairs whose evolutionary relationship are best described by the corresponding alignment parameters used. That is to say, probabilistic alignment with weight parameters $\{w(a,b),\nu\}$ best detect the sequence pairs whose evolution parameters $\{w_c(a,b), \nu_c\}$ being $w_c(a,b) = w(a,b)$ and $\nu_c = \nu$. Interested readers can find more details in [10].

Let us now discuss the relation between a substitution scoring matrix (used in optimal alignment) and the corresponding substitution weight matrix (used in hybrid alignment). We again illustrate with the linear gap function. Readers interested in knowing the affine-gap version can find detailed documentation in [32]. The possible input parameters to (32) are the substitution matrix $w(a,b)$ and the linear gap parameter ν.

As described in [10], the hybrid algorithm is defined in a certain subspace of the above scoring parameter space. These parameters can be represented more succinctly in terms of the usual input to a Smith-Waterman-type algorithm, which contains a substitution score $s(a,b)$ (e.g, the PAM or BLOSUM scoring matrix) and a linear gap cost, say δ. Affine gap version can be found in [32]. In terms of $s(a,b)$ and δ, the parameters become

$$\nu = y \exp(-\delta) \tag{40}$$
$$w(a,b) = y^2 \exp[\lambda_{ug} s(a,b)] \tag{41}$$

where $y = \sqrt{1+\exp(-2\delta)} - \exp(-\delta)$ plays the role of fugacity and λ_{ug} is the unique positive root of the equation

$$\sum_{a,b} e^{\lambda_{ug} s(a,b)} p(a) p(b) = 1,$$

for a given model of amino acid background frequency $p(a)$.

References

1. T.F. Smith, M.S. Waterman: J. Mol. Biol. **147**, 195 (1981)
2. R. Hughey, A. Krogh: CABIOS **12**, 95 (1996)
3. A. Dembo, S. Karlin, O. Zeitouni: Ann. Prob. **22**, 2022 (1994)
4. S. Karlin, S.F. Altschul: Proc. Natl. Acad. Sci. USA **87**, 2264 (1990)
5. E.J. Gumbel: *Statistics of Extremes* (Columbia University Press, New York 1958)
6. S.F. Altschul, W. Gish, W. Miller, E.W. Myers, D.J. Lipman: J. Mol. Biol. **215**, 403 (1990)
7. M.O. Dayhoff, R.M. Schwartz, B.C. Orcutt: 'A Model of Evolutionary Change in Proteins'. In *Atlas of Protein Sequence and Structure*, Ed. by M.O. Dayhoff, R.V. Eck, **5** supp. 3, pp. 345–358 (1978), Natl. Biomed. Res. Found.
8. S. Henikoff, J.G. Henikoff: Proc. Natl. Acad. Sci. USA **89**, 10915 (1992)
9. S.B. Needleman, C.D. Wunsch : J. Mol. Biol. **48**, 433 (1970)
10. Y.-K. Yu, T. Hwa: J. Comp. Biol. **8**, 249 (2001)
11. R. Arratia, P. Morris, M.S. Waterman: J. Appl. Probab. **25**, 106 (1988)
12. S. Karlin, S.F Altschul: Proc. Natl. Acad. Sci. USA **90**, 5873 (1993)
13. S. Karlin, A. Dembo: Adv. Appl. Prob. **24**, 113 (1992)
14. M.S. Waterman, L. Gordon, R. Arratia: Proc. Natl. Acad. Sci. U.S.A. **84**, 1239 (1987)

15. R. Arratia, M.S. Waterman: Ann. Appl. Prob. **4**, 200 (1994)
16. R. Bundschuh, T. Hwa: in *RECOMB99 – Proceedings of the Third Annual International Conference on Computational Molecular Biology*, Ed. by S. Istrail, P. Pevzner, M. Waterman, (ACM Press, New York 1999) pp. 25-32
17. T.F. Smith, M.S. Waterman, C. Burks: Nucleic Acids Research **13**, 645 (1985)
18. J.F. Collins, A.F.W. Coulson, A. Lyall: CABIOS **4**, 67 (1988)
19. R. Mott: Bull. Math. Biol. **54**, 59 (1992)
20. M.S. Waterman, M. Vingron: Stat. Sci. **9**, 367 (1994)
21. M.S. Waterman, M. Vingron: Proc. Natl. Acad. Sci. USA **91**, 4625 (1994)
22. S.F. Altschul, W. Gish: Methods in Enzymology **266**, 460 (1996)
23. R. Olsen, R. Bundschuh, T. Hwa: 'Rapid Assessment of Extremal Statistics for Gapped Local Alignment'. in *Proceedings of The Seventh International Conference on Intelligent Systems for Molecular Biology (ISMB99)*, Ed. by T. Lengauer (AAAI Press, Menlo Park 1999) pp. 211-222
24. S. Eddy, G. Mitchison, R.Durbin: J. Comp. Biol. **2**, 9 (1995)
25. A. Milosavljevic, J. Jurka: CABIOS **9**, 407 (1993)
26. C. Barret, R. Hughey, K. Karplus: CABIOS **13**, 191 (1997)
27. R. Bundschuh: 'An Analytic Approach to Significance Assessment in Local Sequence Alignment with Gaps' in *Proceedings of the Fourth Annual International Conference on Computational Molecular Biology*, Ed. by R. Shamir, S. Miyano, S. Istrail, P. Pevzner, M. Waterman (ACM press, New York 2000) pp. 86-95
28. R. Mott, R. Tribe: J. Comp. Biol. **6**, 91 (1999)
29. D. Siegmund, B. Yakir: Ann. Stat. **28**, 657 (2000)
30. D.S. Fisher, D.A. Huse: Phys. Rev. B **43**, 10728 (1991)
31. T. Halpin-Healy, Y.-C. Zhang: Phys. Rep. **254**, 215 (1995)
32. Y.-K. Yu, R. Bundschuh, T. Hwa: Bioinformatics **18**, 865 (2002)
33. Y.-k. Yu, R. Bundschuh, T. Hwa: 'Statistical Significance and Extreme Ensemble of Gapped Local Hybrid Alignment', in *Biological Evolution and Statistical Physics* (Lecture Notes in Physics, vol 585), Ed. by M. Lassig, A. Valleriani (Springer Verlag, Berlin 2002) pp. 3–21
34. A.G. Murzin, S.E. Brenner, T. Hubbard, C. Chothia: J. Mol. Biol. **47**, 536 (1995)
35. S.E. Brenner, C. Chothia, T.J.P. Hubbard: Proc. Natl. Acad. Sci. USA **95**, 6073 (1998)
36. M. Gribskov, N.L. Robinson: Comput. Chem. **20**, 25 (1996)
37. M.J. Bishop, E.A. Thompson: J. Mol. Biol. **190**, 159 (1986)
38. Rigorously speaking, the weight $\mathcal{W}[\mathbf{a}, \mathbf{b}]$ is given by

$$\mathcal{W}[\mathbf{a}, \mathbf{b}] = [2\nu + w(a_M, b_N)] \cdot \mathcal{W}_{M-1,N-1}$$
$$+ \sum_{m=1}^{M-1} [\nu + w(a_m, b_N)] \cdot \mathcal{W}_{m-1,N-1}$$
$$+ \sum_{n=1}^{N-1} [\nu + w(a_M, b_n)] \cdot \mathcal{W}_{M-1,n-1}.$$

39. J.L. Thorne, H. Kishino, J. Felsenstein: J. Mol. Evol. **33**, 114 (1991)
40. J.L. Thorne, H. Kishino, J. Felsenstein: J. Mol. Evol. **34**, 3 (1992)

Resolution of Some Paradoxes in B-Cell Binding and Activation: A Computer Study

Gyan Bhanot

This is a description of work done in collaboration with Yoram Louzoun and Martin Weigert at Princeton University. In our computer study, we explain certain puzzling aspects of the binding and activation of B-Cells with a new hypothesis about the rate of endocytosis of receptors. The first such puzzle stems from experiments by Dintzis et al which suggest that the binding and activation of B-Cells is sensitive to the valence (number of binding sites or haptens) on antigen. For valence less than 10-20, B-Cells fail to activate for any concentration of antigen. For larger values of valence, they activate only in a narrow range of concentration. Another puzzle has to do with the non-immunogenicity of chimerical B-Cells, which present receptors with erroneous light chain arrangements. We performed a computer experiment to model the B-Cell surface with embedded receptors diffusing in the surface lipid layer. We presented these surface receptors with antigen with varying concentration and valence. Using experimentally reasonable values for the binding and unbinding probabilities for the binding sites on the antigens, we simulated the dynamics of the binding process. Using the single hypothesis that the rate of endocytosis of bound receptors is significantly higher than that of unbound receptors, and that this rate varies inversely as the square of the mass of the bound, connected receptor complex, we are able to reproduce all the qualitative features of the Dintzis experiment. We were also able to generate some testable predictions on how chimeric B-Cells might be non-immunogenic.

1 Introduction

This paper is a description of work done in collaboration with Yoram Louzoun and Martin Weigert at Princeton University [1]. I begin with a brief introduction to the human immune system and the role of B and T Cells in it [2]. Next, I describe the B-Cells receptor/antibody and how errors in the coding for the light chains on these receptors can result in chimerical B-Cells with different light chains on the same receptor or different types of receptors on the same cell. After this, I describe the Dintzis experiments [3–5] and the efforts to explain these experimental results using the concept of an Immunon [6,7]. There is also analytic work by Perelson [8] using rate equations to model the binding and activation process. This is followed by a description of our computer modeling experiment, its results and conclusions [1].

2 Brief Description of Human Immune System

The human immune system [2], on encountering pathogen, has two distinct but related responses. There is an immediate response, called the Innate Response and there is also a slower, dynamic response, called the Adaptive Response. The Innate Response, created over aeons by the slow evolutionary process, is the first line of defense against bacterial infections, chemicals and parasites. It comes into effect immediately and acts mostly by phagocytosis (engulfment). The Adaptive Response is evolving even within an individual, is slower in its action (with a latency of 4-7 days) but is much more versatile. This Adative Response is created by a complex process involving cells called lymphocytes. A single microliter of fluid in the body contains about 2500 lymphocytes.

All cellular components of the Immune System arise in the bone marrow from hematopoietic stem-cells, which differentiate to produce the other more specialized cells of the immune system. Lymphocytes derive from a lymphoid progenitor cell and differentiate into two cell types called the B-Cell and the T-Cell. These are distinguished by their site of differentiation, the B-Cells in the bone marrow and the T-Cell in the thymus. B- and T-Cells both have receptors on their surface that can bind to antigen (pieces of chemical, peptides, etc.) An important difference between B- and T-Cell receptors is that B-Cell receptors are bivalent (have two binding areas) while T-Cell receptors are monovalent (with a single binding area). In the bone marrow, B-Cells are presented with self antigen, eg. pieces of the body's own molecules. Those B-Cells that react to such self antigen are killed. Those that do not are released into the blood and lymphatic systems. T-Cells on the other hand are presented with self antigen in the thymus and are likewise killed if they react to it.

Cells of the body present on their surface pieces of protein from inside the cell in special structures called the MHC (Major Histocompatibility Complex) molecules. MHC molecules are distinct between individuals and each individual carries several different alleles of MHC molecules. T-Cells are selected in the thymus to bind to some MHC of self but not to any self peptides that are presented on these MHC molecules. Thus, only T-Cells that might bind to foreign peptides presented on self MHC molecules are released from the thymus. There are two types of T-Cells, distinguished by their surface proteins. They are called CD8 T-Cells (also called killer T-Cells) and CD4 T-Cells (also called helper T-Cells).

When a virus infects a cell, it uses the cell's DNA/RNA machinery to replicate itself. However, while this is going on, the cell will present on its surface pieces of viral protein on MHC molecules. CD8 T-Cells in the surrounding medium are programmed to bind strongly to such MHC molecules presenting non-self peptides. After they bind to the MHC molecule, they send a signal to the cell to commit suicide (apoptose) and then unbind from the infected cell. Also, once activated in this way, the CD8 T-Cell will replicate aggressively and seek out other infected cells to send them the suicide signal. The CD4 T-Cells on the other hand, recognize viral peptides on B-cells and macrophages (specialized cells which phagocytose or engulf pathogens, digest them and present their peptide pieces on MHC molecules). The role of the CD4 T-Cell, when it binds in this way, is to signal the B-Cell and macrophages to activate and proliferate.

B-Cell that are non-reactive to self antigens in the bone marrow are released into the blood and secondary lymphoid tissue. They have a life time of about three days unless they successfully enter lymphoid follicles, germinal centers or the spleen and get activated by binding to antigen presented to them there. Those that have the correct antibody receptors to bind strongly to viral peptide (antigen), will become activated and will start to divide, thereby producing multiple copies of themselves with their specific high affinity receptors. This process is called 'clonal selection' as the clone which is fittest (binds most strongly to presented antigen) is selected to multiply. The B-Cells that bind to antigen will also endocytose their own receptors with bound antigen and present it on their surface on MHC-II molecules for an activation signal from CD4 T-Cells. Once a clone is selected, the B-Cells also mutate and proliferate to produce variations of receptors to achieve an even better binding specificity to the presented antigen. B-Cells whose mutation results in improved binding will receive a stronger activation signal from the CD4 T-Cells and will out-compete the rest. This process is called 'affinity maturation'. Once the optimum binding specificity B-cells are produced, they are released from the germinal centers. Some of these differentiate into plasma cells which release large numbers of antibodies (receptors) with high binding affinity for the antigen. These antibodies mark the virus for elimination by macrophages. Some B-Cells go into a latent phase (become memory B-Cells) from which they may be activated if the infection recurs.

It is clear from the above discussion that there are two competing pressures in play when antigen binds to B-Cells. One pressure is to maximize the number of surface bound receptors, until a critical threshold is reached when the B-Cell is activated and will proliferate. The other pressure is to endocytosis the receptor-antigen complex followed by presentation of the antigen peptide on MHC-II molecules, binding to CD4 T-Cells and an activation signal from that binding. To function optimally, the immune system must carefully balance these two processes of binding and endocytosis.

Unbound receptors on the surface of B-Cells are endocytosed at the rate of about one receptor every half hour. However, the binding and activation of B-Cells happens in a time scale of a few seconds to a minute (for references to many of the details of the numerical values used in this paper, refer to the references in [1].) If endocytosis is to compete with activation, as it must for the process described above to work, then bound receptors must be endocytosed much more frequently than once every half hour. Since there is no data available on the exact rate of endocytosis for bound receptors, we made the assumption in our simulation that the probability of endocytosis of a single B-Cell receptor bound to antigen is of the same order of magnitude as the probability of binding of antigen to the receptor. There is a strong probability that multiple receptors are linked by bound antigen before they are endocytosed. We make the reasonable assumption that the probability of endocytosis of the receptor-antigen cluster is inversely proportional to the square of the mass of the cluster.

Let us now discuss, in a very simplified way, the structure of the B-Cell receptor/antibody. The B-Cell receptor is a Y shaped molecule consisting of three equal sized segments, connected by disulfide bonds. The antigen binding sites are at the tip of the arms of the Y. These binding sites are made up of two strands (heavy and light) each composed of two regions, one which is constant and another which is highly variable, called the constant and variable regions respectively. The process that forms the antibody first creates a single combination of the heavy and light chains (H,L) sections and then combines two such (H,L) sections by disulfide bonds to create the Y shaped antibody. In diploid species, such as humans, whose DNA strands come from different individuals, there are four ways to make the (H,L) combinations using genes from either of the parent DNA strands. Thus if the parent types make H1, L1, and H2, L2 respectively, in principle, it would be possible to make four combinations: (H1,L1), (H2,L2), (H1,L2) and (H2,L1). The classical dogma in immunology is allelic exclusion, which asserts that, in a given B-Cell, when two strands of (H,L) fuse to form a receptor, only the same (H,L) combination is always selected. This will ensure that for a given B-Cell, all the receptors are identical. However, sometimes this process does not work and B-Cells are found with both types of light chains in receptors on the same cell [9].

It turns out that there are two distinct types of light chains, called κ and λ. Normally, in humans, the ratio of B-Cells with κ or λ chains is 2:1 with each cell presenting either a $\kappa\kappa$ or a $\lambda\lambda$ light chain combination. However, as mentioned above, sometimes allelic exclusion does not work perfectly and B-Cells present $\kappa\lambda$ receptors or the same cell presents receptors of mixed type - a combination of some which are $\kappa\kappa$, some which are $\lambda\lambda$ and some which are $\kappa\lambda$. A given antigen will bind either to the λ or the κ chain, or to neither, but not to both. Thus a $\kappa\lambda$ B-Cell receptor is effectively monovalent. Furthermore, a B-Cell with mixed $\kappa\kappa$ and $\lambda\lambda$ receptors would effectively have fewer receptors available for a given antigen.

It is possible to experimentally enhance the probability of such genetic errors and study the immunogenicity of the resulting B-Cells. This has been done in mice. The surprising result from such experiments is that chimerical B-Cells are non-immunogenic [9]. We shall attempt to explain how this may come about as a result of our assumption about endocytosis.

3 The Dintzis Experimental Results and the Immunon Theory

Dintzis et al. [3–5] did an in-vivo (mouse) experiment using five different fluoresceinated polymers as antigen (Ag). The results of the experiment were startling. It was found that to be immunogenic, the Ag mass had to be in a range of $10^5 - 10^6$ Daltons (1 Dalton = 1 Atomic Mass Unit) and have a valence (number of effective binding sites) greater than 10-20. Antigen with mass or valence outside this range elicited no immune response for any concentration. Within this range of mass and valence, the response was limited to a finite range of antigen concentration.

A model based on the concept of an Immunon was proposed to explain the results [6,7]. The hypothesis was that the B-Cell response is quantized, ie. to trigger an immune response, it is necessary that a minimum number of receptors be connected in a cluster cross linked by binding to antigen. This linked cluster of receptors was called an Immunon and the model came to be called the 'Immunon Model'. However, a problem immediately presents itself: Why are low valence antigens non immunogenic? Why can one not form large clusters of receptors using small valence antigen? The Immunon model had no answer for this question.

Subsequently, Perelson et al. [8] developed mathematical models (rate equations) to study the antigen-receptor binding process. Assuming that B-Cell response is quantized, they were able to show that at low concentration, because of antigen depletion (too many receptors, too little antigen), an Immunon would not form. However, the rate equations made the flaws in the Immunon model apparant. They were not able to explain why large valence antigen were necessary for an immune response nor why even such antigen was tolerogenic (non-immunogenic) at high concentration.

4 Modeling the B-Cell Receptor Binding to Antigen: Our Computer Experiment

The activation of a B-cell is the result of local surface processes leading to a cascade of events that result in release of antibody and/or presentation of antigen. The local surface processes are binding, endocytosis and receptor diffusion. Each of these is governed by its own time and length scales, some of which are experimentally known.

To model B-cell surface dynamics properly, the size of the modeled surface must be significantly larger than the largest dynamic length scale we wish to model and the time steps used must be smaller than the smallest dynamic time scale. Further, the size of the smallest length scale on the modeled surface must be smaller than the smallest length scale in the dynamics. The size of a B-cell receptor is 3 nm and this is the smallest surface feature we will model. The size of a typical antigen in our simulation is 5-40 nm. The diffusion rate of receptors is of the order of $D = 10 \text{x} 1 - 5^{-10} \text{cm}^2/\text{s}$ and the time scale for activation of a cell is of the order of a few seconds to a few tens of seconds ($\tau \sim 100s$). Hence the linear size of the surface necessary in our modeling is $L > \sqrt{D\tau} \sim 1 - 2\mu m$. This is the maximum distance that a receptor will diffuse in a time of about $100s$.

We choose a single lattice spacing to represent a receptor. The linear size of our surface was chosen to be 1000 lattice units which approximately represents a physical length of $3 - 4\mu m$. The affinity of receptor-hapten binding is 10^5M^{-1} for a monovalent receptor. The affinity of a bivalent receptor depends on the valence of the antigen and on the distribution of haptens on the antigen. The weight of a single hapten is a few hundred Daltons. Hence, the ratio of the on-rate to the off-rate of a single receptor hapten pair is $\sim 100 - 1000$. We choose an on-rate of 0.2 and an off rate of 0.001 in dimensionless units. Our unit of time was set to 0.05 milli second. This was done by choosing D in dimensionless units to be 0.1 which means that the effective diffusion rate is $(0.1 \text{x} (3 \text{nm})^2 / (0.05 \text{ms}) \sim 2.0 \text{x} 10^{-10} \text{cm}^2/\text{s}$. The affinity of chimeric B-Cell receptors was set lower because they bind to DNA with a lower affinity. For them, we used an on-rate of 0.1 and an off-rate of 0.01.

The cell surface was chosen to have periodic boundary conditions as this simplifies the geometry of the modeling considerably. The size of our cell surface is equivalent to 20% of a real non-activated B-cell. A B-cell typically has 50000 receptors on its surface. Hence, we modeled with 10000 receptors initially placed on random sites of the lattice. In each time step, every receptor was updated by moving it to a neighboring site (if the site was empty) with a probability $D = 0.1$. Receptors that are bound to antigen were not allowed to move. At every time step, receptors which have free binding sites can bind to other haptens on the antigen or to any other antigen already present on the surface. They can also unbind from hapten to which they are bound. Once an antigen unbinds from all receptors, it is released within 5 time steps on average. Once every 20 time steps, the receptors were presented with new antigen at a constant rate which was a measure of the total antigen concentration. We varied this concentration rate in our modeling.

The normal rate of endocytosis of unbound receptors is once every 1/2 hour. If this is the rate of endocytosis for bound receptors also, it will be too small to play a role in antigen presentation. Thus we must assume that a bound receptor has an higher probability of being endocytosed compared to an unbound receptor. A receptor can bind to two haptens and every antigen can bind to multiple receptors. This cross linking leads to the creation of large complexes. We assume that the probability to enodcytose a receptor-antigen complex is inversely proportional to the square of its mass. The mass of the B cell receptor is much higher than the mass of the antigens, so, when computing the mass of the complex we can ignore the mass of the antigen. We thus set the endocytosis rate only as a function of the number of bound receptors. The rate of endocytosis for the entire complex was chosen to be inversely proportional to the square of the number of receptors in the complex. More specifically, we set the probability to endocytose an aggregate of receptors to be 0.0005 divided by the square of the number of receptors in the aggregate. For chimeric B-cells we reduced the numerator in this probability by a factor of 100.

5 Results

The results of our computer study are shown in Fig. 1, where the solid line shows the number of bound receptors after 10 seconds of simulation as a function of antigen valence. The data are average values over several simulations with different initial positions for receptors and random number seeds. The dashed line shows the number of endocytosed receptors. One observes a clear threshold below which the number of bound surface receptors stays close to zero followed by a region where the number of bound receptors increases and flattens out. This establishes that we can explain the threshold in antigen valence in the Dintzis experiment.

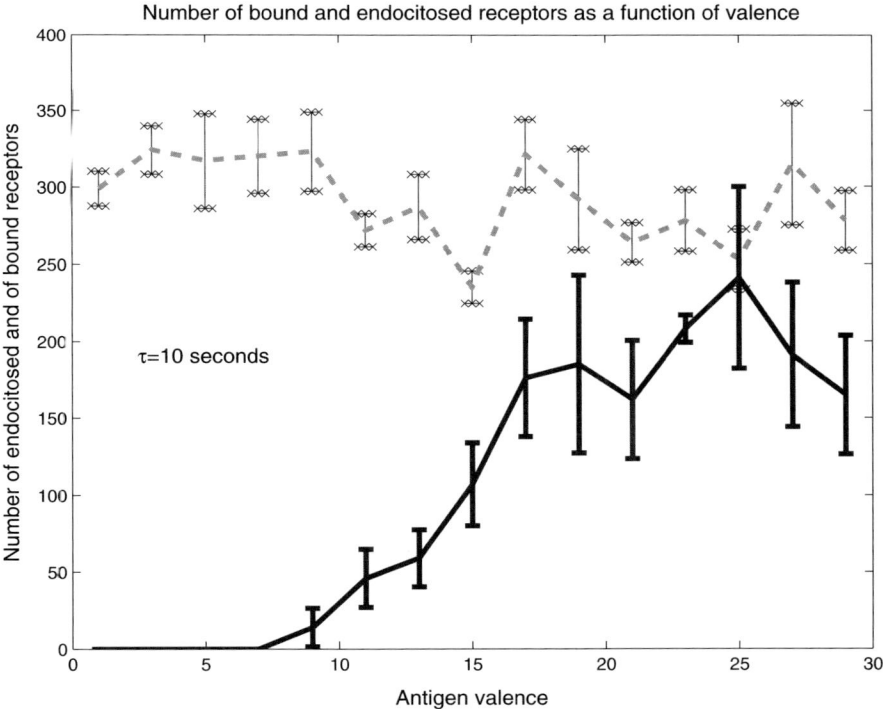

Fig. 1. Dependence of the number of bound receptors and number of endocytosed receptors on Antigen Valence for medium levels of concentration after 10 seconds

The reason for the threshold is easy to understand qualitatively. Once an antigen binds to a receptor, the probability that its other haptens bind to the other arm of the same receptor or to one of the other receptors present in the vicinity is an exponentially increasing function of the number of haptens. Also, once an antigen is multiply bound in a complex, the probability of all the haptens unbinding is an exponentially decreasing function of the number of bound haptens. Given that receptors once bound may be endocytosed, low valence antigen bound once will most likely be endocytosed or will unbind before it can bind more than once (ie. before it has a chance to form an aggregate and lower its probability to endocytose.) As the valence increases, the unbinding probability decreases and the multiple binding probability increases until it overcomes the endocytosis rate. Finally, for high valence, one will reach a steady state between the number of receptors being bound and the number endocytosed in a given unit of time.

In Figs. 2 and 3, we show the number of bound receptors (solid line) and endocytosed receptors (dashed line) as a function of the antigen concentration for two different values of valence. Figure 2 has data for high valence

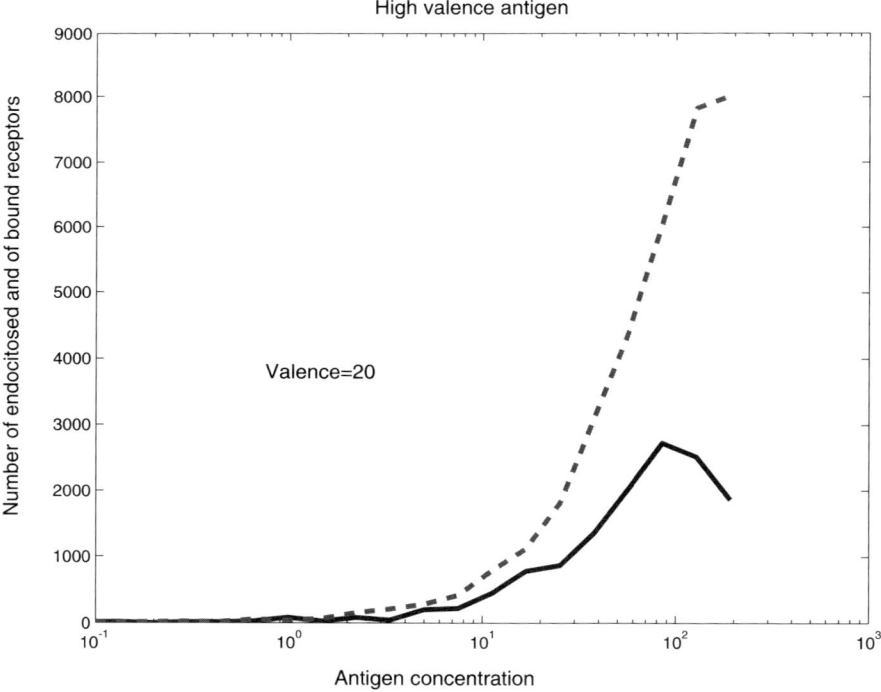

Fig. 2. Dependence of the number of bound receptors and number of endocytosed receptors on antigen concentration for high valence antigens after 10 seconds of simulation

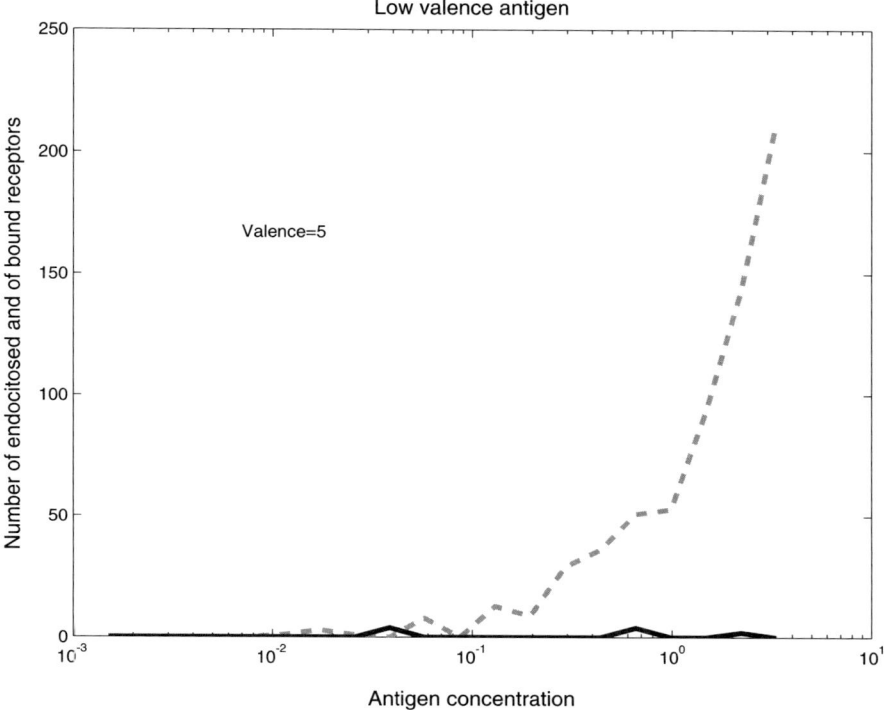

Fig. 3. Dependence of the number of bound receptors and number of endocytosed receptors on antigen concentration for low valence antigens after 10 seconds of simulation

(20) and Fig. 3 for low valence (5). It is clear that for high valence, there is a threshold in concentration below which there are no bound receptors (no immune response) followed by a range of concentration where the number of bound receptors increases followed by a region where it decreases again. The threshold at low concentration is easy to understand. It is caused by antigen depletion (all antigen that binds is quickly endocytosed). The depletion at high concentration comes about because of too much endocytosis, which depletes the pool of available receptors. For low valence (Fig. 3), there is no range of concentrations where any surface receptors are present. The reason is that the valence is too low to form aggregates and lower the rate of endocytosis and is also too low to prevent unbinding events from happening fast enough. Thus all bound receptors get quickly endocytosed. The high rate of endocytosis for high concentration probably leads to tolerance, as the cell will not survive such a high number of holes on its surface.

Figures 1, 2 and 3 are the major results of our modeling. They clearly show that the single, simple assumption of an increased rate of endocytosis for bound receptors and reasonable assumptions about the way this rate depends upon the mass of the aggregated receptors is able to explain both the low concentration threshold for immune response as well as the high concentration threshold for tolerogenic behavior in the Dintzis experiment. It can also explain the dependence of activation on valence, with a valence dependent threshold (or alternately, a mass dependent threshold) for activation.

Now consider the case of chimeric B-Cells. It turns out that these cells bind to low affinity DNA but are not activated [9]. DNA has a high valence and is in high concentration when chimeric B-Cells are exposed to it. To model the interaction of these B-Cells, we therefore used a valence of 20 and lowered the binding rate. We considered two cases:

Case 1: The B-cell has $\kappa\kappa$ and $\lambda\lambda$ receptors in equal proportion. This effectively halves the number of receptors since antigen will bind either to the $\kappa\kappa$ receptor or the $\lambda\lambda$ receptor but not to both. Figure 4 shows the results of our modeling for this case. Note that the total number of bound receptors is very low. This is due to the low affinity. However, the endocytosis rate is high, since receptors once bound will be endocytosed before they can bind

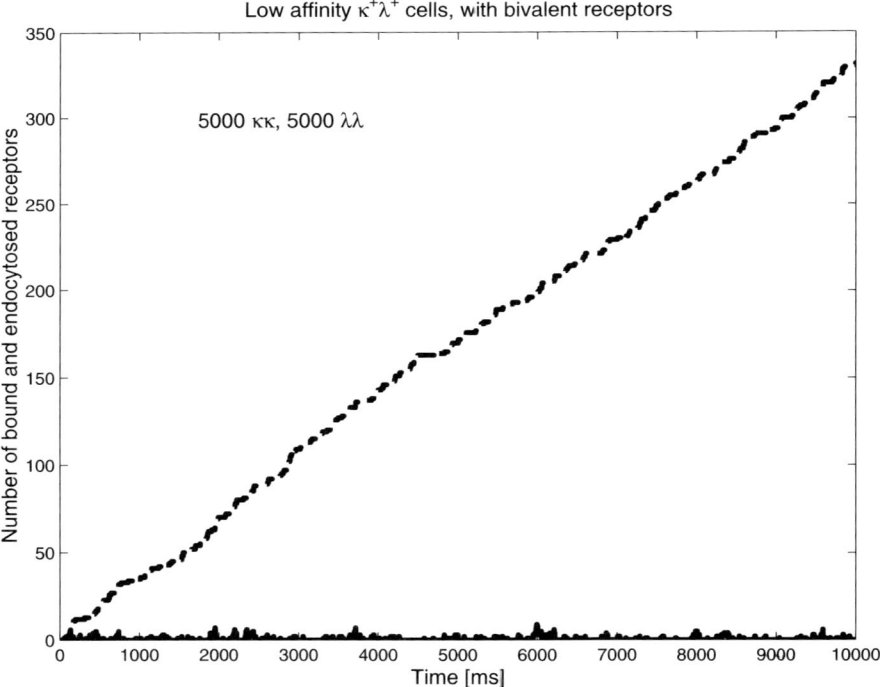

Fig. 4. The number of bound and endocytosed receptors for a cell with 50% $\kappa\kappa$ and 50% $\lambda\lambda$ receptors. These cells would be non-immunogenic because of low levels of activation from the low binding

again and lower their probability of endocytosis. Thus in this case we would expect tolerogenic behavior because of the low number of bound receptors. Case 2: The κ and λ are on the same receptor. This means that the receptor is effectively monovalent since antigen that binds to one of the light chains will not, in general, bind to the other. In a normal bivalent receptor, the existence of two binding sites creates an entropy effect where it becomes likely that if one of the sites binds, the other binds as well. The single binding site on the $\kappa\lambda$ receptors means that antigen binds and unbinds, much like the case of the T-Cell. Thus, although the number of bound receptors at any given time reaches a stady state, the endocytosis rate is low, since receptors do not stay bound long enough to be endocytosed. Figure 5 shows the results of the modeling which are in agreement with this qualitative picture. The non-immunogenicity of $\kappa\lambda$ cells would come about because of the low rate of endocytosis and consequent lack of T-Cell help.

Our modeling thus shows that for chimeric receptors, non-immunogenicity would arise from subtle dynamical effects which alter the rates of binding and

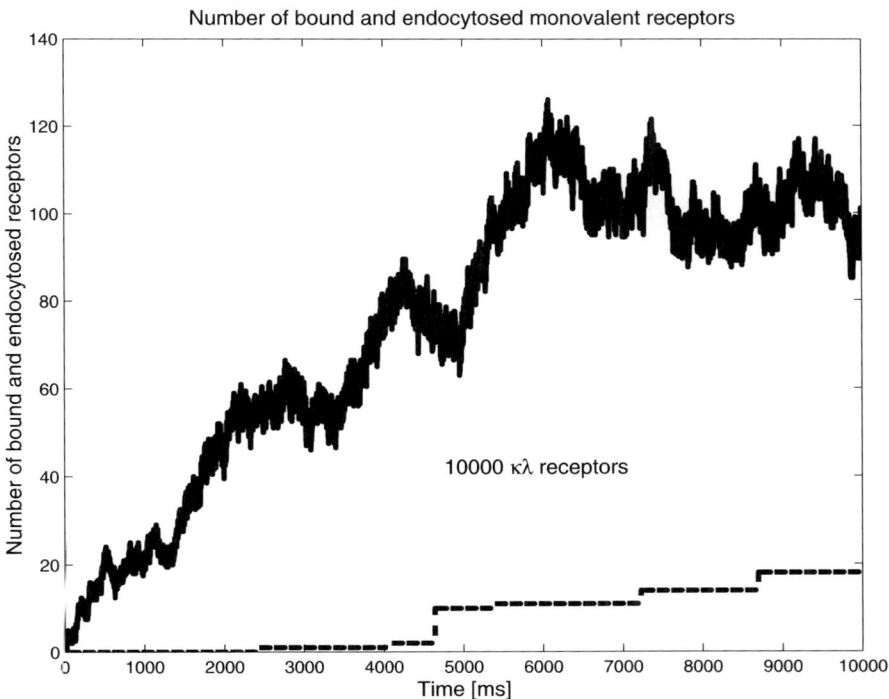

Fig. 5. The number of bound and endocytosed receptors for a cell with only $\kappa\lambda$ receptors. These cells would be non-immunogenic because of low levels of endocytosis and consequent lack of T-Cell help

endocytosis so that either activation or T-Cell help would be compromised. These predictions could be tested experimentally.

Dedication and Acknowledgement

I would like to dedicate this paper to the memory of my friend Phil Seiden who invited me to contribute to this volume and promised to show me the lifestyle of Southern Florida but who passed away suddenly before he could do so. Phil was a colleague at IBM Research, a wonderful human being and a brilliant scientist. I will miss his humor, intelligence and humanity.

I also thank Professor Luc Wille for inviting me to write this paper and I thank my collaborators at Princeton University, Martin Weigert and Yoram Louzoun, for their patient help in my efforts to learn a little biology. Finally, I thank the Computational Biology Center at IBM Research for allowing me leave to work at Princeton University.

References

1. Y. Louzoun, G. Bhanot, M. Weigert: Bull. Math. Biol. **65** 535 (2003).
2. C.A. Janeway, P. Travers, M. Walport, J.D. Capra: *Immunobiology – The Immune System in Health and Disease* (Elsevier Science, London, and Garland Publishing, New York 1999)
3. R.Z. Dintzis, M. Okajima, M.H. Middleton, G. Greene, H.M. Dintzis: J. Immunol. **143**, 4 (1989)
4. J.W. Reim, D.E. Symer, D.C. Watson, R.Z. Dintzis, H.M. Dintzis: Mol. Immunol. **33**, 17 (1996)
5. R.Z. Dintzis, M.H. Middleton, H.M. Dintzis: J. Immunol. **131**, 2196 (1983)
6. B. Vogelstein, R.Z. Dintzis, H.M. Dintzis: Proc. Natl. Acad. Sci. USA **79**, 2 (1982)
7. H.M. Dintzis, R.Z. Dintzis, B. Vogelstein: Proc. Natl. Acad. Sci. USA **73**, 2619 (1976)
8. B. Sulzer, A.S. Perelson: Math. Biosci. **135**, 187 (1996); Mol. Immunol. **34**, 63 (1997)
9. Y. Li, H. Li, M. Weigert: J. Exp. Med. **195**, 181 (2002)

Proliferation and Competition in Discrete Biological Systems

Yoram Louzoun and Sorin Solomon

The three most basic processes in biological systems are proliferation, death and competition. The interplay between these processes was analyzed for more than 200 years in the framework of Ordinary Differential Equations (ODEs) and Partial Differential Equations (PDEs). However, recently it has been shown that many systems in chemistry, biology, finance and social sciences present emerging features which are not easy to guess from the elementary interactions of their microscopic individual components. It turns out that taking into account the actual individual/discrete character of the microscopic components of these systems is crucial for explaining their macroscopic behavior. We model the behavior of prototype biological systems, containing proliferation and death, taking into account the discrete distribution of the agents in such systems. The emerging population dynamics is different from the one predicted by PDEs and ODEs. The total population of the agents increases indefinitely in situations where the classical lore predicted a decreasing population. We analyze the behavior of these systems using a renormalization group analysis to show that the discrete distribution of the agents leads to a localized aggregation of agents with surprising emerging behavior. We then enlarge this analysis to include the effects of different regimes of competition.

1 Introduction

Biological systems were analyzed in the last two hundred years mainly using differential equations [1,2]. This type of equations is valid only under very strict limitations, which apply to most physical systems, but is not valid in most biological systems. In order to test these equations, we analyze prototype biological systems. These systems are composed of the main elements in any biological system: proliferation, death and competition. We show the failure of differential equations even in these very simple system. We then propose a modeling methodology, which has the capacity to express much of the successes of the other modeling methods while transcending their limitations and in particular the limitations which prevented the appropriate understanding of the central issue: the emergence of complex deterministic macroscopic functions out of simple stochastic microscopic interactions.

The methodology is based on recognizing the discreteness and spatial inhomogeneity of the biological units [3,4] and tailoring a specific model for each biological situation by representing the objects and interactions appropriate for each scale.

We start by presenting a very simple generic model which contains proliferating (and dying) individuals. We show that in reality it behaves very differently from its representation in terms of continuum density distributions: in situations where the continuum equations predict population extinction, the individuals self-organize in spatio-temporally localized adaptive patches, which insure their survival and development. This phenomenon is due to the emergence of large, macroscopic structures from an apparently uniform background [39] due to the amplification of small, microscopic fluctuations which originate in the individualized character of the elementary components of the system. This mechanism insures in particular that on large enough two-dimensional surfaces, even if the average growth rate is negative (due to a very large death rate), adaptive structures always emerge and flourish.

The remainder of this chapter is organized as follows. In Sect. 2 we describe in detail the model system, which contains proliferating and dying cells. We then develop the appropriate PDE and solve it. The results of the PDE are compared to a precise microscopic simulation (MS) of this system, and we show that the proliferation rate predicted by the MS is many orders of magnitude larger than the one expected from the PDE. In particular one obtains proliferation under conditions in which the naive continuum limit predicts decay. In Sect. 3 we identify the source of the difference between MS and PDE and discuss some methodological aspects of this difference. In Sect. 4 we describe the trade-off between the emergence of local structure through enhancement of microscopic fluctuations and the smoothing of these structures through competition. In Sect. 5 we analyze systematically the behavior of our system, and describe the different regimes the system can have. In Sect. 5 we introduce additional reactions, which limit the growth of the proliferating entities. We show that under certain conditions the macroscopic inhomogeneity is even increased by the effect of competition.

Summarizing, in this chapter we are opening the way to a wide front re-evaluation of the dynamics that is supposed to emerge out of the known microscopic elementary interactions of biological systems. We suggest that the main effects be re-scrutinized and provide the particular methods to do so. Paradoxically, the new methods that we propose are closer to the intuition of the practitioners in biology. Indeed the objects which MS manipulates are the very same that biologists encounter in their experimental work: cells, molecules, etc. (rather than abstract approximations such as differential equations).

2 Dynamics of Discrete Proliferating Agents

Proliferating entities can be described as agents A that duplicate whenever they meet a homogeneously distributed stimulating signal S. These agents have in general also a typical death rate d. Some examples of such agents are: in immunology – immune cells reacting to an antigen [6]; in ecology – animals proliferating whenever they find food.

This system can be described as a two-reactions system (Fig. 1). The first reaction is that whenever an A agent meets a stimulating signal S it multiplies with a probability λ. We will denote this as:

- $A + S \rightarrow A + A + S.\ \lambda$

The second mechanism is the death of an agent with a probability μ:

- $A \rightarrow \Phi.\ \mu$.

The S and A agents diffuse with a rate of D (which can be different for A and for S). Moreover, we let both the A and the S agents diffuse for a long enough time to assure the homogeneity of their distribution. Only then do we start the A reactions.

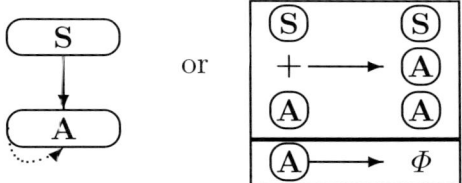

Fig. 1. The reactions taking place in a system of proliferating cells. 1) Whenever an A agent meets an S agent, a new A agent is created. $A + S \rightarrow S + A + A$, with a rate of λ 2) Each A agent has a constant probability μ of dying. $A \rightarrow \Phi$. The Partial Differential Equation (PDE) approximation in the continuum limit for these 2 reactions is: $\dot{A} = (<S>\lambda - \mu)A + D\nabla^2 A$. In the left figure full arrow represent activation, and dotted arrows represent destruction

The naive lore would expect the growth rate of A to be proportional to the number of A and S pairs. Thus the number of new A agents produced in a time interval Δt would be:

$$\Delta A = SA\lambda \Delta T. \tag{1}$$

The number of A deaths is proportional to the total A population:

$$\Delta A = -\mu A \Delta t. \tag{2}$$

Thus the total change in A is:

$$\Delta A = (S\lambda - \mu)A\Delta t + \text{diffusion}. \tag{3}$$

This equation becomes a PDE [7] when we take the limit $\Delta t \to 0$:
$$\dot{A} = (S\lambda - \mu)A + D\nabla^2 A. \tag{4}$$
The solution of this PDE with homogeneous initial conditions is [7]:
$$A(x) = e^{(S\lambda - \mu)t}, \tag{5}$$
showing the expected exponential behavior.

This A population will decay if the average growth rate is lower than the decay rate $((S\lambda - \mu) > 0)$ [1]. On the other hand, if the average growth rate is higher than the death rate $((S\lambda - \mu) > 0)$ the A concentration grows with an exponential growth rate of $(S\lambda - \mu)$. This exponential growth will end when other mechanisms, discussed in the following sections, become important.

In order to test the prediction of the PDE, we simulated the system using discrete agents moving on a lattice. We first let all the agents diffuse freely, until they reach a homogeneous distribution. We then start the interactions of the A agent. We use a regime in which the birth rate is much lower than the death rate. Following the PDE one would assume the total A population will decrease to zero. However, the result is quite the opposite: instead of the 'death' predicted by the PDE, one finds abundant 'life' under a wide range of conditions. The A population increases exponentially, instead of decreasing to zero (Fig. 2). This simulation was performed using both asynchronous and synchronous simulation tools, in order to check for artifacts of the simulation, but the results were similar in both simulations.

The PDE treatment fails in this case, since it ignores the discrete distribution of the S agent. Indeed, if the signaling agent (S) is a random discrete agent, its distribution is uniform over large scales but locally its density will be granular. The difference between the large scale S uniformity and its microscopic granularity is the origin of the macroscopic differences between the A dynamics predicted by the PDE and by the MS treatment. The discreteness of S induces automatically microscopic fluctuations. The macroscopic distribution of S will always be homogeneous, but it will always have microscopic fluctuations. The diffusion of the S will change the conformation of these microscopic fluctuations, but their distribution will not change.

In the PDE treatment, one assumes that the S fluctuations are smoothed due to the diffusion, and that the time limited fluctuations have no effect on the average behavior of the system. These two assumptions are wrong. The fluctuations in S cannot be smoothed since they are due to the inherent discreteness of the agents. The effects of the fluctuations cannot be averaged since the S and the A agents are strongly correlated. The local dynamics of the As at a point i in space are:
$$\dot{A}_i(t) = (S_i(t)\lambda - \mu)A_i(t) + \text{diffusion}. \tag{6}$$
In the PDE treatment this dynamics is averaged to:
$$\begin{aligned}\langle \dot{A} \rangle_r &= \lambda \langle S \rangle_r \langle A \rangle_r(x) - \mu \langle A \rangle_r + D\nabla^2 \langle A \rangle_r \\ &= (\lambda S - \mu)\langle A \rangle_r + D\nabla^2 \langle A \rangle_r,\end{aligned} \tag{7}$$

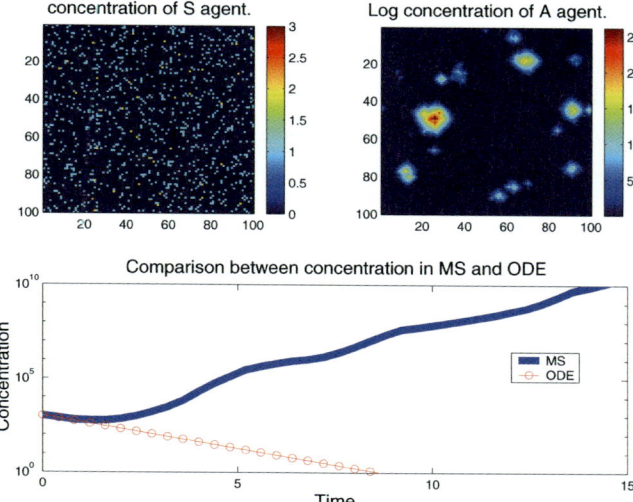

Fig. 2. Comparison between the MS and PDE description of the system described in Fig. 1. The A drawing is the local concentration of S agents at every lattice site. The B drawing represents the log of the A agent concentration at every site. We use a log representation in order to show the large variance in the local concentration of the A agents. The C drawing shows the average A agent concentration in the MS (interrupted lines) and in the differential equation model (circles)

where r represents a region over which the distribution S is smooth enough to be treated by the classical analytical methods. Eq. (6) transforms into Eq. (7) if we can separate $\langle S_i A_i \rangle_r$ into $\langle S_i \rangle_r \langle A_i \rangle_r$. Thus in order for Eq. (7) to hold S and A have to be independent variables [8] at least in the small region r. Eq. (7) is the very basis of any PDE approach. Yet, it is wrong in the case we present here, since A_i is high precisely on the sites where S_i is high. Thus S and A are in fact strongly correlated.

One could argue that the diffusion of both A and S will average the local fluctuations of the A population and weaken the effect of the correlation term. We will show in the next sections that this is not the case, but first let us discuss some general implication of our results.

3 How Well Do Different Methods Deal with Discreteness?

The MS helped us uncover two basic features:
- The population increase in the MS is much higher than that computed in the PDE treatment: the solution of the PDE is exponentially decaying while that of the MS is exponentially growing.

- Diffusion by itself is not enough to smooth the A concentration distribution.

These phenomena admit multiple interpretations in various fields:

- If the individuals are interpreted as interacting molecules, the resulting chemical system self-organizes into spatial patches of high density which evolve adaptively in a way similar to the first self-sustaining systems that might have anticipated living cells.
- If the individuals are the carriers of specific genotypes represented in a genetic space, the patches can be identified with species, which rather than becoming extinct, evolve between various genomes (locations in the genetic space) by abandoning regions of low viability in favor of more viable regions. This adaptive speciation behavior emerges in spite of the total randomness we assume for the individual motions in the genetic space (mutations).
- In immunology these agents represent the activation of immune cells in the presence of minute amounts of antigen. These cells will be activated and create new clones, instead of dying.

The crucial role of microscopic discretization in the emergence of macroscopic inhomogeneity might be unfamiliar and even strange for the reader. After all, microscopic corrections to the differential equations should lead to microscopic corrections in their solutions. That this is however not the case is known in the theory of electric conductivity since the late 1950's. Indeed, by solving the Schrödinger equation for periodic potentials one obtains electronic wave functions that spread out over the entire space (which implies electric conductivity) [9]. However, at least in two dimensions, the slightest random uniform noise leads to spatially localized wave functions (insulators) [39]. Similarly, here the presence of microscopic discretization of S induces intrinsically microscopic spatial inhomogeneities in the S distribution. Thus we see that discretization implies fluctuations: the MS predicts 'life' where the PDE predicted 'death'.

4 Single S Analysis [10]

In order to understand the 'source of life' in this system one has to concentrate on the microscopic conditions around individual S agents, rather than looking at the local average growth rate $\lambda S - \mu$. To illustrate this point, Fig. 3 represents the evolution of the A cloud following a *single S* agent as it jumps around randomly. As can be seen there, the A concentration tracks the S agent as the latter performs a random walk. Clearly, the colony does not decay to extinction. Instead, it seems to follow the single S agent, 'trying' to keep its center of mass at the S agent's location. As shown in the inset, each jump of the single agent is followed by a momentary decrease in the height of

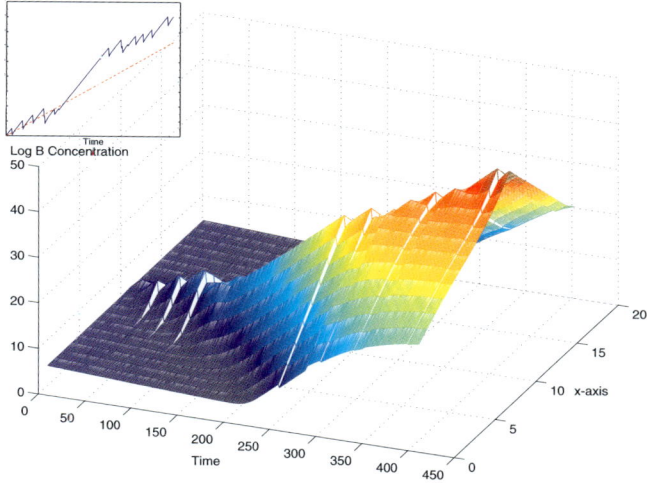

Fig. 3. The profile of an A island as a function of time as it follows the random motion of a single S agent. The cross-section of the island is taken through the current location of the S agent. The inset shows the time evolution of the height of the A concentration at the point at which S is currently located (solid blue line). The A colony is seen to grow, although the average growth rate over the entire space is negative (S is extremely low since there is only one agent in the whole simulation space, thus $\lambda S - \mu \approx -\mu$. The dashed red line shows the exponential growth with coefficient $\epsilon_0 - 2dD_S\kappa$ where $\kappa = Log_e(\lambda/D_A)$, is the slope of the island (this slope is exhibited in the main graph, and can be derived from a simple approximate calculation). Likewise, ϵ_0 is $\lambda - \mu - 2dD_A$ which can be derived similarly

the A concentration, but due to the multiplicative process there is an overall increase [10].

Let us first consider the simplest situation of a single S agent jumping randomly (with diffusion coefficient D_S) between the locations in an infinite d-dimensional space. In between S-jumps the A-density at the S location grows exponentially [11] as $n_A(t) \sim n_A(0)e^{(\lambda-\mu-2dD_A)t}$. Here λ, μ and $2dD_A$ stand for the proliferation, death, and the loss due to diffusion, respectively. The estimation is made by neglecting the flow of A's from neighboring sites to the S site, which is justified when the A concentration in the neighboring sites is much lower than on the S site. In the same limit, the ratio between the height of the A density at the S location and the height of the A density on a neighboring site is easily estimated to be equal to: λ/D_A. Consequently, each S jump corresponds to a sudden decrease by a factor of λ/D_A in the height of the A hill. As there are in average $2dD_S$ such jumps per unit time, the net effect of proliferation, diffusion and death, gives the A concentration at the S site as a function of time:

$$n_A(t) = n_A(0)e^{(\lambda-\mu-2dD_A-2dD_S \ln(\lambda/D_A))t}. \tag{8}$$

This approximation is in good agreement with the simulation shown in Fig. 4. The slope of the island, on a log-scale, is indeed seen to be $\ln(\lambda/D_A)$, and the time dependence of the height of the A island in between S jumps is indeed given approximately by an exponent, $(\lambda - \mu - 2dD_A)t$. Consequently the dashed line (in the inset), which represents Eq. (3), follows closely the actual growth seen in the simulation (blue line). The difference between the analytical estimate and the simulation is mainly due to cases where two or more S jumps follow each other rapidly. In such cases the island's shape does not stabilize before another S jump is made and corrections need to be made to the estimates. These rather rare events modify somewhat the actual result.[1]

One may ask what happens in cases when single colonies are unstable (i.e where the exponent in Eq. (3) is negative). One possibility is that in such a situation the continuum approximation is valid and the A concentration decays to zero. Another possibility is that, although single isolated colonies are unstable, global effects such as islands growing, joining and splitting give us back the survival feature. In particular, since large colonies are more stable than small colonies, one may expect the typical size of an "active" colony to grow with time. This behavior is demonstrated in Fig. 4 which shows the active clusters in a two-dimensional system developing in time. Evidently, the small clusters either decay or merge into larger and larger clusters.

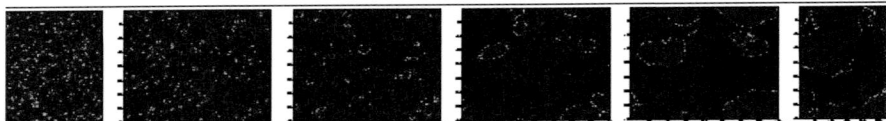

Fig. 4. The spatial distribution of A's for different times (earlier times to the left), where concentrations greater than 10 are colored red and concentrations below 10 are color coded

5 RG Analysis [10]

In order to obtain a more rigorous bound on the parameter range in which life overcomes the gloomy prognosis of the macroscopic analysis we used the renormalization group (RG) analysis which indicates that on large enough surfaces 'life always wins'. For higher dimensions, the domination of life still extends to arbitrary low n_S densities, but a minimal finite λ value is required.

In RG, the collective behavior of the system is identified by integrating out the small length scale and short time fluctuations, leaving us with an

---

[1] The above analysis becomes invalid if $D_A = 0$, where the spatial dimensions of the island do not grow at all. We do not consider this singular case in this paper.

effective theory for the large scale objects. Here, these are the large, stable islands shown in Fig. 4. The new, effective theory is characterized by renormalized coupling constants, i.e., modified numerical values of the effective rates (growth rate, dearth rate, hopping, etc.) on the larger length scale. The process of decimating small fluctuations is then iterated again and again, giving us flow lines which reflect the evolution of the effective values of the coupling constants as one integrates over larger and larger scales l.[2]

Figure 5 shows the flow lines of m ($m = \mu - \lambda n_S$) and λ due to the iteration of the decimation process[3]. The flow is given by the equations:

$$\frac{dm}{dl} = 2m - \frac{\lambda^2 n_S}{2\pi D}$$
$$\frac{d\lambda}{dl} = \lambda[2 - d + \frac{\lambda}{2\pi D}] \qquad (9)$$

For $d \leq 2$ we see that for large length and time scales (that is, after many iterations of the decimation process), λ grows without limit while m eventually becomes negative. This implies that on the large scale, the system actually behaves as if $\lambda n_S > \mu$, and life always wins.

In higher dimensions ($d > 2$) Fig. 5 indicates a dynamical phase transition where for part of the parameter space the system flows to negative m (life) and for another part the system flows to positive m (death).

It should be noted that the flow portrayed in Fig. 5 is associated with larger and larger length scales. For a finite system, the flows should be truncated and the size of the system may be crucial: simulations with parameters identical to that of Fig. 4, lead to extinction when carried out on a system size four times smaller.

6 Mechanisms Limiting Population Growth

The proliferation of the A agents in the models of Sect. 2 must be limited by some saturation mechanism. We have shown that a high death rate (even a death rate higher than the birth rate) is not enough in order to limit the population proliferation. The required mechanisms can be either an external regulator [13], or a limit on the resources available to the proliferating agent [14].

---

[2] The details of this RG analysis, which involves the presentation of the exact Master equation of the process as a field integral and the ϵ-expansion around the critical dimension $d_c = 2$ are beyond the scope of this paper and can be found elsewhere. [12]
[3] $D \equiv D_A + D_S$ is the effective diffusion constant

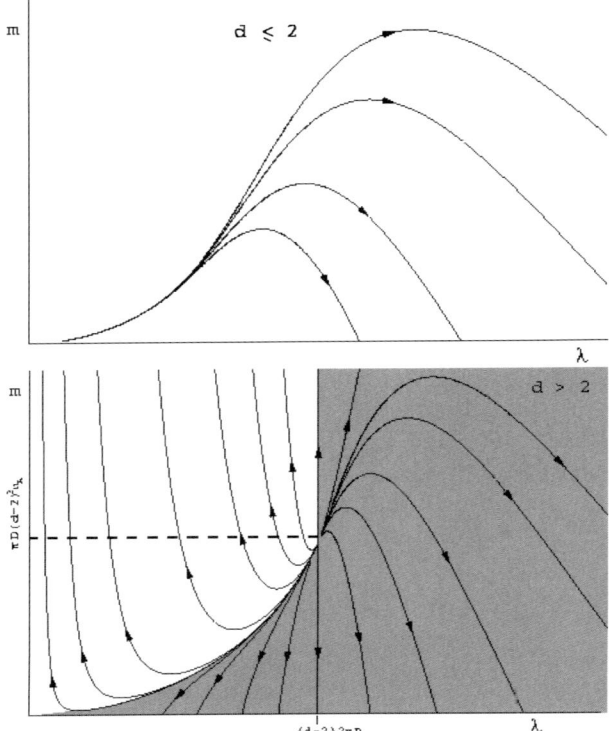

Fig. 5. Lower panel shows flow lines for $d > 2$. Shaded region flows to negative mass ('life'). Upper panel shows flow lines for $d \leq 2$, the whole parameter space flows to negative mass

6.1 Local Competition

Many of the mechanisms limiting the A proliferation can be described as competition. Competition for a resource means that whenever two A agents need the resource either to survive or to proliferate only one of them will get it and the other will die (or fail to proliferate). In ecology competition can be over food or water, while in the immune system competition can be over access to an antigen (Fig. 6).

Such a system is sometimes called a Lotka-Voltera system [15,16] and can be classically described using a local differential equation:

$$\dot{A}(r) = \lambda S(r)A(r) - \mu A(r) - \gamma A(r)^2 + D\nabla^2 A(r) \tag{10}$$

For low values of A this system will behave in a similar way to the system described in Sect. 2, where we have already shown the failure of the PDEs. As the A population increases the saturation term becomes important and the A population will be limited by the saturation term.

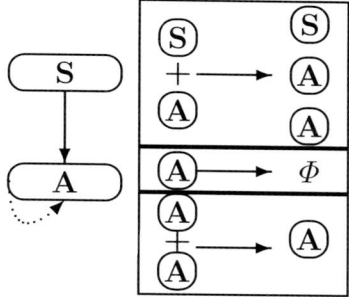

Fig. 6. Reaction scheme for a system containing local competition. This system has constant proliferation and death rates as in Fig. 1. However it has an extra mechanism of local competition: Whenever two A agents meet there is a probability that one of them will be destroyed $A + A \to A$

On can analyze two cases. If $(\lambda S - \mu) > 0$ the total population will increase at almost any point in space until limited by the saturation, and then will stabilize around a steady state at: $A = \gamma/(\lambda S - \mu)$. The discreteness of S induces only a low level of fluctuations around this steady state. The more interesting case is when $(\lambda S - \mu) < 0$. We have shown in the previous section that, even in this case, the point $A = 0$ is not a stable steady state (in contrast to the PDE result). The steady state occurs when the A population in the regions of high A population is saturated by the competition term. The only difference between this case and the one presented in the first section is the limited size of the high A population zones (Fig. 7).

Paradoxically the conditions in which PDE and the MS show similar features are seldom realized in nature. They represent extreme cases in which a certain agent is capable to fill all the available space. In the animal world only a few species enjoy such a status (for example, humans and their parasites). In immunology this regime corresponds to a cell type filling all the blood and immune system (systemic diseases). Except for these extreme cases the situation in which $(\langle S \rangle \lambda - \mu) < 0$ seems to be more appropriate for realistic systems.

6.2 Global Competition

The proliferation of the A agent and the competition between different A agents can occur over very different spatial scales. The proliferation is a local mechanism, while the factor leading to competition can be a global one. In ecological systems the global character of the limiting factor may be related to the fact that the predators move much faster than their prey and that their population is proportional to the total prey population. Therefore the predator population constitutes an indirect interaction between distant prey locations. We will show that in such cases the effects of inhomogeneity are even more important than in the previous sections. As an aside let us note

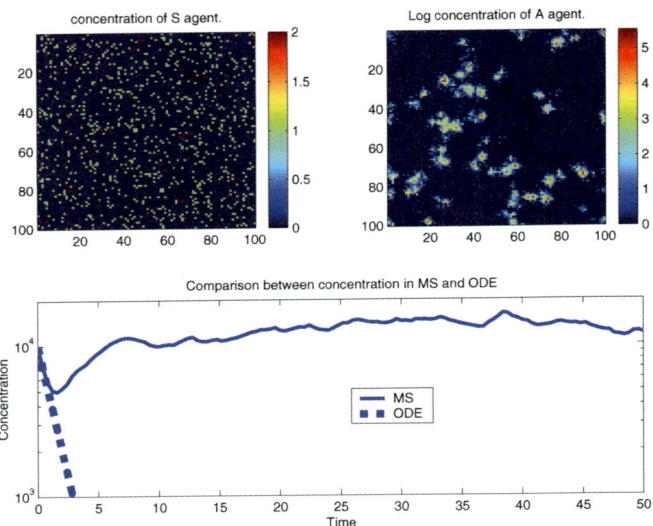

Fig. 7. Comparison between PDE and MS for the system defined by Fig. 6. This system has an average negative growth rate yielding an exponential decay in the ODE description as in the previous section. In the MS we observe a limited growth of the A islands, which is due to the competition between As. Thus even regions in which the local A concentration would grow are limited to a finite concentration

that this kind of reactions although formally local (a lion can eat a prey only if they meet) are very effective in inducing top down effects; that is, the regulation of the local prey population is enforced by their total population.

MS can describe in a natural way only local interactions. It is non-trivial to describe in MS factors that are acting on widely different scales. The solution in this case would be to create a new type of hybrid models (HM), which will allow us, in particular, to express the interaction between agents and their averages over various ranges.

A simple example for such system is obtained by modifying the regulation mechanism of the system presented in the previous section (Fig. 8). There the two mechanisms of proliferation and competition were local interactions. We can now replace the local competition by a global competition. Even though from a microscopic point of view the non-locality might look unnatural this is the only way that non-linearities can arise at the macroscopic level. Indeed, a term of the form $\langle A \rangle^2$ can appear only as an average over terms of the type $A(r)\langle A \rangle$. Therefore the modification of Eq. (6) would be

$$\dot{A}(r) = \lambda S(r) A(r) - \mu A(r) - \gamma A(r) \langle A \rangle + D \nabla^2 A(r) \tag{11}$$

That is, we replaced the local competition in the system described in Sect. 4 by a global competition term. We find that passing from local to global reactions has crucial implications. By comparing the behavior of the system

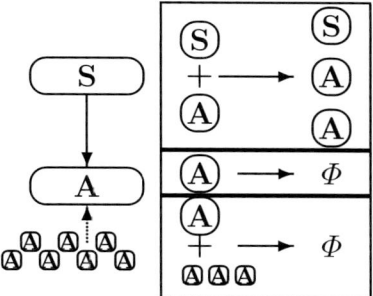

Fig. 8. Global competition reaction scheme. The mechanisms of proliferation and death are similar in this system to those described in Fig. 1. The extra mechanism in this system is a global competition: Each A agent has an extra death rate proportional to the average concentration of the A's

Eq. (11) with the system Eq. (10) one find that they are dramatically different in all the parameter regions. Even in the region in which the system Eq. (10) is described correctly by a PDE, the PDE are still incapable of describing properly the system (11). Indeed in the system (11) (for most parameter sets) the entire space is empty of As except for a single island centered around the point at which the S concentration is maximal (Fig. 9).

We can understand this system by ignoring the diffusion term and simplifying the equation to:[4]

$$\dot{A} = S(r)A(r) - A(r)\langle A \rangle = A(r)(S(r) - \langle A \rangle) \qquad (12)$$

One can see that in every point in which $S(r)$ is lower than S_{max} the $A(r)$ density will vanish, while around the point r_{max} in which $S(r) = S_{max}$ a region of high A density will appear. Suppose that one starts with all the $A(r)$ very small. As long as the average concentration of the A's ($\langle A \rangle$) is lower than S_{max}, the local concentration $A(r)$ at the point r_{max} will rise (according to Eq. (12)). This in turn will raise the value of $\langle A \rangle$. Eventually $\langle A \rangle$ will reach the value of S_{max}. What will be the equilibrium value of the $A(r)$'s at every other point? In every point where $S(r) < S_{max}$, one will have $A(r)(S(r) - \langle A \rangle) = A(r)(S(r) - S_{max}) < 0$, which according to 12 means that the concentration at those points will decay to 0. It is clear, therefore, that the entire A population responsible for $\langle A \rangle = S_{max}$ is due to a very dense island situated around r_{max}. The equality $\langle A \rangle = S_{max}$ is not affected by the A diffusion, which only has the effect of spreading the distribution of the $A(r)$'s in a limited region around r_{max}. One sees that the dynamics is dominated in this case by the most extreme stochastic local fluctuation in the $S(r)$ distribution, and not by the global or average properties of the $S(r)$'s.

---

[4] We rescaled the equation to $\lambda = 1$, $\gamma = 1$ and added the death term μ into the S density. This can be done without any loss of generality.

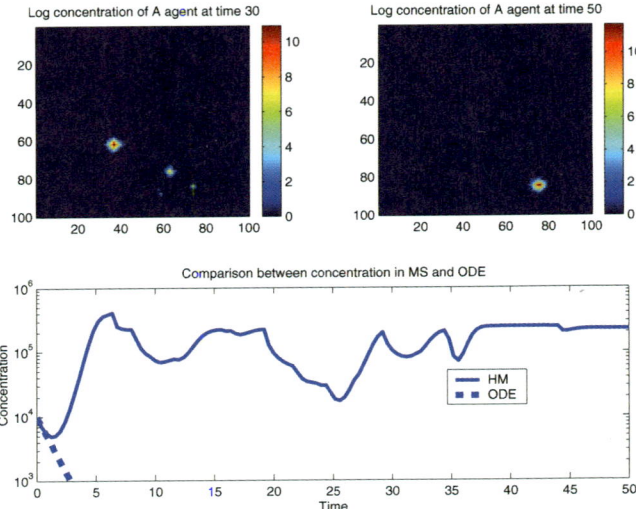

Fig. 9. Comparison between PDE and HM for a system with an average negative growth rate and a global competition term (as defined in Fig. 8). In this model a single large island with a very high A concentration appears. The competition with this island destroys all the other small A islands. The top figures show the log of the concentration of A agents at different times. The bottom part shows the comparison between the ODE and the HM

At this stage, one can re-introduce the S diffusion term, and show that two regimes can exist. In most of phase space all the A population is concentrated in a single island, but in a small part of the phase space, multiple islands can exist for a long time (Fig. 10). Note again that the diffusion of the S does not lead to a smoothing of the A fluctuations. It only leads to a movement of the high A population zone. This zone does not disappear, since its lifespan is much longer than the lifespan of a single A. (It is the time it takes to destroy a microscopically large number of As.) Within this lifespan a new large fluctuation of S will appear within this zone and the zone will restart to grow.

6.3 Emergence of Complexity

In the previous sections it has been shown that the discrete character of the elementary agents S leads to unexpectedly rich and complex behavior of the A population. The complex spatio-temporal emergent structure of the A population cannot be predicted by a PDE approach. However a system which has only proliferation and death (linear) terms leads eventually to a divergent A population. In the present section we introduced few types of competition terms. Local competition limits the size and maximum concentration of each macroscopic A island. However, rather than interfering with the emergence

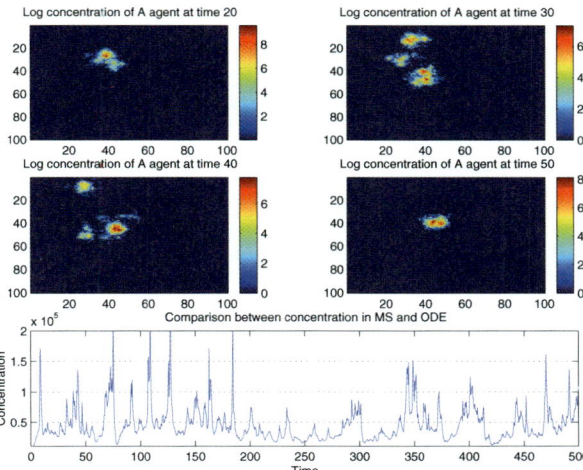

Fig. 10. The model defined in Fig. 8 in a regime with high S diffusion rate. When the S diffusion rate is very high, a few large A islands are created. As opposed to Fig. 8, the largest A island does not dominate. The various islands exhibit adaptive behavior. They look for the maxima of the S concentration, split, merge and die. The top windows shows different snapshots of the A distribution. The small islands are destroyed while the large islands grow. The bottom window is the long term evolution of this system and shows intermittent fluctuations

of macroscopic features, the competition terms turn out to enhance the local and complex character of the A islands. Global competition can endow these macroscopic islands with an emergent social (or sometimes anti-social) behavior. The evolving macroscopic objects may now have apparent 'goals': seeking food, multiplying, dying, or destroying the competing islands. The lifespan of these emergent islands is many orders of magnitude longer than the lifespan of their components (which of course remain non-adaptive and totally mechanical during the entire history of the system).

7 Discussion

Dynamic biological systems have traditionally been described using PDEs. PDEs were invented and proved useful in simple physical systems that fulfilled certain restrictions:

- a very large density of interacting particles at every point in space;
- a high reaction rate;
- a diffusion rate high enough to smooth local fluctuations;
- an immediate result of each interaction. i.e. no delay between the interaction and its result.

All of these assumptions are usually false in biological systems. For such systems:

- The density of interacting particles is small. For example, the number of interacting cells at every point is on the order of tens.
- The reaction rate is low. The interaction between cells is a long process involving a long chain of molecular sub-processes.
- The diffusion rate can in many cases be low. Moreover, the motion of cells cannot be described as a diffusion process. The motion is best described as a slow directional motion directed by many chemical signals.
- Most biological processes require the production of new molecules or the division of cells. Each of these process can take hours or days. Thus there is a delay between an action and its result.

In this work we showed that even in simple cases PDEs fail to describe appropriately the dynamics. This is as well. Systems that were too simple to present complex emergent features in the framework of PDEs, when studied appropriately, turned out to present a wide range of experimentally documented features.

7.1 Dimensionality

Our results suggest that the dimensionality of the system and its size are crucial features for its capability to emerge and sustain life. This may explain the fact that most of the ecological systems are two-dimensional. Reinterpreting in the genome space, the present results provide the conceptual link between the atomized structure of the life building blocks and the explosive Darwinian tandem, noise + proliferation.

7.2 Inter-Scale Information Flow

Autocatalysis results in the flow of information from small scale to large-scale features [17]; from the microscopic details of the system to its macroscopic behavior. The rapid evolution of small-scale features to a macroscopic scale affects the distribution of the macroscopic variables. In particular the macroscopic variables that are dominated by the sum of a small number of large values (instead of on the sum of a large number of small values) have a very specific (fractal-power law) space-time behavior. Such effects are enhanced when there is a direct effect of the macroscopic values on the microscopic mechanisms. Such a dialog between different scales is the obvious reality in biological systems [18].

Many biological systems evolve at time scales much longer than their fundamental microscopic time scales. In PDEs the emergence of long time scales cannot occur naturally, while in MS such scales can emerge spontaneously. In MS, the agents self-organize [19] in collective structures with emergent adaptive properties and with dynamical space-time scales of their own. The effective dynamics of these emerging objects is much slower than that of their components (and this fact constitutes their very definition [20]). Analyzing the system at the level of the emerging objects requires automatically the transition to longer time scales and larger spatial scales. Each level corresponds to a new space-time scale in a hierarchy of complementary representations of the same complex macroscopic system.

We found in this chapter that one of the core mechanism for the emergence of collective macroscopic objects is the interplay between autocatalysis and the discrete microscopic structure of the components of the system. This mechanism is totally missed by the mean field description, which predicts uniform extinction in conditions in which the real system emerges actually very lively collective adaptive objects.

References

1. T.R. Malthus: *An Essay on the Principle of Population* (printed for J. Johnson in St Paul's Churchyard, London 1798); (reprinted Macmillan and Co, London 1894)
2. P.F. Verhulst, Correspondence Mathématique et Physique publié par A. Quetelet **10**, 113 (1838); (English translation in D. Smith, N. Keyfiz: *Mathematical Demography* (Springer, New York 1977), pp. 333–337; Mem. Acad. Royale Bruxelles **28**, 1 (1844)
3. P.E. Seiden, F. Celada: J. Theor. Biology **158**, 329 (1992)
4. Y. Louzoun, S. Solomon, H. Atlan, I.R. Cohen: Bull Math Biol. **65**, 375 (2003)
5. Mott N.F.: Adv. Phys. **16**, 49 (1967); Y. Nagaoka, H. Fukuyama: *Anderson Localization*, Springer Series in Solid State Sciences, Vol. 39, (Springer Verlag, Berlin, New York 1982)
6. M. Croft: Current Opinion in Immunology **6**, 431 (1991)
7. W.E. Boyce, R. C. DiPrima: *Elementary Differential Equations and Boundary Value Problems*, 4th ed. (Wiley, New York 1986)
8. A. Papoulis: *Probability, Random Variables, and Stochastic Processes*, 2nd ed. (McGraw-Hill, New York, 1984) pp. 144-145
9. B.I. Shkolvskii, A.L. Efros: *Electronic Properties of Doped Semiconductors* (Springer Verlag, New York 1984)
10. N. Shnerb, Y. Louzoun, E. Bettelheim, S. Solomon: Proc. Natl. Acad. Sci. USA **97** 10322 (2000)
11. D. R. Nelson, N. M. Shnerb: Phys. Rev. **E 58**, 1383 (1998)
12. N. Shnerb, E. Bettelheim, Y. Louzoun, O. Agam, S. Solomon: Phys. Rev. E **63**, 021103 (2001)
13. H. Jiang, L. Chess: Annu. Rev. Immunol. **18** 185 (2000)
14. C. Darwin: *The Origin of Species* (John Murray, London 1859)
15. A.J. Lotka: *Elements of physical biology* (Williams and Wilkins Co., Baltimore 1925)
16. V. Volterra: *Variazioni e Fluttuazioni del Numero d'Individui in Specie Animali Conviventi* Atti della R. Accademia nazionale dei Lincei, Memorie della Classe di scienze fisiche, matematiche e naturali, (VI) **2**, 31 (1926)

17. A. Turing: Philos. Trans. R. Soc. London **B237**, 37 (1952)
18. H. Atlan: Thesis Eleven **52**, 5 (1998)
19. H. Atlan: Physica Scripta **36**, 563 (1987)
20. S. Solomon: 'The Microscopic Representation of Complex Macroscopic Phenomena: Critical Slowing Down - A Blessing in Disguise'. In *Annual Reviews of Computational Physics II*, ed. by D. Stauffer (World Scientific, Singapore 1995) pp. 243-294

Privacy and Data Exchanges

Bernardo A. Huberman

This chapter reviews cryptographic solutions to two problems relevant to the processing of large data sets such as those encountered in bioinformatics. Both center on issues of data sharing while maintaining privacy. The first problem is that of deciding whether two data sets residing at distinct sites share common information, without revealing their contents. The second is that of surveying a population of individuals and, if necessary contacting them again depending on the answers given, without revealing their identity or answers. After a brief tutorial on public key cryptography, solutions to both of these problems are provided and discussed.

1 Introduction

The advent of the word wide web has enabled information exchanges on a global scale that were inconceivable a few years ago. Citizens of virtually any country can have, at the click of the mouse, access to timely information and services that used to be restricted to a few privileged individuals. A striking example of this phenomenon is seen in the new ways by which the scientific community communicates new results. Not long ago, few scientists were fortunate to be part of some inner circle that sent to each other advance notice of breakthroughs and new results in their fields which would later appear in print. The rest of the scientific community had to wait and learn of these new results and discoveries through the mailing of reports and papers, or, more generally but even slower, their publication in scientific journals. This created a class system in science, which was largely defined by who had access to timely information.

All this has changed with the advent of the Internet. Now any scientist with access to a computer and the network can instantaneously receive new results and exchange ideas with the authors and colleagues, while contributing to their spreading throughout the community. Equally important, access to data that used to be hard to obtain, such as results from experiments or reprints of papers published in costly journals, is now available for free through a number of reliable web sources. Thus information, which used to be scarce and therefore expensive, is now plentiful and nearly free, while the scarce resource – and therefore the expensive one – is the knowledge required to process that information.

As with many other technological breakthroughs that are first exploited in the scientific domain, the ability to exchange information on a global scale mediated by the Internet quickly migrated to the public at large, and in the process generated both commercial value and public uses. This widespread use did not arrive without creating new problems for users, ranging from the protection of valuable data from unauthorized access, to issues of individual privacy rights that are becoming central to the public discourse around the use of the Internet.

Given that many of the papers presented in this volume deal with biological problems that also straddle the scientific, commercial and public arenas, I think that it is appropriate to describe some novel mechanisms that we have recently invented to deal with two pervasive problems [1]. The first problem is that of discovering whether or not data sets have common characteristics while keeping them private, while the second problem relates to the finding of individuals with given characteristics and contacting them repeatedly while safeguarding their privacy. Any time a researcher has a data set that she wants to compare with another one at a remote site (for the purpose of identifying a meaningful sequence, for example), she faces the issue of making sure that the remote site owner, or other users of the remote site, do not discover what her set contains. The mechanism that I'll describe below makes this possible without resorting to a third party, a serious constraint in the global information infrastructure.

The second problem that we addressed arises in data collection for biological and health research. The very ability to gather and disseminate fine-grained data in the medical and behavioral fields through the Internet has led to expressions of concern about privacy issues and to public reactions that in some cases have translated into laws [2,3]. In the case of some European Community nations, strong restrictions have already been placed on the ability of those who collect personal data to release it without explicit individual consent. The saliency of privacy issues on the Internet stems from the different and conflicting needs that users and providers have when it comes to the use of information. While individuals would like certain guarantees about how their data is obtained and used, few entities provide them. On the other hand, many institutions, both public and private, feel they have a right to access this data, whose analysis can also have positive effects for individuals, such as better products and cures for particular diseases.

Once again, it would seem that that all that is needed in order to protect individual privacy is a trusted party or entity that would act as an intermediary between the subjects and the researchers while protecting their privacy. The difficulty with this alternative is that it is hard to find someone or an institution that everyone likes. Worse, it provides a single point of failure, for if this entity were compromised all data files could suddenly become public. Even with legal protections, citizens might anticipate that laws could change with time, as in the case of adoption rights, where today it is possible to obtain the identity of parents who gave children away for adoption at a time when the legal standard offered them anonymity for life.

As I show below, there is a technical solution to this problem that allows for investigators to have access to individual data and to contact them with further questions while at the same time preserving their full privacy. And this is done without resorting to any trusted third party.

Since the solution to both problems involves zero knowledge cryptographic techniques, I will first provide a lightning survey of such techniques, to be followed by a simple exposition of the mechanisms that solve these problems. The interested reader can read the references to our original paper for a more detailed explanation of what is involved in setting up these mechanisms. Next I will describe how to find out if several data sets are similar or not without revealing their content, and then I will move on to the mechanism that allows to conduct surveys of people who can reveal data while their privacy is protected.

2 A Lightning Review of Cryptographic Techniques

The mechanisms that I'll describe rely on a variety of cryptographic techniques, which in turn exploit the use of two fundamental cryptographic primitives: hash functions and public key systems. For the benefit of the reader unfamiliar with this field I will now describe the general properties of these two primitives.

In general, cryptographic functions operate on inputs such as "messages" and "keys", and produce outputs such as "cipher texts" and "signatures". It is common to treat all of these inputs and outputs as large integers according to some standardized encoding. Throughout this exposition I will assume that any value involved in a cryptographic function is a large integer, no matter what it may be called.

A cryptographic hash function, H, is a mathematical transformation that takes a message m of any length, and computes from it a short fixed-length message, which we'll call $H(m)$. This fixed length output has the important property that there is no way to find what message produced it short of trying all possible messages by trial and error. Equally important, even though there may exist many messages that hash to the same value, it is computationally infeasible to find even two values that "collide". This practically guarantees that the hash of a message can "represent" the message in a way which is very difficult to cheat. An even stronger property that we will require is that the output of a cryptographic hash function cannot be easily influenced or predicted ahead of time. Thus someone who wanted to find a hash with a particular pattern (beginning with a particular prefix, say) could do no better than trial and error. In practice, hash functions such as MD5 (Message Digest Algorithm) and SHA (Secure Hash Algorithm) are often assumed to have these properties.

Public key encryption (or signature) relies on a pair of related keys, one secret and one public, associated with each individual participating in a communication. The secret key is needed to decrypt (or sign), while only the public key is needed to encrypt a message (or verify a signature). A public key is generated by those wishing to receive encrypted messages, and broadcast so that it can be used by the sender of the message to encode it. The recipient of this message then uses his own private key in combination with his public key to decrypt the message. While slower than secret key cryptography, public key systems are preferable when dealing with networks of people that need to be reconfigured fairly often. Popular public key systems are based on the properties of modular arithmetic.

3 Secret Matching of Data Sets

It is often the case that a researcher needs to compare a private data set with one residing at some remote site in order to identify it. While the Internet and several bioinformatics companies provide mechanisms for doing so, they do not remove the disincentive inherent in having to disclose the private content of the data set to an unknown service who could also disclose it to competitors. What one needs is a mechanism that circumvents this problem. In what follows I'll describe such a mechanism using the proverbial crypto characters Alice and Bob, Alice representing a researcher on one end and Bob the service or another researcher at the other end. Assume for the sake of illustration that Alice has a list containing a and b, which could be long protein sequences, and Bob's server has a and d. What they first have to do is illustrated in Fig. 1.

Alice's list: a,b a,b Bob's list: a,d

Alice generates a secret key x Bob generates a secret key y

a, b, c, x, y are integers.
Alice and Bob agree on a common prime number p.
All computations are done modulo p.

Fig. 1. Schematic representation of Public Key Cryptography (see text)

Next, both Alice and Bob proceed to go through the following set of operations.

1. Alice computes $(a)^x$ and $(b)^x$ and sends them to Bob, while Bob computes $(a)^y$ and $(d)^y$ and sends it to Alice.
2. Next Alice takes the data she just received from Bob, which she cannot decipher because she does not know y, and computes $(a^y)^x$ and $(d^y)^x$ and sends them back to Bob. Bob does the same with Alice's message, and he computes $(a^x)^y$ and $(b^x)^y$ which he then proceeds to send back to Alice.

Now, since $a^{xy} = a^{yx}$, they both know they have a but Bob doesn't know Alice has b and vice versa.

Notice that the whole security of the operation would be compromised if Alice or Bob were able to compute either x or y from the data they initially sent to each other. But that is almost impossible because of the intractability of the discrete logarithm problem: given integers a and b and prime p, it is computationally hard to find integer n such that

$$b^n = a(\bmod p). \tag{1}$$

This method, which I illustrated using only two inputs from both Alice and Bob, generalizes to any large set of data and thus allows for either party to find whether or not they have a set in common without revealing what the data is.

4 Private Surveys in the Public Arena

I will now describe a technique for conducting surveys that allows investigators to have access to individuals and to contact them with further questions while at the same time preserving their full privacy. This solution relies on zero knowledge cryptographic techniques developed in the context of secure distributed computation. Our scheme allows a researcher to issue a survey to a number of individuals who can answer in what effectively amounts to anonymous fashion, while they can still be tracked over time and queried on additional items without the researcher learning the identity of the subjects. Moreover, the solution we propose does not even require a trusted third party, which for the reasons state above is not a suitable solution.

The "magic trick" behind this solution can be explained in simple terms by resorting to a physical analogy. We first describe this analogy and then explain how to implement it computationally.

Consider a bulletin board where survey questions are posted for all members of a community to see. For the sake of simplicity in the exposition, we'll assume that the answers to these questions are of the form "yes" and "no", although the mechanism is much more general. Each subject answers the question by effectively anonymously "placing" on the bulletin board two unlocked boxes, labeled "yes" and "no" with locks designed in such a way that the subject only has the key to the one corresponding to his answer. This is shown in Fig. 2.

Fig. 2. Private surveys in the public arena: schematic representation of surveying method guaranteeing participants' privacy (see text)

Since no one else knows which of the two keys the subject has, others, including the researcher, cannot tell how a given subject responded. And yet, he can contact each of the respondents that answered the question in a given way by creating a box that can be unlocked only by members of the selected group, as shown in Fig. 3.

Placing messages in this box and then locking it, allows communication with members of this group, defined by their answer to the question. Thus the researcher can ask group members further questions. This mechanism need not be restricted to the researcher: it can also allow members of the group to communicate with each other (e.g., as a chat forum) without them learning the identities of others in the group. All of this occurs in full view of the whole community, but with decrypting abilities possessed only by those who answered in a given fashion.

This technique provides a potential solution to the dilemma of protecting privacy or making it public. Notice that it does not require a trusted third party, although the underlying implementation, which we discuss below, does require the users to trust standard and tested cryptographic protocols. This trust is no different from that we put on a locksmith when asking for a copy of our household key, or on the manufacturer of a garage door opener.

A simple application of this technique counts individuals with a given property. All that is required is to post a message with a key requesting an acknowledgment from all members using that key. The number of an-

Fig. 3. Private surveys in the public arena: participants that gave a specific answer ("yes" in this case) can be contacted again without having their identity revealed (see text)

swers compared to the whole population yields a useful frequency. Another form of panel research would follow a group over time, effectively conducting prospective surveys by simply adding more questions to the bulletin board and watching what happens to the frequencies. This would also allow looking for correlations among members of different groups. That is accomplished by repeating the original procedure in a more refined fashion.

This physical metaphor can actually be implemented and automated in a transparent fashion by using public key cryptographic systems [1]. These systems rely on a pair of related keys, one secret and one public, associated with each individual participating in a communication. The secret key is needed to decrypt (or sign), while only the public key is needed to encrypt a message (or verify a signature). A public key is generated by those wishing to receive encrypted messages, and broadcast so that it can be used by the sender of the message to encode it. The recipient of this message then uses his own private key in combination with his public key to decrypt the message. In our particular application we use the additional property that by constraining the product of two or more public keys to be equal to a specific large number, it is only possible to generate a set of such keys in which only one of the keys has a corresponding private key. This provides the computational basis for the analogy of the two locks described above: each person answers the question by posting two public keys, constrained so that their product matches a value given as part of the question. The person can only have a private key for one of the posted public keys, and selects the private key corresponding to the answer.

The security of the full system requires addressing additional issues. For instance, to what extent do laws protect people from having to reveal their secret keys? Another issue is the size and diversity of the group, enabling people to effectively hide among other members. In some cases, incentives for participation and correct answers can be important and some possible answers have been proposed, like markets for secrets [5].

This mechanism provides a third alternative to the dilemma of having to choose between privacy and the public interest. While these two have been part of the public discourse for many years, the new developments in genetic research and information systems raise them to heightened concern. While the social benefits of novel privacy mechanisms are not usually considered in policy discussions of the use of cryptography, they illustrate an important opportunity for allowing widespread use of these technologies.

5 Conclusion

The ease of communication and the ensuing exchanges of data mediated by the Internet have raised interesting problems concerning both the security of the data exchanged and the need to keep it private in certain situations. Two examples of relevance to the bioinformatics community are the problem of having to match data sets with a remote server without revealing the content of the data, and how to conduct repeated surveys of particular individuals in a population without compromising their privacy. In both cases, the standard answers are either to resort to a trusted third party or to desist in having the data exposed to manipulations that could actually reveal its nature or the identity of the target.

As I have shown, it is possible to use zero knowledge techniques to solve these problems in ways that ensure privacy without having to resort to trusted third parties. Moreover, these mechanisms can be implemented and deployed on desktop machines. While a few years ago it required large machines to perform modular arithmetic operations at a suitable speed, the availability of crypto accelerators make these techniques quite feasible and potentially useful.

References

1. B.A. Huberman, M. Franklin, T. Hogg: 'Enhancing Privacy and Trust in Electronic Communities', in *Proceedings of the ACM Conference on Electronic Commerce (EC99)*, (ACM Press, New York 1999) pp. 78-86
2. H. Watzman: Nature **394**, 214 (1998)
3. D. Adam: Nature **411**, 509 (2001)
4. P.A. Roche, G.J. Annas: Nature Genetics **2**, 392 (2001)
5. E. Adar, B.A. Huberman, 'A Market for Secrets', FirstMonday, August 2001. See: http://www.firstmonday.org/issues/issue6_8/adar/index.html

Part IV

Pattern Recognition

Statistical Physics and the Clustering Problem

Sebastiano Stramaglia, Leonardo Angelini, Carmela Marangi,
Luigi Nitti, and Mario Pellicoro

This chapter reviews statistical approaches to the clustering problem, i.e. the task of partitioning data-sets in classes in such a way that points in the same class are more similar to one another than to those in other classes. Although this is technically an ill-posed problem, it is of great importance in a wide range of applications and numerous methods have been proposed to tackle it. This paper reviews mainly the coupled maps approach to clustering which performs a non-parametric classification without any assumptions on the distribution of clusters or the number of classes. The technique is illustrated on a biological example (reconstruction of phylogenetic trees) and one from coding theory. The merits of various approaches and the remaining challenges are also discussed.

1 Introduction

Clustering methods aim at partitioning a set of data-points in classes such that points that belong to the same class are more *similar* than points belonging to different classes [1]. These classes are called clusters and their number may be preassigned or can be a parameter to be determined by the algorithm. Clustering is an ill-posed problem because a given data-set can be partitioned in many ways without a clear criterion to prefer one clustering over another. In particular the clustering output depends on the resolution at which data-points are processed; many clustering algorithms provide as output the full hierarchy of classes which is obtained as the resolution is varied.

There exist applications of clustering in such diverse fields as pattern recognition [2], astrophysics [51], high energy physics [4], satellite data analysis [5], communications [6], biology [7], business [8], macro-molecular crystallography [9], speech recognition [10], analysis of dynamical systems [12], buried land-mine detection by IR imaging [11] and many others. Two main approaches to clustering can be identified: parametric and non-parametric clustering.

Non-parametric approaches make few assumptions about about the data structure and, typically, follow some local criterion for the construction of clusters. Typical examples of the non-parametric approach are the agglomerative and divisive algorithms that produce dendrograms [13]. In the last few years non-parametric clustering algorithms have been introduced employing the statistical properties of physical systems. The Super-Paramagnetic approach by Domany and coworkers [14] exploits the analogy to a model granular magnet: the spin-spin correlation of a Potts model, living on the datapoints lattice and with pair-couplings decreasing with the distance, is used to partition points in clusters. The synchronization properties of a system of coupled chaotic maps are used in [15] to produce hierarchical clustering.

In parametric methods clustering is treated as a problem of unsupervised learning, i.e. learning a probability distribution from a set of samples (examples). Some assumptions about the underlying data structure must be made, for example in generative mixture models [16] data are viewed as coming from a mixture of probability distributions, each representing a different cluster. In other words, the probability density function $P_0(x)$, x being the generic point in the data space, is formed from a linear combination of basis functions, the number q of basis functions representing the number of clusters:

$$P_0(x) = \sum_{\alpha=1}^{q} p(x|\alpha)p(\alpha), \tag{1}$$

where $p(\alpha)$ are the mixing parameters, and the component densities, in the Gaussian case, read:

$$p(x|\alpha) = \frac{1}{(2\pi\sigma_\alpha^2)^{d/2}} \exp\left(-\frac{|x-\mu_\alpha|^2}{2\sigma_\alpha^2}\right), \tag{2}$$

where d is the number of components of the vector x, μ_α is the prototype associated to cluster α and σ_α is the corresponding spread. The parameters of this distribution ($\{p(\alpha)\}$, $\{\mu_\alpha\}$ and $\{\sigma_\alpha\}$) are adjusted to achieve a good match with the distribution of the input data. This can be obtained by maximizing the data likelihood (ML) [16] or the posterior (MAP) if additional prior information on the parameters is available [16].

It is worth stressing that the most important problem in this frame is that *one risks finding what one is searching for*. The goal is to discover genuine structures present in a given data-set and to discard those structures which are due to chance, as realizations of the randomness. The competition between genuine structures and noise, in the case of very high dimensional input data, give rise to the phenomenon termed *retarded learning*: below a critical size of the data-set no information at all about the underlying structure can be extracted. We refer the reader to [17], and references therein, for an excellent discussion on retarded learning.

Many parametric clustering methods are based on a cost function: the best partition of points in clusters is assumed to be the one with minimum cost. Often cost functions incorporate the loss of information incurred by the clustering procedure when trying to reconstruct the original data from the compressed cluster representation: the most popular algorithm to optimize a cost function is K-means [16].

It is important to stress the difference between *central* clustering, where it is assumed that each cluster can be represented by a prototype [18], and *pairwise* clustering where data are indirectly characterized by pairwise comparisons instead of explicit coordinates [19]; pairwise algorithms require as input only the matrix of dissimilarities. Obviously the choice of the measure of dissimilarity is not unique and it is crucial for the performance of any pairwise clustering method. It is worth remarking that it often happens that the dissimilarity matrix violates the requirements of a distance measure, i.e. the triangular inequality does not necessarily hold.

In the next section a biological application of non-parametric clustering will be described: reconstruction of phylogenetic trees from human DNA sequences. In Sect. 3 we will describe an interesting frame to derive cost functions for clustering, the autoencoder frame.

2 Hierarchical Clustering for Phylogeny Reconstruction

In many applications a hierarchical approach to clustering allows one to better describe the underlying structure of the data distribution. The optimal partition of the data set is then extracted from the full hierarchy of solutions according to problem dependent optimality criteria or stability requirements with respect to statistical fluctuations. For a class of problems, however, the solution is itself inherently hierarchical, as it is the case of cluster analysis of DNA sequences for phylogeny reconstruction in the framework of molecular biology studies of human evolution. The study of the patterns of genetic diversity in modern human populations can help unravel the puzzle of the human evolutionary story as due to ancient migrations, population expansions and bottlenecks. Relationships among populations can be assessed on the basis of differences in nucleotide composition of specific DNA sequences, adopted as molecular markers of biological diversity. Even if the choice of the appropriate genetic analysis is still on debate, a preeminent role has been achieved by analysis of mitochondrial DNA (mtDNA), mainly due to its high mutation rate [21,22]. Moreover the mtDNA exists in a large number of copies in each cell and it is maternally inherited and not recombinant, i.e. it is inherited as a single block or haplotype. According to a suitably defined genetic distance among haplotypes, the cluster analysis performed here provides a phylogenetic tree from samples of a given geographical area. As a further step a stability criterion is applied to determine extensive macro classes (haplogroups) for comparison with anthropological classifications of the same area.

2.1 Coupled Map Clustering (CMC) Algorithm

Following the idea of natural clustering arising from collective behavior of complex dynamical systems, a new clustering algorithm has been recently proposed, which is based on the cooperative behavior of an inhomogeneous lattice of coupled chaotic maps leading to the formation of clusters of almost synchronized maps with very correlated chaotic trajectories. The clusters' structure is biased by the architecture of the couplings among the maps and a full hierarchy of clusters can be achieved using the mutual information between pairs of maps as a similarity index. The algorithm (henceforth called CMC) performs a non-parametric partition of a data without prior assumptions about the number of classes and geometric distribution of clusters. In the original formulation [15], CMC is a clustering tool to process an input data set of arbitrary nature. Obviously, since we are dealing here with a specific application to sequence classification, a version of CMC tailored to the peculiar data is designed in order to improve its performances. Before entering the specific application to the field of human evolution studies, we briefly review the CMC algorithm. Let us consider a set of N points (representing here DNA sequences) in a D-dimensional space (with D equal to the number of variant sites). We assign a real dynamical variable [-1,1] to each point and define pair-interactions $J_{ij} = \exp\left(-d_{ij}/2\alpha^2\right)$, where α is the local length scale and d_{ij} is a suitable measure of distance between points i and j in our D-dimensional space. The time evolution of the system is given by:

$$x_i(t+1) = \frac{1}{C_i} \sum_{j \neq i} J_{ij} f\left(x_j(t)\right), \tag{3}$$

where $C_i = \sum_{j \neq i} J_{ij}$, and we choose the map $f(x) = 1 - 2x^2$. Due to the choice of the function f, equations (3) represent the dynamical evolution of chaotic maps x_i coupled through pair interactions J_{ij}. The lattice architecture is fully specified by fixing the value of α as the average distance of k-nearest neighbors pairs of points in the whole system (our results are quite insensitive to the particular value of k). For the sake of computational economy, we consider only interactions of a map with a limited number of maps whose distance is less than 3α, and set all other J_{ij} to zero. Starting from a random initial configuration of x, equations (3) are iterated until the system attains its stationary regime, corresponding to a macroscopic attractor which is independent of the initial conditions. To study the correlation properties of the system, we consider the mutual information I_{ij}, between pairs of variables whose definition is as follows [23]. If the state of element i is $x_i(t) > 0$ then it is assigned a value 1, otherwise it is assigned 0: this generates a sequence of bits, in a certain time interval, which allows the calculation of the Shannon entropy H_i for the i-th map. In a similar way the joint entropy H_{ij} is calculated for each pair of maps and finally the mutual information is given by $I_{ij} = H_i + H_j - H_{ij}$. The mutual information is a good measure of correlations and it is practically precision independent, due to the coarse graining of the dynamics. If maps i and j evolve independently then $I_{ij} = 0$; if the two neurons are exactly synchronized then the mutual information achieves its maximum value, in the present case $\ln 2$, due to our choice of the function

f. The algorithm identifies clusters with the linked components of the graph obtained by drawing a link between all the pairs of maps whose mutual information exceeds a threshold θ; θ controls the resolution at which data are clustered. Hierarchical clustering is obtained repeating this procedure for an increasing sequence of θ-values: each clustering level is extracted as a partition of data with a finite stability region in the θ parameter. The most stable solution identifies the optimal partition of the given data set.

We note that since clustering is performed in a hierarchical way, this algorithm provides an effective tool for phylogeny reconstruction. Results at different hierarchical levels can be identified as different branch levels of a phylogenetic tree, where the stable clustering solution would represent the terminal branching of the tree. We limit ourselves to consider a valid phylogenetic tree whenever we are dealing with homologous sequences verifying stationary conditions on the stochastic evolutionary process, i.e when one compares processes having the same type of dynamics on different lineages, as it is the case of intraspecies evolution.

Let us spend a few words on selection of the algorithm parameters, namely the number of neighboring maps k and the resolution θ. We stress here that the algorithm is a deterministic one, being the dependence from the initial random configuration of the maps wiped out because of the peculiar dynamics of chaotic systems . That implies a dependence of final results on the particular choice of the external parameter, even though, as already tested in several contexts of application [15], clustering solutions provided by the CMC algorithm display robustness against quite a rough tuning of k and θ.

2.2 Distance Measures

The main advantage of this algorithm is that it is a deterministic algorithm of general applicability, since no a priori knowledge of cluster structure is to be assumed. However, in order to tailor the method to the specific application of direct comparison of DNA segments, we need to provide a suitable definition of genetic distance measure. Let us remind that a DNA sequence can be represented by a one-dimensional strings of letters taken from a 4 symbol alphabet {A,C,G,T}, namely the four nucleotide bases (adenine, cytosine, guanine, thymine) which the DNA is composed from.

For example, in Table 1 we depict a set of 13 short pieces (16 sites long) of human mtDNA, each belonging to a different individual. Note that all the sequences are referred to a reference sequence: only variations with respect to the reference are reported.

A genetic distance between two individuals is commonly defined as the number of differences of nucleotides at corresponding sites in the selected DNA segment. The major drawback of such a definition is that it does not take into account heterogeneous variation rate at different sites. For population divergence time estimate, the grouping of haplotypes is usually performed on the basis of shared patterns at relatively rapidly changing sites.

Table 1. Example: a data-set of mtDNA sequences.

Reference	G	A	A	C	T	T	T	C	G	G	T	G	C	T	A	A
$1 - UNA_{115}$	-	-	-	-	-	-	-	-	-	A	C	-	T	-	-	-
$2 - AWYN_{320}$	-	-	-	-	-	-	C	-	-	A	C	-	T	-	-	-
$3 - CITAK_{352}$	-	-	-	-	-	-	-	-	-	A	C	-	T	-	G	-
$4 - LANI_{17}$	-	-	-	-	-	C	-	-	-	A	C	-	T	-	-	-
$5 - DANI_{33}$	-	-	-	-	-	C	-	-	-	A	C	-	T	-	-	-
$6 - LANI_{8}$	C	-	-	-	-	C	-	-	-	A	C	-	T	-	-	-
$7 - DANI_{24}$	-	-	-	-	-	C	-	-	-	A	C	-	T	C	-	-
$8 - UNA_{94}$	-	-	-	-	-	-	C	-	-	-	-	-	-	-	-	T
$9 - UNA_{98}$	-	-	-	-	-	-	-	-	-	-	-	-	-	-	-	T
$10 - UNA_{89}$	-	-	-	-	-	-	-	-	A	-	-	-	-	-	-	T
$11 - UNA_{78}$	-	-	G	-	-	-	-	-	-	-	-	-	-	-	-	-
$12 - CITAK_{343}$	-	-	G	-	C	-	-	-	-	-	-	-	-	-	-	-
$13 - CITAK_{291}$	-	-	G	-	-	-	-	-	-	-	-	-	-	-	-	-

Discrimination occurs at highly variant sites which should then give the most relevant contribution to cluster identification; in other words, highly variant sites are more *informative*. Accordingly, we propose to weight the ordinary distance between sequences by introducing an *entropic* term [20]. We define the distance between two sequences i and j of length S as follows:

$$d_{ij} = \sum_{s=1}^{S} \delta_{ij}^s E^s, \qquad (4)$$

where δ_{ij}^s is 1 if the $\{ij\}$ pair exhibits different nucleotides at site s, otherwise it is 0. The term E^s is expressed as an entropy:

$$E^s = -\sum_{l=1}^{4} p_l^s \log p_l^s, \qquad (5)$$

where the index l runs over the number of different nucleotides and p_l^s represents the frequency of the nucleotide l at site s within the data-set at hand. An appealing feature of this *entropic* distance is the lack of bias by any biological model of genetic distance, although, depending on the data-set, the information provided, without any complementary assumption on the sequence generating processes, might not fully solve all the sequences classification ambiguities. Moreover the entropic distance is strictly related to the specific context of haplogroup discrimination.

In order to clarify what the entropic distance elicits from data, let us discuss how it would work on the example depicted in Table 1. The most informative sites are the 3rd, the 6th, the 10th, the 11th, the 13th, and the 16th site; indeed they are the most *varying*. As a consequence of the distance measure we adopted, we would expect that the sequences $\{1, 2, 3, 4, 5, 6, 7\}$ form a cluster, because they have the same nucleotide on the informative 10th, 11th and 13th locations. Analogously, the sequences $\{8, 9, 10\}$ would form another cluster because they coincide on the 16th location which is varying, whereas $\{11, 12, 13\}$ would form a cluster since they have the same nucleotide on the varying third site. At a lower resolution, we would also expect that sequences $\{1, 2, 3\}$ and $\{4, 5, 6, 7\}$ would form two subclusters because of the sixth location.

Another interesting measure of the distance between mtDNA sequences has been recently proposed: it incorporates a weight which depends on a measure of different evolution rate on sites, called *site variability* [24]. The distance is given by

$$D_{ij} = \sum_{s=1}^{S} \nu^s, \qquad (6)$$

where the site variability ν^s is defined as

$$\nu^s = 1/\nu_m \sum_{i<j}^{N} \delta_{ij}^s / K_{ij}, \qquad (7)$$

K_{ij} being an estimate of the overall genetic distance of the pair $\{ij\}$ as determined by the model of the evolution process described in [25,26] and employing a Stationary Markov Process. The site variability is then normalized to its maximum value ν_m in the whole data set. Our simulations show that the entropic weights and the site variabilities, evaluated on the same data set, are very correlated and practically equivalent for the clustering purpose [20].

2.3 Experiment

As an application of the CMC method, we report the analysis of a sample of 202 subjects from the Pacific area, known to generate 89 haplotypes and which have been also thoroughly studied by anthropological point of view (results of the anthropological analysis can be found in [27]). The data set consists of 89 sequences of 71 variant sites taken from the mtDNA hypervariable region HVRI. The CMC results are shown in Fig. 1 as a dendrogram illustrating the temporal evolution where the time scale is expressed in terms of the resolution parameter. New clusters originate at each branching point, their cardinality being described by the radius of the corresponding symbol.

The final classification has been obtained using the site variability weighted distance D (similar results are obtained using the entropic distance d), $k = 10$ and $\theta = 0.35$. The optimal value of θ have been selected by applying the resampling technique, described in Sect. 3.4, on 100 different subsets of 67 sequences randomly extracted from the original data set. The optimal value of θ, corresponding to the terminal branching of the dendrogram, represents the final classification to be compared with results from known techniques. Our clustering results are consistent with anthropological data. The same sample has been investigated with two other methods widely used for DNA sequence classification, the Neighbor Joining (NJ) method [28] by applying the NEIGHBOR program available through the PHYLIP package [29], and the Reduced Median Network (RMN) [30]. The NJ method generates a tree starting from an estimate of the genetic distance matrix, here calculated by Stationary Markov Model of [25,26]. In agreement with results from CMC, the NJ tree evidences three main subdivisions: the largest group of sequences (49 haplotypes) - identified as Group I - is clearly distinguished from the two other clusters, Group II and Group III. These three clusters coincide with the big clusters found by CMC (see Fig. 1). The RMN method was used for deeper insight of haplotype genetic relationships. It generates a network which harbors all most parsimonious trees. The network (data not shown) results are quite complex, as a consequence of the high number of haplotypes considered in the analysis, while its reticulated structure reflects the high rate of homoplasy in the dynamic evolution of the mtDNA HVRI region. The topological structure of the RMN also evidences three major haplotype clusters, which reflects the same *haplotype composition* shown by the CMC tree.

2.4 Discussion

We have described a method for sequence classification that is based on a hierarchical clustering algorithm and employs a biologically motivated definition of distance. The main advantage of this approach relies in high effectiveness and low computational cost which make it suitable for analysis of large amounts of data. Moreover, thanks to the general applicability of the algorithm, any prior biological information may be coded in an ad-hoc distance definition, thus improving the reliability of sequence grouping. This is the rationale for the introduction of distance measures weighted by entropic or site variability terms that account for the dependency of classification on the different rates of variation occurring on sites. Performances obtained by applying both distance definitions on a population data set of a given geographical area have been compared with classification obtained using NJ and RMN. We find that this method performs at least as well as these two well-known techniques but at lower complexity and computational costs. Moreover, compared to RMN, the method has the main advantage of providing an easy reading and interpretation of results regardless the data set size.

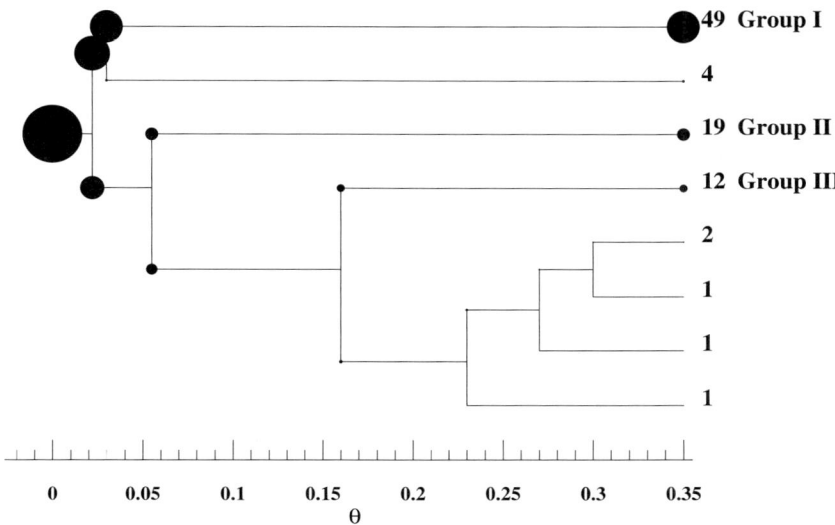

Fig. 1. The tree of clusters found by the clustering algorithm on the genetic data-set described in the text. The horizontal axis represents the resolution parameter θ and can be interpreted as the time axis. The most stable partition and the cardinality of clusters, found by a resampling technique, is described on the left and corresponds to $\theta = 0.35$. The three clusters, denoted as Group I, II and III, coincide with those found by other methods, namely NJ and RMN (see text)

3 The Auto-encoder Frame

In this section we describe the Probabilistic Autoencoder approach to clustering, where Folded Markov chains are used to derive cost functions [31]. Some examples of two-stage folded Markov chains, and the corresponding algorithms for clustering and topographic mapping [32], are thoroughly analyzed in [33], where it is also shown that the cost function for pairwise clustering, introduced in [19], may be seen as a consequence of Bayes' theorem and the requirement of minimal average distortion in a probabilistic autoencoder. We introduce a new class of cost functions for pairwise clustering which can be obtained, in the autoencoder frame, by requiring *maximal similarity* instead of minimal distortion. We show that the cost functions here introduced provide a non-hierarchical clustering of points where dense connected regions of points in the data space are recognized as clusters.

3.1 Cost Functions

We will limit our discussion to autoencoders described by one-stage folded Markov chains. Let us consider a point x, in a data space, sampled with probability distribution $P_0(x)$; a code index $\alpha \in \{1, \ldots, q\}$ is assigned to x

according to conditional probabilities $P(\alpha|x)$. A reconstructed version of the input, x', is then obtained by use of the Bayesian decoder:

$$P(x'|\alpha) = \frac{P(\alpha|x') P_0(x')}{P(\alpha)}. \tag{8}$$

The joint distribution of x, x' and α, describing this encoding-decoding process, is

$$P(x, x', \alpha) = P_0(x) P(\alpha|x) P(x'|\alpha); \tag{9}$$

owing to (8), the joint distribution reads:

$$P(x, x', \alpha) = \frac{P_0(x) P_0(x') P(\alpha|x) P(\alpha|x')}{P(\alpha)}. \tag{10}$$

The conditional probabilities $\{P(\alpha|x)\}$ are the free parameters that must be adjusted to force the autoencoder to emulate the identity map on the data space.

Let $d(x, x')$ be a measure of the distortion between input and output of the autoencoder. The average distortion is then given by:

$$\mathcal{D} = \sum_{\alpha=1}^{q} \int dx \int dx' \frac{P_0(x) P_0(x') P(\alpha|x) P(\alpha|x')}{P(\alpha)} d(x, x'). \tag{11}$$

Moreover, let $s(x, x')$ be a measure of the similarity between input and output; the average similarity is then given by

$$\mathcal{S} = \sum_{\alpha=1}^{q} \int dx \int dx' \frac{P_0(x) P_0(x') P(\alpha|x) P(\alpha|x')}{P(\alpha)} s(x, x'). \tag{12}$$

It is natural to postulate a one-to-one mapping between values of distortion and similarity, $s = F(d)$, with F a strictly decreasing function. A good autoencoder is obviously characterized by a low value of $\mathcal{D}$ and high value of $\mathcal{S}$. However we remark that the two requirements $Min(\mathcal{D})$ and $Max(\mathcal{S})$, for reasonable choices of F, are not generically equivalent.

Now we turn back to the clustering problem. Given a data-set $\{x_i\}$ of cardinality N, partitioning these points in q classes corresponds, in this frame, to designing an autoencoder, with q code indices, acting on the data space. We choose the encoder to be deterministic:

$$P(\alpha|x) = \delta_{\alpha\ \sigma(x)}, \tag{13}$$

$\sigma(x) \in \{1, \ldots, q\}$ being the code index associated to x. The estimate for the average distortion (11), based on the data-set at hand, is given by $\hat{\mathcal{D}} = N H_d[\sigma]$, where we introduce the Hamiltonian H_d for the Potts variables $\{\sigma_i\}$:

$$H_d[\sigma] = \sum_{\alpha=1}^{q} \frac{\sum_{i,j=1}^{N} \delta_{\alpha\sigma_i} \delta_{\alpha\sigma_j} d_{ij}}{\sum_{k=1}^{N} \delta_{\alpha\sigma_k}}, \tag{14}$$

where $\sigma_i = \sigma(x_i)$, $d_{ij} = d(x_i, x_j)$. It turns out that H_d is equivalent to the cost function for pairwise clustering, influential in the clustering literature, introduced in [19].

The estimate for the average similarity is, similarly, given by $\hat{S} = -NH_s[\sigma]$, where we introduce the Hamiltonian H_s:

$$H_s[\sigma] = -\sum_{\alpha=1}^{q} \frac{\sum_{i,j=1}^{N} \delta_{\alpha\sigma_i}\delta_{\alpha\sigma_j} s_{ij}}{\sum_{k=1}^{N} \delta_{\alpha\sigma_k}}. \tag{15}$$

If we choose the autoencoder by minimizing the average distortion, then the best partition of the data-set in q classes corresponds to the ground state of H_d. If we choose it by maximizing the average similarity, then the ground state of H_s must be sought for, instead. Since both $\{d_{ij}\}$ and $\{s_{ij}\}$ may be taken positive, it follows that H_d is characterized by antiferromagnetic couplings between the Potts variables, while H_s is made of ferromagnetic couplings. Denominators in both H_d and H_s serve to enforce the coherence among the q clusters. In particular, without the denominator the ground state of H_s would correspond to a single big cluster.

The form of the function F, determining the relation between s and d, has to be specified. In what follows we consider two forms of this relation. A scale-free relation

$$s_{ij} = F_\gamma(d_{ij}) = \left(\frac{d_{ij}}{\langle d \rangle}\right)^{-\gamma}, \tag{16}$$

depending on the exponent γ, and a scale-dependent relation

$$s_{ij} = F_a(d_{ij}) = \exp\left(-\frac{1}{2a^2}\left(\frac{d_{ij}}{\langle d \rangle}\right)^2\right), \tag{17}$$

dependent on the scale a. In the formulas above, $\langle d \rangle$ is the average dissimilarity over all the pairs of data-set points. The exponent γ will be restricted to assume small values so as to characterize the corresponding Potts model by long-range ferromagnetic couplings; the scale parameter a will be bounded in $[0, 1]$.

At this point it is worth stressing that minimization of the distortion and maximization of the similarity yield, in the autoencoder frame, different cost functions. The Hamiltonian H_d embodies the requirement that pairs of distant points (large d_{ij}) should belong to different clusters. On the other hand, the Hamiltonian H_s, for reasonable choices of F, concentrates on pairs of close points (small d) and forces them to belong to the same cluster. In other words, H_s may be seen to implement the idea that clusters should be searched for as dense connected regions in the data space.

3.2 Deterministic Annealing

We consider two optimization algorithms to find the configuration of minimum cost: simulated annealing [34] and mean-field annealing [35]. Both approaches associate a Gibbs probability distribution to the functional to be optimized. Simulated annealing is a Monte-Carlo technique which samples the Gibbs distribution as the temperature is reduced to zero, while mean-field annealing attempts to track an approximation, to the mean of the distribution, known as *mean field* approximation [36]. We remark that an efficient mean-field annealing algorithm for cost function (14), based on the Expectation-Maximization scheme [37], is described in [19]: the generalization of that algorithm to (15) is straightforward.

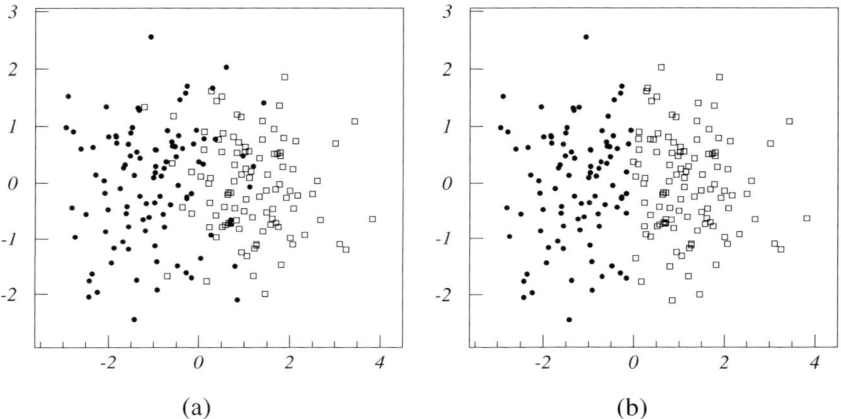

Fig. 2. (**a**) An artificial data set made of two Gaussian distributed clusters, each consisting of 100 points. Empty squares and black circles refer to the two different clusters. (**b**) Clustering result obtained by minimization of H_d (see text)

3.3 Experiments

We describe now the application of the variational criteria for clustering, described above, to some artificial and real data-sets. In many cases cost functions H_d and H_s have very close global minima. For example in Fig. 2a we depict an artificial data-set generated by two overlapping isotropic Gaussian distributions. In this case the natural measure of dissimilarity is Euclidean metrics, and we use $q = 2$. In Fig. 2b the corresponding ground state of H_d [38] is depicted: it is very close to the Bayesian solution, i.e. the solution obtained drawing the symmetry plane for the centers of the two Gaussians. A similar partition is obtained minimizing, by simulated annealing, H_s. As a measure of the difference between two partitions $\{\sigma_i\}$ and $\{\eta_i\}$, we evaluate the following quantity:

$$\epsilon = \frac{1}{N(N-1)} \sum_{i=1}^{N} \sum_{j=1, j\neq i}^{N} \left(\delta_{\sigma_i \sigma_j} - \delta_{\eta_i \eta_j}\right)^2 \qquad (18)$$

which counts the number of pairs of points upon which the two partitions disagree. Using the scale-dependent F_a, we find the ground state of H_s to differ from those of H_d by $\epsilon < 0.01$ varying a in $[0.05, 1]$. Analogously, using the scale-free F_γ, with $\gamma \in [0.1, 1.5]$, we find $\epsilon < 0.02$ when we compare the ground state of H_s with those of H_d. Hence, on this data set, the cost functions introduced above work similarly within wide ranges of γ and a values. We find a similar behavior with respect to the famous IRIS data of Anderson [39]. This data set has often been used as a standard for testing clustering algorithms: it consists of three clusters (Virginica, Versicolor and Setosa) and there are 50 objects in $\mathbf{R}^4$ per cluster. Two clusters (Virginica, Versicolor) show considerable overlap. The clustering result, with $q = 3$ and minimizing H_d, consists of three clusters of 61, 39 and 50 points respectively, with a 90% correct classification percentage. We obtain exactly the same partition by minimizing H_s using a scale-free F (with $\gamma \in [0.15, 1.45]$), and using a scale-dependent F (with $a \in [0.25, 1]$). For $a \in [0.1, 0.25]$ we obtain, in the scale-dependent case, a slightly different partition with clusters' sizes 58, 42, 50 and correct classification percentage 93.3%. These results show that also in the IRIS case the pairwise clustering procedures by distortion minimization and similarity maximization are almost equivalent.

A typical situation resulting in different answers from H_d and H_s is depicted in Fig. 3a. This two-dimensional data-set is made up of an elongated cluster and a Gaussian distributed circular one. It is evident that two dense connected regions are present, and that the farthest pairs of points belong to the same connected region. This is the type of data-set for which minimizing the distortion is not equivalent to maximizing the similarity. In fig. 3b the partition we obtain by minimizing H_d is depicted: it fails to recognize the structure in the data-set. Let us now consider the ground state of H_s with the scale-dependent F. For $a < 0.7$ the ground state, depicted in Fig. 3c, recognizes with 99% accuracy the data structure. At $a \sim 0.7$ a transition phenomenon occurs: the configuration depicted in Fig. 3c ceases to be the global minimum, the new ground state (Fig. 3d) being very close to the solution by H_d. In Fig. 4a we depict the efficiency of the classification versus the resolution parameter a, for the scale dependent F, while in Fig. 4b we consider a sequence of a-values and we plot the ϵ between partitions corresponding to adjacent values of a. The peak at $a = 0.7$ is the indicator of the transition between global minima. Finally, in Fig. 4c the size of the two clusters, versus a, is depicted. Concerning the scale-free F, in Fig. 5 the same plots as in Fig. 4 are depicted, showing that the *good* minimum is stable for a wide range of γ-values. The choice of the optimization algorithm deserves a

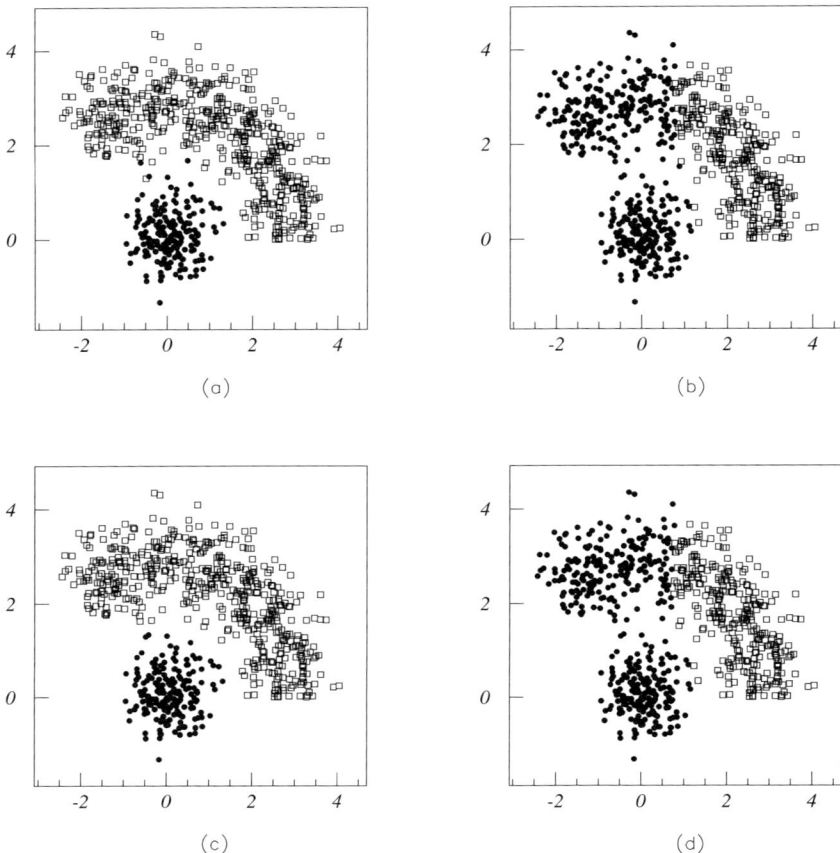

Fig. 3. (a) An artificial data set made of an elongated cluster of 500 points (empty circles) and a circular cluster of 200 points (black circles). (b) Partition by minimizing H_d. (c) Partition by minimizing scale dependent H_s with $a < 0.7$. (d) Partition by minimizing scale dependent H_s with $a > 0.7$

comment. All the results described above are obtained by simulated annealing; we also apply the mean-field annealing scheme, described in [19], and we always find a configuration very close to the one from simulated annealing, while spending less computational time. This confirms that optimization algorithms rooted in mean-field theory yield quickly a good solution on these problems [35].

3.4 Resampling Technique for Unsupervised Estimation of the Number of Classes

In many practical situations the number of classes present in the data set at hand is unknown: much work [13] has been done on methods for assess-

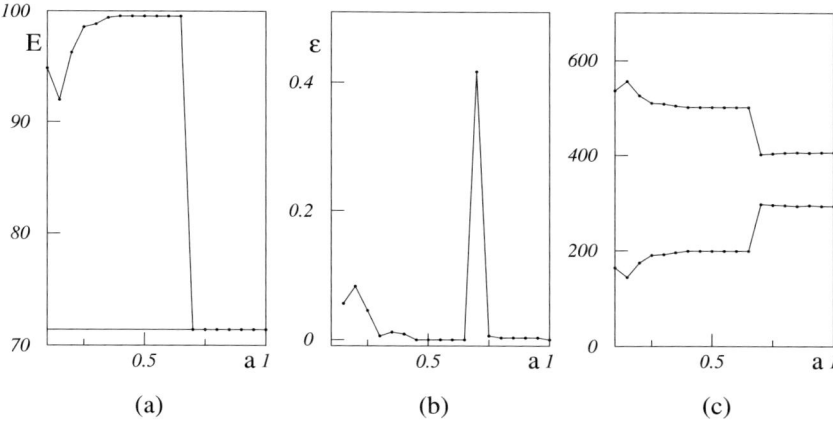

Fig. 4. (a) The efficiency (percentage of correctly classified points) versus a, obtained on the data-set depicted in Fig. 2 by minimizing H_s with scale-dependent F. The dashed line is the efficiency obtained by minimization of H_d. (b) The ϵ parameter, (see the text) between partitions corresponding to adjacent values of a, is plotted versus a. (c) The size of the two output clusters versus a

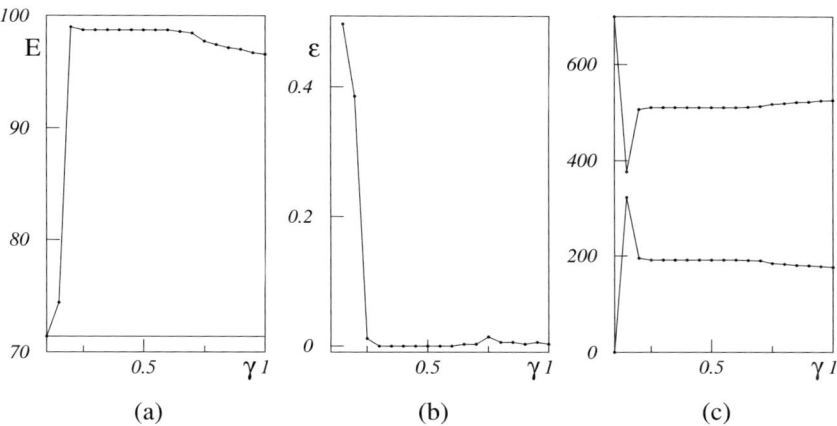

Fig. 5. (a) The efficiency (percentage of correctly classified points) versus γ, obtained on the data-set depicted in Fig. 2 by minimizing H_s with scale-independent F. The dashed line is the efficiency obtained by minimization of H_d. (b) The ϵ parameter, (see the text) between partitions corresponding to adjacent values of γ, is plotted versus γ. (c) The size of the two output clusters versus γ

ing the probable number of clusters or structures within unknown data sets making use of nothing more than the available data itself. In a recent paper [40], a resampling method for unsupervised estimation of cluster validity was proposed to select, within the same algorithm, the parameters corresponding to the optimal partition. In this approach, subsets of the data at hand are randomly formed and the clustering algorithm is applied to each subset: a figure of merit is proposed to identify the stable clustering solutions, which are less likely to be due to fluctuations or chance. Here we propose the use of a resampling technique, similar to those introduced in [40], to fix the number of classes q in the pairwise clustering algorithms described in the previous subsections. Let us describe the resampling technique adopted here. Given a data set of cardinality N, the clustering algorithm for q clusters described above (for example minimizing average distortion of the autoencoder) is applied to the full data-set and to k resamples of the original data set. A resample is obtained by selecting randomly a subset of fN points from the N points, $f \in [0, 1]$ being the dilution factor. A figure of merit $F_M(q)$ is evaluated as follows:

$$F_M(q) = \left\langle \frac{N_r}{N_f} \right\rangle_k ; \qquad (19)$$

N_f is the number of pairs $\{i, j\}$ of points which were belonging to a same cluster when partitioning the full data-set and have both survived the resampling, while N_r is the number of pairs, among these N_f, which are still in a same cluster when partitioning the resample. The average $\langle \cdot \rangle_k$ is over the k different resamples. The figure of merit is evaluated for several values of q. Clearly, $F_M(q) \in [0, 1]$ with $F_M = 1$ for perfect score. Let us now describe our results, obtained on some artificial data sets. In what follows we employ the clustering algorithm corresponding to distortion minimization, and we find that the results are stable against variations of f in the interval $[0.6, 0.7]$. In Fig. 6a a two-dimensional data set, made up of three separated clusters, is shown. The figure of merit, depicted in Fig. 6b, clearly shows a maximum at $q = 3$. In Fig. 6c, the three clusters are almost overlapping; nevertheless F_M still shows a clear maximum at $q = 3$ (Fig. 6d). The data set depicted in Fig. 7a is two-dimensional and consists of six clusters. In this case (Fig. 7b) the figure of merit has two peaks, one for $q = 2$ and the other in $q = 6$: both the two clusters solution and the (correct) six clusters one appear to be stable in this case. Finally, in Fig. 7c three clusters in six dimensions are shown: F_M shows a clear peak in $q = 3$ (Fig. 7d). These results show that inspection of the curve F_M versus q may help to infer the number of classes present in a data set.

3.5 Discussion

In this section we address non-hierarchical pairwise clustering and, working in the probabilistic autoencoder frame, we introduce a class of cost functions

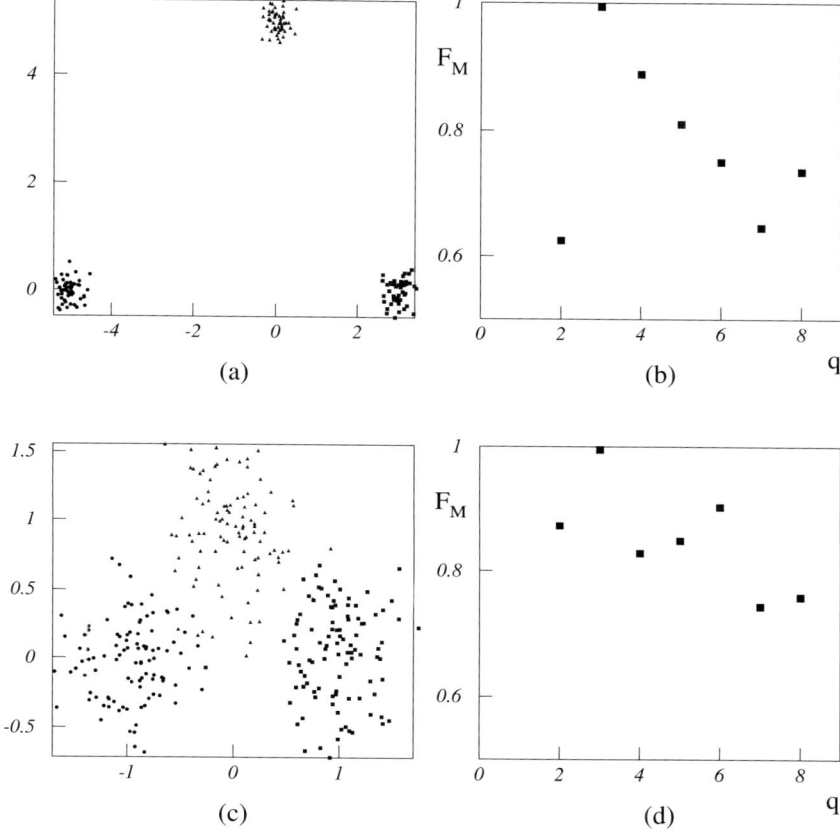

Fig. 6. (a) A two-dimensional data-set, made of three separated clusters, is represented. (b) The figure of merit, evaluated over 50 resamples of the data set shown in (a) and using $f = 0.7$, is plotted versus the number of classes q. (c) A two-dimensional data-set, made of three almost overlapping clusters, is represented. (d) The figure of merit, evaluated over 50 resamples of the data set shown in (c) and using $f = 0.7$, is plotted versus the number of classes q

arising from the request of maximal average similarity between the input and the output of the autoencoder. Our simulations show that the partition provided by these new cost functions corresponds to extract dense connected regions in data space, and that a relevant discrepancy with the partition provided by the cost function introduced in [19] is to be expected in case of non-trivial geometry of clusters. We note that the approach to clustering here described has some similarities with the method in [14]: indeed in both cases clustering is mapped onto a ferromagnetic Potts model with couplings decreasing with the distance. In the superparamagnetic approach, however, q is not related to the number of classes present in the data-set and one obtains

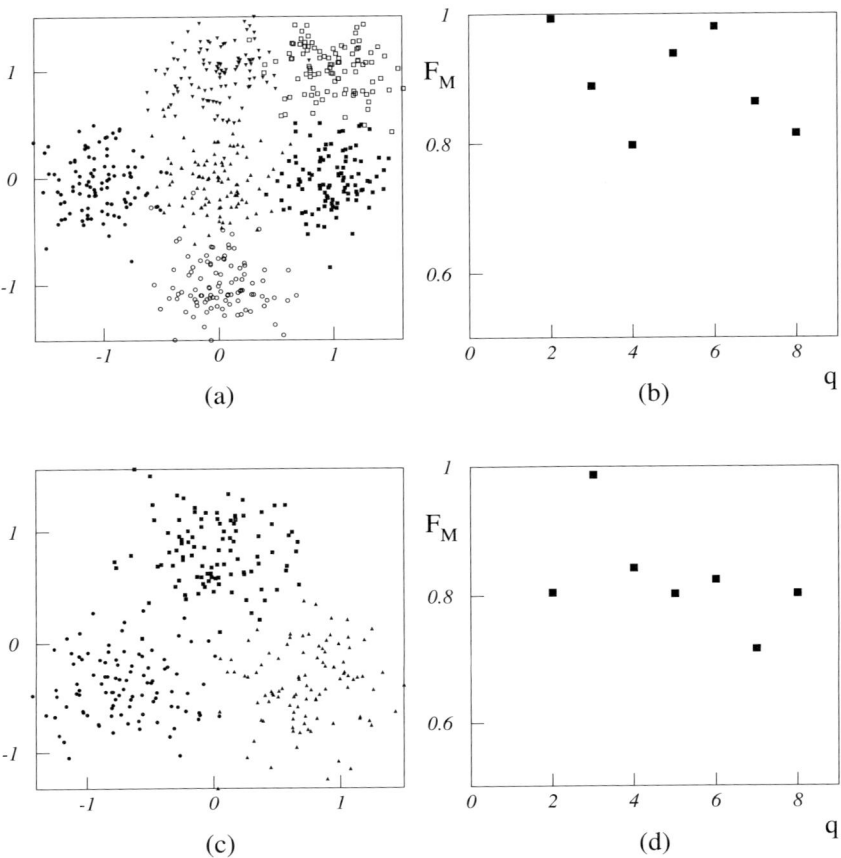

Fig. 7. (a) A two dimensional data-set, made of six clusters, is represented. (b) The figure of merit, evaluated over 50 resamples of the data set shown in (a) and using $f = 0.7$, is plotted versus the number of classes q. (c) A four dimensional data-set, made of three clusters, is represented (first two principal components). (d) The figure of merit, evaluated over 50 resamples of the data set shown in (c) and using $f = 0.7$, is plotted versus the number of classes q. Compare this plot with that in Fig. 6d

hierarchical clustering as the temperature of the Potts model is varied. In the present case q is the number of classes, which is supposed to be known (non-hierarchical clustering), and the denominators in the Hamiltonian, ensuring clusters' coherence, lead to a non-trivial ground state which reflects data structure. We consider two classes of cost function. Scale-free cost functions depend on the exponent γ, while scale-dependent ones depend on the scale-parameter a. Varying a, i.e. changing the resolution at which the data-set is processed, may give rise to transitions between different partitions, as one intuitively might expect; in the scale-free case, the clustering output is fairly stable, with respect to γ, in a wide range. We also propose a resampling technique as a means to choose q in situations where the number of classes is ambiguous: some encouraging application results are reported.

4 Conclusions

In the last few years huge amounts of data are becoming accessible in many fields, thus making it necessary to develop new tools for dealing with the avalanche of available data. Many innovative algorithms for data analysis are rooted on Statistical Physics. We have described some novel approaches to clustering, one of the most effective methodologies to compress and/or extract meaningful information from data.

This research is an activity of the Center of Innovative Technologies for Signal Detection and Processing, Bari.

References

1. B.D. Ripley: *Pattern Recognition and Neural Networks.* (Cambridge University Press, Cambridge 1996)
2. R.O. Duda, P.E. Hart: *Pattern Recognition and scene analysis.* (Wiley, New York 1973)
3. A. Dekel, M.J. West: Astrophys. J. **228**, 411 (1985)
4. P. Chiappetta, P. Colangelo, P. De Felice, G. Nardulli, G. Pasquariello: Phys. Lett. B **322**, 219 (1994)
5. A. Baraldi, P. Blonda, F. Parmiggiani, G. Satalino: Optical Engineering **39**, 907 (2000)
6. Y. Linde, A. Buzo, R.M. Gray: IEEE Trans. on Communications **28**, 84 (1980)
7. U. Alon, N. Barkai, D.A. Notterman, K. Gish, S. Ybarra, D. Mack, A.J. Levine: Proc. Natl. Acad. Sci. USA **96**, 6745 (1999)
8. L. Kullmann, J. Kertesz, R.N. Mantegna: Physica A **287**, 412 (2000)
9. C. Giacovazzo: *Direct Phasing in Crystallography.* (Oxford University Press, Oxford 1998)
10. T. Kosaka, S. Sagayama: 'Tree-structured speaker clustering for fast speaker adaptation'. In: *Proceedings of the 1994 IEEE International Conference on Acoustic, Speech and Signal Processing - Vol. 1*, (IEEE, New York 1994) pp. 245–248
11. C. Marangi, L. Angelini, F. De Carlo, G. Nardulli, M. Pellicoro, S. Stramaglia: 'Clustering by inhomogeneous chaotic maps in landmine detection'. In: *Proceedings of SPIE - Vol. 4170*, (SPIE 2001) pp. 122–132
12. A. Hutt, M. Svensen, F. Kruggel, R. Friedrich: Phys. Rev. E **61**, R4691 (2000)
13. A.K. Jain, R.C. Dubes: *Algorithms for Clustering Data.* (Prentice Hall, New York 1988)
14. M.Blatt, S. Wiseman, E. Domany: Phys. Rev. Lett. **76**, 3251 (1996)
15. L. Angelini, F. De Carlo, C. Marangi, M. Pellicoro, S. Stramaglia: Phys. Rev. Lett. **85**, 554 (2000)
16. C.M. Bishop: *Neural Networks for Pattern Recognition.* (Oxford University Press, Oxford 1995)

17. A. Engel, C. Van den Broeck: *Statistical Mechanics of Learning*. (Cambridge University Press, Cambridge 2001)
18. K. Rose, E. Gurewitz, G.C. Fox: Phys. Rev. Lett. **65**, 945 (1990)
19. T. Hofmann, J.M. Buhmann: IEEE Trans. P.A.M.I. **19**, 1 (1997)
20. L. Angelini, M. Attimonelli, M. De Robertis, M. Mannarelli, C. Marangi, L. Nitti, M. Pellicoro, G. Pesole, C. Saccone, S. Stramaglia, M. Tommaseo: "CMC: a novel clustering method for human sequence classification". (Submitted)
21. M. Pagel: Nature **401**, 877 (1999)
22. M. Ingman, H. Kaesmann, S. Paabo, U. Gyllensten: Nature **408**, 708 (2000)
23. S. Wiggins: *Introduction to Applied Nonlinear Dynamical Systems and Chaos*. (Springer, Berlin 1990)
24. G. Pesole, C. Saccone: Genetics **157**, 859 (2001)
25. C. Lanave, G. Preparata, C. Saccone, G. Serio: Jour. Mol. Evol. **20**, 86 (1984)
26. C. Saccone, C. Lanave, G. Pesole, G. Preparata: Meth. Enzymol. **183**, 570 (1990)
27. M. Tommaseo, M. Attimonelli, M. De Robertis, F. Tanzariello, C. Saccone: Am. J. Phys. Anthropol. **117**, 49 (2002)
28. N. Saitou, M. Nei: Mol. Biol. Evol. **4**, 406 (1987)
29. J. Felsenstein: *PHYLIP, Phylogeny Inference Package*. (Genetics Dept., University of Washington, Seattle)
30. H.J. Bandelt, P. Forster, C.S. Bryan, M.B. Richards: Genetics **141**, 743 (1995)
31. S.P. Luttrel: Neural Computation **6**, 767 (1994)
32. C.M. Bishop, M. Svensen, C.K.I. Williams: Neural Computation **10**, 215 (1997)
33. T. Graepel: *Statistical Physics of clustering algorithms*, Diplomarbeit, FB Physik, Institut für Theoretische Physik, Technische Universität Berlin (1998)
34. S. Kirkpatrick, C.D. Gelatt, M.P. Vecchi: Science **220**, 671 (1983)
35. A.L. Yuille, J.J. Kosowsky: Neural Computation **6**, 341 (1994)
36. G. Parisi: *Statistical Field Theory*. (Addison Wesley, Reading 1988)
37. A.P. Dempster, N.M. Laird, D.B. Rubin: Jour. Royal Stat. Soc. **39**, 1 (1977)
38. In the text we use an operational definition of *ground state* as the best output over a number (10-50) of simulated annealing runs. The true ground state might be found only by an unpractical exhaustive search.
39. E. Anderson: Bull. Amer. Iris Soc. **59**, 2 (1935)
40. E. Levine, E. Domany: Neural Computation **13**, 2573 (2001)

The Challenges
of Clustering High Dimensional Data

Michael Steinbach, Levent Ertöz, and Vipin Kumar

Cluster analysis divides data into groups (clusters) for the purposes of summarization or improved understanding. For example, cluster analysis has been used to group related documents for browsing, to find genes and proteins that have similar functionality, or as a means of data compression. While clustering has a long history and a large number of clustering techniques have been developed in statistics, pattern recognition, data mining, and other fields, significant challenges still remain. In this chapter we provide a short introduction to cluster analysis, and then focus on the challenge of clustering high dimensional data. We present a brief overview of several recent techniques, including a more detailed description of recent work of our own which uses a concept-based clustering approach.

1 Introduction

Cluster analysis [23,29] divides data into meaningful or useful groups (clusters). If meaningful clusters are the goal, then the resulting clusters should capture the natural structure of the data. For example, cluster analysis has been used to group related documents for browsing, to find genes and proteins that have similar functionality, and to provide a grouping of spatial locations prone to earthquakes. However, in other cases, cluster analysis is only a useful starting point for other purposes, e.g., data compression or efficiently finding the nearest neighbors of points. Whether for understanding or utility, cluster analysis has long been used in a wide variety of fields: psychology and other social sciences, biology, statistics, pattern recognition, information retrieval, machine learning, and data mining.

In this chapter we provide a short introduction to cluster analysis, and then focus on the challenge of clustering high dimensional data. We present a brief overview of several recent techniques, including a more detailed description of recent work of our own which uses a concept-based approach. In all cases, the approaches to clustering high dimensional data must deal with the "curse of dimensionality" [5], which, in general terms, is the widely observed phenomenon that data analysis techniques (including clustering), which work well at lower dimensions, often perform poorly as the dimensionality of the analyzed data increases.

2 Basic Concepts and Techniques of Cluster Analysis

2.1 What Cluster Analysis Is

Cluster analysis groups objects (observations, events) based on the information found in the data describing the objects or their relationships. The goal is that the objects in a group should be similar (or related) to one another and different from (or unrelated to) the objects in other groups. The greater the similarity (or homogeneity) within a group and the greater the difference between groups, the better the clustering.

The definition of what constitutes a cluster is not well defined, and in many applications, clusters are not well separated from one another. Nonetheless, most cluster analysis seeks, as a result, a crisp classification of the data into non-overlapping groups. Fuzzy clustering [22] is an exception to this, and allows an object to partially belong to several groups.

To illustrate the difficulty of deciding what constitutes a cluster, consider Fig. 1, which shows twenty points and three different ways that these points can be divided into clusters. If we allow clusters to be nested, then the most reasonable interpretation of the structure of these points is that there are two clusters, each of which has three subclusters. However, the apparent division of the two larger clusters into three subclusters may simply be an artifact of the human visual system. Finally, it may not be unreasonable to say that the points form four clusters. Thus, we again stress that the notion of a cluster is imprecise, and the best definition depends on the type of data and the desired results.

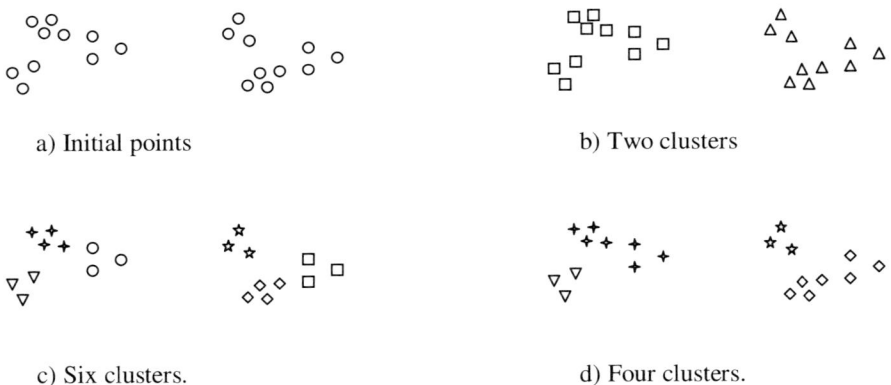

Fig. 1. Different clusterings for a set of points

2.2 What Cluster Analysis Is Not

Cluster analysis is a classification of objects from the data, where by "classification" we mean a labeling of objects with class (group) labels. As such, clustering does not use previously assigned class labels, except perhaps for verification of how well the clustering worked. Thus, cluster analysis is sometimes referred to as "unsupervised classification" and is distinct from "supervised classification", or more commonly just "classification", which seeks to find rules for classifying objects given a set of pre-classified objects. Classification is an important part of data mining, pattern recognition, machine learning, and statistics (discriminant analysis and decision analysis).

While cluster analysis can be very useful, either directly or as a preliminary means of finding classes, there is more to data analysis than cluster analysis. For example, the decision of what features to use when representing objects is a key activity of fields such as data mining, statistics, and pattern recognition. Cluster analysis typically takes the features as given and proceeds from there. Thus, cluster analysis, while a useful tool in many areas, is normally only part of a solution to a larger problem that typically involves other steps and techniques.

2.3 The Data Matrix

Objects (measurements, events) are usually represented as points (vectors) in a multi-dimensional space, where each dimension represents a distinct attribute (variable, measurement) describing the object. For simplicity, it is usually assumed that values are present for all attributes. (Techniques for dealing with missing values are described in [23,29].) Thus, a set of objects is represented (at least conceptually) as an m by n matrix, where there are m rows, one for each object, and n columns, one for each attribute. This matrix has different names, e.g., pattern matrix or data matrix, depending on the particular field.

The data is sometimes transformed before being used. One reason for this is that different attributes may be measured on different scales, e.g., centimeters and kilograms. In cases where the range of values differs widely from attribute to attribute, these differing attribute scales can dominate the results of the cluster analysis, and it is common to standardize the data so that all attributes are on the same scale. A simple approach to such standardization is, for each attribute, to subtract of the mean of the attribute values and divide by the standard deviation of the values. While this is often sufficient, more statistically "robust" approaches are available, as described in [29].

Another reason for initially transforming the data is to reduce the number of dimensions, particularly if the initial number of dimensions is large. We defer this discussion until later in this chapter.

2.4 The Proximity Matrix

While cluster analysis sometimes uses the original data matrix, many clustering algorithms use a similarity matrix, S, or a dissimilarity matrix, D. For convenience, both matrices are commonly referred to as a proximity matrix, P. A proximity matrix, P, is an m by m matrix containing all the pairwise dissimilarities or similarities between the objects being considered. If x_i and x_j are the ith and j^{th} objects, respectively, then the entry at the ith row and j^{th} column of the proximity matrix is the similarity, s_{ij}, or the dissimilarity, d_{ij}, between x_i and x_j. For simplicity, we will use p_{ij} to represent either s_{ij} or d_{ij}. Figures 2a, 2b, and 2c show, respectively, four points and the corresponding data and proximity (distance) matrices. (Different types of proximities are described in Sect. 2.7.)

For completeness, we mention that objects are sometimes represented by more complicated data structures than vectors of attributes, e.g., character strings or graphs. Determining the similarity (or differences) of two objects in such a situation is more complicated, but if a reasonable similarity (dissimilarity) measure exists, then a clustering analysis can still be performed. In particular, clustering techniques that use a proximity matrix are unaffected by the lack of a data matrix.

2.5 The Proximity Graph

A proximity matrix defines a weighted graph, where the nodes are the points being clustered, and the weighted edges represent the proximities between points, i.e., the entries of the proximity matrix (see Fig. 2c). While this proximity graph can be directed, which corresponds to an asymmetric proximity matrix, most clustering methods assume an undirected graph. Relaxing the symmetry requirement can be useful in some instances, but we will assume undirected proximity graphs (symmetric proximity matrices) in our discussions.

From a graph point of view, clustering is equivalent to breaking the graph into connected components (disjoint connected subgraphs), one for each cluster. Likewise, many clustering issues can be cast in graph-theoretic terms, e.g., the issues of cluster cohesion and the degree of coupling with other clusters can be measured by the number and strength of links between and within clusters. Also, many clustering techniques, e.g., single link and complete link (see Sec. 2.10), are most naturally described using graph representations.

2.6 Some Working Definitions of a Cluster

As mentioned above, the term, cluster, does not have a precise definition. However, several working definitions of a cluster are commonly used and are given below. There are two aspects of clustering that should be mentioned in conjunction with these definitions. First, clustering is sometimes viewed

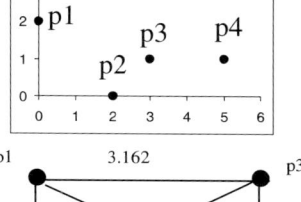

point	x	y
p1	0	2
p2	2	0
p3	3	1
p4	5	1

	p1	p2	p3	p4
p1	0.000	2.828	3.162	5.099
p2	2.828	0.000	1.414	3.162
p3	3.162	1.414	0.000	2.000
p4	5.099	3.162	2.000	0.000

Fig. 2. Four points (top left), their proximity graph (bottom left), and their corresponding data (top right) and proximity (distance) matrices (bottom right)

as finding only the most "tightly" connected points while discarding "background" or noise points. Second, it is sometimes acceptable to produce a set of clusters where a true cluster is broken into several subclusters (which are often combined later, by another technique). The key requirement in this latter situation is that the subclusters are relatively "pure", i.e., most points in a subcluster are from the same "true" cluster.

1. **Well-Separated Cluster Definition:** A cluster is a set of points such that any point in a cluster is closer (or more similar) to every other point in the cluster than to any point not in the cluster. Sometimes a threshold is used to specify that all the points in a cluster must be sufficiently close (or similar) to one another.

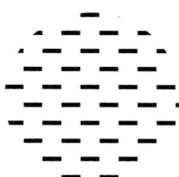

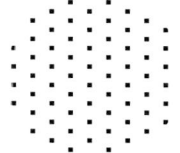

Fig. 3. Three well-separated clusters of two-dimensional points

However, in many sets of data, a point on the edge of a cluster may be closer (or more similar) to some objects in another cluster than to objects in its own cluster. Consequently, many clustering algorithms use the following criterion.

2. **Center-based Cluster Definition:** A cluster is a set of objects such that an object in a cluster is closer (more similar) to the "center" of a cluster, than to the center of any other cluster. The center of a cluster is often a centroid, the average of all the points in the cluster, or a medoid, the "most representative" point of a cluster.

Fig. 4. Four center-based clusters of two-dimensional points

3. **Contiguous Cluster Definition (Nearest Neighbor or Transitive Clustering):** A cluster is a set of points such that a point in a cluster is closer (or more similar) to one or more other points in the cluster than to any point not in the cluster.

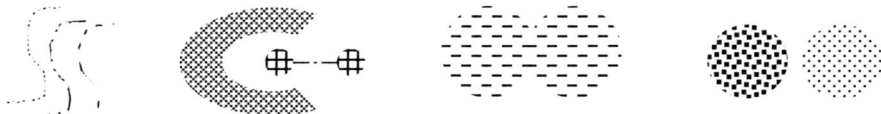

Fig. 5. Eight contiguous clusters of two-dimensional points

4. **Density-based Definition:** A cluster is a dense region of points, which is separated by low-density regions, from other regions of high density. This definition is more often used when the clusters are irregular or intertwined, and when noise and outliers are present. Notice that the contiguous definition would find only one cluster in Fig. 6. Also note that the three curves don't form clusters since they fade into the noise, as does the bridge between the two small circular clusters.
5. **Similarity-based Cluster Definition:** A cluster is a set of objects that are "similar", and objects in other clusters are not "similar". A variation on this is to define a cluster as a set of points that together create a region with a uniform local property, e.g., density or shape.

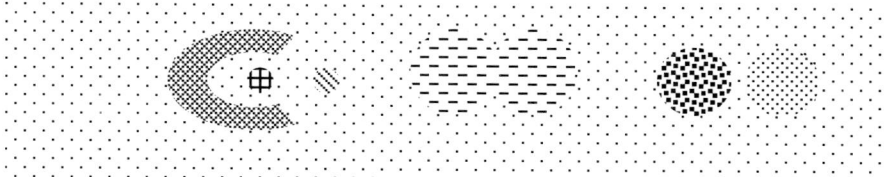

Fig. 6. Six dense clusters of two-dimensional points

2.7 Measures (Indices) of Similarity and Dissimilarity

The notion of similarity and dissimilarity (distance) seems fairly intuitive. However, the quality the quality of a cluster analysis depends critically on the similarity measure that is used and, as a consequence, hundreds of different similarity measures have been developed for various situations. The discussion here is necessarily brief.

Attribute Types and Scales

The proximity measure (and the type of clustering used) depends on the attribute type and scale of the data. The three typical types of attributes are shown in Table 1, while the common data scales are shown in Table 2.

Table 1. Different attribute types.

Binary	Two values, e.g., true and false.
Discrete	A finite number of values, or integers, e.g., counts.
Continuous	An effectively infinite number of real values, e.g., weight.

Table 2. Different attribute scales.

Qualitative	Nominal	The values are just different names, e.g., colors or zip codes.
	Ordinal	The values reflect an ordering, nothing more, e.g., good, better, best.
Quantitative	Interval	The difference between values is meaningful, i.e., a unit of measurement exists. For example, temperature on the Celsius or Fahrenheit scales.
	Ratio	The scale has an absolute zero so that ratios are meaningful. Examples are physical quantities such as electrical current, pressure, or temperature on the Kelvin scale.

Euclidean Distance and Some Variations

The most commonly used proximity measure, at least for ratio scales (scales with an absolute 0) is the Minkowski metric, which is a generalization of the distance between points in Euclidean space.

$$p_{ij} = \left(\sum_{k=1}^{d} |x_{ik} - x_{jk}|^r \right)^{1/r}, \qquad (1)$$

where r is a parameter, d is the dimensionality of the data object, and x_{ik} and x_{jk} are, respectively, the kth components of the ith and j^{th} objects, x_i and x_j.

For $r = 1$, this distance is commonly known as the L_1 norm or city block distance. If $r = 2$, the most common situation, then we have the familiar L_2 norm or Euclidean distance. Occasionally one might encounter the $L_{\max}$ norm (L_∞ norm), which represents the case $r \to \infty$. Figure 7 gives the proximity matrices for the L_1, L_2 and L_∞ distances, respectively, using the data matrix from Fig. 2.

The r parameter should not be confused with the dimension, d. For example, Euclidean, Manhattan and supremum distances are defined for all values of d, 1, 2, 3, $\cdots$, and specify different ways of combining the differences in each dimension (attribute) into an overall distance.

point	x	y
p1	0	2
p2	2	0
p3	3	1
p4	5	1

L2	p1	p2	p3	p4
p1	0.000	2.828	3.162	5.099
p2	2.828	0.000	1.414	3.162
p3	3.162	1.414	0.000	2.000
p4	5.099	3.162	2.000	0.000

L1	p1	p2	p3	p4
p1	0.000	4.000	4.000	6.000
p2	4.000	0.000	2.000	4.000
p3	4.000	2.000	0.000	2.000
p4	6.000	4.000	2.000	0.000

L_∞	p1	p2	p3	p4
p1	0.000	2.000	3.000	5.000
p2	2.000	0.000	1.000	3.000
p3	3.000	1.000	0.000	2.000
p4	5.000	3.000	2.000	0.000

Fig. 7. Data matrix and the corresponding L_1, L_2, and L_∞ proximity matrices

Finally, note that various Minkowski distances are metric distances. In other words, given a distance function, *dist*, and three points **a**, **b**, and **c**, these distances satisfy the following three mathematical properties: reflexivity (*dist*(**a**, **a**) = 0), symmetry (*dist*(**a**, **b**) = *dist*(**b**, **a**)), and the triangle inequality (*dist*(**a**, **c**) ≤ *dist*(**a**, **b**) + *dist*(**b**, **a**)). Not all distances or similarities are metric, e.g., the Jaccard measure of the following section. This introduces potential complications in the clustering process since in such cases, **a** similar (close) to **b** and **b** similar to **c**, does not necessarily imply **a** similar to **c**. The concept based clustering, which we discuss later, provides a way of dealing with such situations.

Similarity Measures Between Binary Vectors

These measures are referred to as similarity coefficients [23], and typically have values between 0 (not at all similar) and 1 (completely similar). The comparison of two binary vectors, **a** and **b**, leads to four quantities:

N_{01} = the number of positions where **a** was 0 and **b** was 1
N_{10} = the number of positions where **a** was 1 and **b** was 0
N_{00} = the number of positions where **a** was 0 and **b** was 0
N_{11} = the number of positions where **a** was 1 and **b** was 1

Two common similarity coefficients between binary vectors are the simple matching coefficient (SMC) and the Jacccard coefficient.

$\text{SMC} = (N_{11} + N_{00}) / (N_{01} + N_{10} + N_{11} + N_{00})$
$\text{Jaccard} = N_{11} / (N_{01} + N_{10} + N_{11})$

For the following two binary vectors, **a** and **b** we get SMC = 0.7 and Jaccard = 0.

a = 1 0 0 0 0 0 0 0 0 0
b = 0 0 0 0 0 0 1 0 0 1

Conceptually, SMC equates similarity with the total number of matches, while J considers only matches on 1's to be important. There are situations in which both measures are more appropriate. For example, if the vectors represent students' answers to a True-False test, then both 0-0 and 1-1 matches are important and these two students are very similar, at least in terms of the grades they will get. If instead, the vectors indicate particular items purchased by two shoppers, then the Jaccard measure is more appropriate since it would be odd to say that the purchasing behavior of two customers is similar, even though they did not buy any of the same items.

2.8 Hierarchical and Partitional Clustering

The main distinction in clustering approaches is between hierarchical and partitional approaches. Hierarchical techniques produce a nested sequence of partitions, with a single, all-inclusive cluster at the top and singleton clusters of individual points at the bottom. Each intermediate level can be viewed as combining (splitting) two clusters from the next lower (next higher) level. (Hierarchical clustering techniques that start with one large cluster and split it are termed "divisive", while approaches that start with clusters containing a single point, and then merge them are called "agglomerative".) While most hierarchical algorithms involve joining two clusters or splitting a cluster into two sub-clusters, some hierarchical algorithms join more than two clusters in one step or split a cluster into more than two sub-clusters.

Partitional techniques create a one-level (unnested) partitioning of the data points. If K is the desired number of clusters, then partitional approaches typically find all K clusters at once. Contrast this with traditional hierarchical schemes, which bisect a cluster to get two clusters or merge two clusters to get one. Of course, a hierarchical approach can be used to generate a flat partition of K clusters, and likewise, the repeated application of a partitional scheme can provide a hierarchical clustering.

There are also other important distinctions between clustering algorithms: Does a clustering algorithm cluster on all attributes simultaneously (polythetic) or use only one attribute at a time (monothetic)? Does a clustering technique use one object at a time (incremental) or does the algorithm require access to all objects (non-incremental)? Does the clustering method allow a cluster to belong to multiple clusters (overlapping) or does it assign each object to a single cluster (non-overlapping)? Note that overlapping clusters are not the same as fuzzy clusters, but rather reflect the fact that in many real situations, objects belong to multiple classes.

2.9 Specific Partitional Clustering Techniques: K-Means

The K-means algorithm discovers K (non-overlapping) clusters by finding K centroids ("central" points) and then assigning each point to the cluster associated with its nearest centroid. (A cluster centroid is typically the mean or median of the points in its cluster and "nearness" is defined by a distance or similarity function.) Ideally the centroids are chosen to minimize the total "error", where the error for each point is given by a function that measures the discrepancy between a point and its cluster centroid, e.g., the squared distance. Note that a measure of cluster "goodness" is the error contributed by that cluster. For squared error and Euclidean distance, it can be shown [4] that a gradient descent approach to minimizing the squared error yields the following basic K-means algorithm. (The previous discussion still holds if we use similarities instead of distances, but our optimization problem becomes a maximization problem.)

Basic K-means Algorithm for finding K clusters.

1. Select K points as the initial centroids.
2. Assign all points to the closest centroid.
3. Recompute the centroid of each cluster.
4. Repeat steps 2 and 3 until the centroids don't change (or change very little).

K-means has a number of variations, depending on the method for selecting the initial centroids, the choice for the measure of similarity, and the way that the centroid is computed. The common practice, at least for Euclidean data, is to use the mean as the centroid and to select the initial centroids randomly.

In the absence of numerical problems, this procedure converges to a solution, although the solution is typically a local minimum. Since only the vectors are stored, the space requirements are $O(m*n)$, where m is the number of points and n is the number of attributes. The time requirements are $O(I*K*m*n)$, where I is the number of iterations required for convergence. I is typically small and can be easily bounded as most changes occur in the first few iterations. Thus, the time required by K-means is efficient, as well as simple, as long as the number of clusters is significantly less than m.

Theoretically, the K-means clustering algorithm can be viewed either as a gradient descent approach which attempts to minimize the sum of the squared error of each point from cluster centroid [4] or as a procedure that results from trying to model the data as a mixture of Gaussian distributions with diagonal covariance matrices [30].

2.10 Specific Hierarchical Clustering Techniques: MIN, MAX, Group Average

In hierarchical clustering the goal is to produce a hierarchical series of nested clusters, ranging from clusters of individual points at the bottom to an all-inclusive cluster at the top. A diagram called a dendogram graphically represents this hierarchy and is an inverted tree that describes the order in which points are merged (bottom-up, agglomerative approach) or clusters are split (top-down, divisive approach). One of the attractions of hierarchical techniques is that they correspond to taxonomies that are very common in the biological sciences, e.g., kingdom, phylum, genus, species, $\cdots$ (Some cluster analysis work occurs under the name of "mathematical taxonomy".) Another attractive feature is that hierarchical techniques do not assume any particular number of clusters. Instead, any desired number of clusters can be obtained by "cutting" the dendogram at the proper level. Finally, hierarchical techniques are thought to produce better quality clusters [23].

In this section we describe three agglomerative hierarchical techniques: MIN, MAX, and group average. For the single link or MIN version of hierarchical clustering, the proximity of two clusters is defined to be minimum of the distance (maximum of the similarity) between any two points in the different clusters. The technique is called single link, because if you start with all points as singleton clusters, and add links between points, strongest links first, these single links combine the points into clusters. Single link is good at handling non-elliptical shapes, but is sensitive to noise and outliers.

For the complete link or MAX version of hierarchical clustering, the proximity of two clusters is defined to be maximum of the distance (minimum of the similarity) between any two points in the different clusters. The technique is called complete link because, if you start with all points as singleton clusters, and add links between points, strongest links first, then a group of points is not a cluster until all the points in it are completely linked, i.e., form a clique. Complete link is less susceptible to noise and outliers, but can break large clusters, and has trouble with convex shapes.

For the group average version of hierarchical clustering, the proximity of two clusters is defined to be the average of the pairwise proximities between all pairs of points in the different clusters. Notice that this is an intermediate approach between MIN and MAX. This is expressed by the following equation:

$$proximity(cluster1, cluster2) = \frac{\sum_{\substack{p_1 \in cluster1 \\ p_2 \in cluster2}} proximity(p_1, p_2)}{size(cluster1) * size(cluster2)} \quad (2)$$

Figure 8 shows a table for a sample similarity matrix and three dendograms, which respectively, show the series of merges that result from using the MIN, MAX, and group average approaches. In this simple case, MIN and group average produce the same clustering.

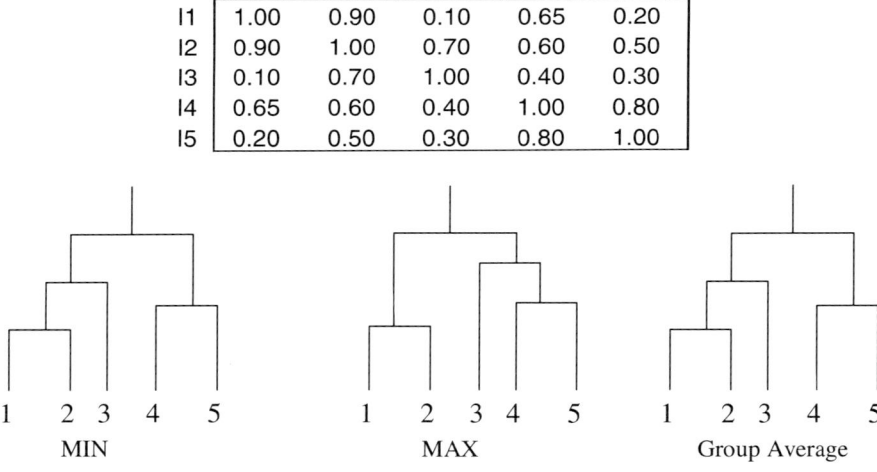

Fig. 8. Dendograms produced by MIN, MAX and group average hierarchical clustering technique

3 The "Curse of Dimensionality"

It was Richard Bellman who apparently originated the phrase, "the curse of dimensionality", in a book on control theory [5]. The specific quote from [5], page 97, is "In view of all that we have said in the forgoing sections, the many obstacles we appear to have surmounted, what casts the pall over our victory celebration? It is the curse of dimensionality, a malediction that has plagued the scientist from the earliest days." The issue referred to in Bellman's quote is the impossibility of optimizing a function of many variables by a brute force search on a discrete multidimensional grid. (The number of grids points increases exponentially with dimensionality, i.e., with the number of variables.) With the passage of time, the "curse of dimensionality" has come to refer to any problem in data analysis that results from a large number of variables (attributes).

In general terms, problems with high dimensionality result from the fact that a fixed number of data points become increasingly "sparse" as the dimensionality increase. To visualize this, consider 100 points distributed with a uniform random distribution in the interval $[0, 1]$. If this interval is broken into 10 cells, then it is highly likely that all cells will contain some points. However, consider what happens if we keep the number of points the same, but distribute the points over the unit square. (This corresponds to the situation where each point is two-dimensional.) If we keep the unit of discretization to be 0.1 for each dimension, then we have 100 two-dimensional cells, and it is quite likely that some cells will be empty. For 100 points and three dimensions, most of the 1000 cells will be empty since there are far more points than cells. Conceptually our data is "lost in space" as we go to higher dimensions.

For clustering purposes, the most relevant aspect of the curse of dimensionality concerns the effect of increasing dimensionality on distance or similarity. In particular, most clustering techniques depend critically on the measure of distance or similarity, and require that the objects within clusters are, in general, closer to each other than to objects in other clusters. (Otherwise, clustering algorithms may produce clusters that are not meaningful.) One way of analyzing whether a data set may contain clusters is to plot the histogram (approximate probability density function) of the pairwise distances of all points in a data set (or of a sample of points if this requires too much computation.) If the data contains clusters, then the graph will typically show two peaks: a peak representing the distance between points in clusters, and a peak representing the average distance between points. Figure 9 shows idealized versions of the data with (left) and without (right) clusters. Also see [6]. If only one peak is present or if the two peaks are close, then clustering via distance based approaches will likely be difficult. Note that clusters of different densities could cause the leftmost peak in the left part of Fig. 9 to actually become several peaks.

There has also been some work on analyzing the behavior of distances for high dimensional data. In [7], it is shown, for certain data distributions, that the relative difference of the distances of the closest and farthest data points of an independently selected point goes to 0 as the dimensionality increases, i.e.,

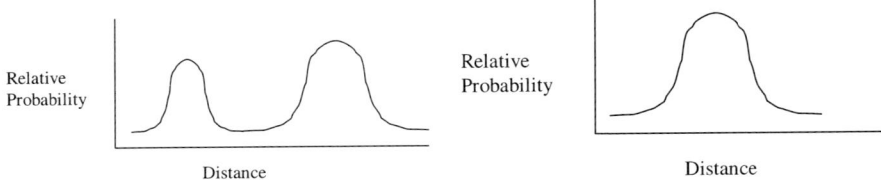

Fig. 9. Plot of interpoint distances for data with (left) and without (right) clusters

$$\lim_{d \to \infty} \frac{\text{MaxDist} - \text{MinDist}}{\text{MinDist}} = 0. \tag{3}$$

For example, this phenomenon occurs if all attributes are i.i.d. (identically and independently distributed). Thus, it is often said, "in high dimensional spaces, distances between points become relatively uniform". In such cases, the notion of the nearest neighbor of a point is meaningless. To understand this in a more geometrical way, consider a hyper-sphere whose center is the selected point and whose radius is the distance to the nearest data point. Then, if the relative difference between the distance to nearest and farthest neighbors is small, expanding the radius of the sphere "slightly" will include many more points.

In [7] a theoretical analysis of several different types of distributions is presented, as well as some supporting results for real-world high dimensional data sets. This work was oriented towards the problem of finding the nearest neighbors of points, but the results also indicate potential problems for clustering high dimensional data.

The work just discussed was extended in [18] to look at the absolute difference, **MaxDist − MinDist**, instead of the relative difference. It was shown that the behavior of the absolute difference between the distance to the closest and farthest neighbors of an independently selected point depends on the distance measure. In particular, for the L_1 metric, **MaxDist − MinDist** increases with dimensionality, for the L_2 metric, **MaxDist − MinDist** remains relatively constant, and for the L_d metric, $d \geq 3$, **MaxDist − MinDist** goes to 0 as dimensionality increases. These theoretical results were also confirmed by experiments on simulated and real datasets. The conclusion is that the L_d metric, $d \geq 3$, is meaningless for high dimensional data.

The previous results indicate the potential problems with clustering high dimensional data sets, at least in cases where the data distribution causes the distances between points to become relatively uniform. However, things are sometimes not as bad as they might seem, for it is often possible to reduce the dimensionality of the data without losing important information. For example, sometimes it is known *a priori* that only a smaller number of variables are of interest. If so, then these variables can be selected, and the others discarded, thus reducing the dimensionality of the data set. More generally, data analysis (clustering or otherwise) is often preceded by a "feature selection" step that attempts to remove irrelevant features. This can be accomplished by discarding features that show little variation or which are highly correlated with other features. (Feature selection is a complicated subject in its own right.)

Another approach is to project points from a higher dimensional space to a lower dimensional space. The idea here is that that often data can be approximated reasonably well even if only a relatively small number of dimensions are kept, and thus, little "true" information is lost. Indeed, such techniques can, in some cases, enhance the data analysis because they are effective in removing noise. Typically this type of dimensionality reduction is accomplished by applying techniques from linear algebra or statistics such as Principal Component Analysis (PCA) [23] or Singular Value Decomposition (SVD) [35].

To make this more concrete we briefly illustrate with SVD. (Mathematically less inclined readers can skip this paragraph without loss.) A singular value decomposition of an m by n matrix, M, expresses M as the sum of simpler rank 1 matrices as follows: $M = \sum_{i=1}^{n} s_i u_i v_i^T$, where s_i, a scalar, is the ith singular value of M, u_i is the ith left singular vector, and v_i is the ith right singular vector. All singular values beyond the first r, where $r = \text{rank}(M)$ are 0 and all left (right) singular vectors are orthogonal to each other and are of unit length. A matrix can be approximated by omitting some of the terms of the series that correspond to non-zero singular values. (Singular values are non-negative and ordered by decreasing magnitude.) Since the magnitudes of these singular values often decrease rapidly, an approximation based on a relatively small number of singular values, e.g., 50 or 100 out of 1000, is often sufficient for a productive data analysis. Furthermore, it is not unusual to see data analyses that take only the first few singular values.

However, both feature selection and dimensionality reduction approaches based on PCA or SVD may be inappropriate if different clusters lie in different subspaces. Indeed, we emphasize that for many high dimensional data sets it is likely that clusters lie only in subsets of the full space. Thus, many algorithms for clustering high dimensional data automatically find clusters in subspaces of the full space. One example of such a clustering technique is "projected" clustering [3], which also finds the set of dimensions appropriate for each cluster during the clustering process. More techniques that find clusters in subspaces of the full space will be discussed in Sect. 4.

In summary, high dimensional data is not like low dimensional data and needs different approaches. The next section presents recent work to provide clustering techniques for high dimensional data. While some of this work is represents different developments of a single theme, e.g., grid based clustering, there is considerable diversity, perhaps because of high dimensional data, like low dimensional data is highly varied.

4 Recent Work in Clustering High Dimensional Data

4.1 Clustering via Hypergraph Partitioning

Hypergraph-based clustering [21] is an approach to clustering in high dimensional spaces, which is based on hypergraphs. (This is also work of one of the authors (Kumar), but not our recent work on clustering referenced earlier, which comes later in this section.) Hypergraphs are an extension of regular graphs, which relax the restriction that an edge can only join two vertices. Instead an edge can join many vertices. Hypergraph-based clustering consists of the following steps:

1. Define the condition for connecting several objects (each object is a vertex of the hypergraph) by a hyperedge.
2. Define a measure for the strength or weight of a hyperedge.
3. Use a graph-partitioning algorithm [28] to partition the hypergraph into two parts in such a way that the weight of the hyperedges cut is minimized.
4. Continue the partitioning until a fixed number of partitions are achieved, or until a new partition would produce a poor cluster, as measured by some fitness criteria.

In [21], the data being clustered is "market basket" data. With this kind of data there are a number of items and a number of "baskets", or transactions, each of which contains a subset of all possible items. (A prominent example of market basket data is the subset of store items (products) purchased by customers in individual transactions hence the name market basket data.) This data can be represented by a set of (very sparse) binary vectors one for each transaction. Each item is associated with a dimension (variable), and a value of 1 indicates that the item was present in the transaction, while a value of 0 indicates that the item was not present.

The individual items are the vertices of the hypergraph. The hyperedges are determined by determining subsets of items that frequently occur together. For example, baby formula and diapers are often purchased together. These subsets of frequently co-occurring items are called frequent itemsets and can be found using relatively simple and efficient algorithms [2].

The strength of the hyperedges is determined in the following manner. If the frequent itemset being considered is of size n, and the items of the frequent itemset are $i_1, i_2 \cdots i_n$, then the strength of a hyperedge is obtained as follows:

1. Consider each individual item, i_j, in the frequent itemset.
2. Determine what fraction of the market baskets (transactions) that contain the other $n - 1$ items also contain i_j. This (estimate of the) conditional probability that i_j occurs when the other items do is a measure of the strength of the association between the items.

3. Average these conditional probabilities together.

More formally the strength of a hyperedge is given by

$$\frac{1}{n}\sum_{j=1}^{N} \mathbf{prob}(\mathbf{i_j}|\mathbf{i_1},\cdots,\mathbf{i_{j-1}},\mathbf{i_{j+1}},\cdots,\mathbf{i_n}). \tag{4}$$

An important feature of this algorithm is that it transforms a problem in a sparse, high dimensional data space into a well-studied graph-partitioning problem that can be efficiently solved.

4.2 Grid Based Clustering Approaches

In its most basic form, grid based clustering is relatively simple:

a) **Divide the space over which the data ranges into (hyper) rectangular cells,** e.g., by partitioning the range of values in each dimension into equally sized cells. See Fig. 10 for a two dimensional example of such a grid.

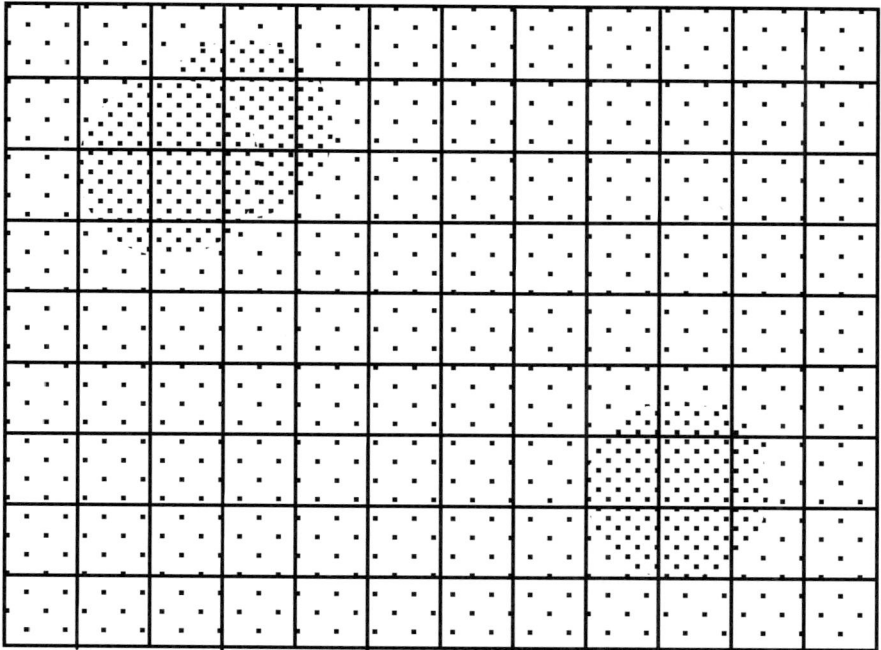

Fig. 10. Two-dimensional grid for cluster detection

b) **Discard Low-density Grid Cells.** This assumes a density based definition of clusters, i.e., that high-density regions represent clusters, while low-density regions represent noise. This is often a good assumption, although density based approaches may have trouble when there are clusters are of widely differing densities.
c) **Combine Adjacent High-density Cells to Form Clusters.** If high-density regions are adjacent, then they are joined to form a single cluster.

Assigning points to cells requires only linear time, i.e., the time complexity is $O(n)$, where n is the number of data points. (However, if the data is high dimensional or some dimensions have a large range, it is necessary to use data structures, e.g., hash tables [8], that do not explicitly store the non-empty cells.) Discarding low-density cells also requires only linear time, at least if only non-empty cells are stored. However, combining dense cells can potentially take $O(n^2)$ time, since it may be necessary to compare each non-empty cell to every other. Nonetheless, if the number of dense grid cells is $O(\sqrt{n})$, then this step will also be linear.

There are a number of obvious concerns about grid-based clustering methods. The grids are square or rectangular and don't necessarily fit the shape of the clusters. This can be handled by increasing the number of grid cells, but at the price of increasing the amount of work, and if the grid size is halved the number of cells increases by a factor of 2^d, where d is the number of dimensions. Also, since the density of a real cluster may vary, making the grid size too small might put "holes" in the cluster, especially with a small number of points. Finally, grid based clustering typically assumes that the distance between points is measure by and L_1 or L_2 distance measure.

Also, despite the appealing efficiency of grid based clustering schemes, there are serious problems as the dimensionality of the data increases. First, the number of cells increases exponentially with increasing dimensionality. For example, even if each dimension is only split in two, there will still be 2^d cells. Given 30 dimensional data, a grid based clustering approach will use, at least conceptually, a minimum of a billion cells. (Again by using algorithms for hash tables or sparse arrays, at most n cells need to be physically represented.) For all but the largest data sets, most of these cells will be empty. More importantly, it is very possible – particularly with a regular grid – that that a cluster might be divided into a large number of cells and that many or even all these cells might have a density less than the threshold.

Another problem is finding clusters in the full-dimensional space. To see this imagine that each point in one of the clusters in Fig. 10 is augmented with many additional variables, but that the values assigned to points in these dimensions are uniformly randomly distributed. Then almost every point will fall into a separate cell in the new, high dimensional space. Thus, as previously mentioned, clusters of points may only exist in subsets of high dimensional spaces. Of course, the number of possible subspaces is also exponential in the dimensionality of the space, yet another aspect of the curse of dimensionality.

CLIQUE

CLIQUE [1] is a clustering algorithm that attempts to deal with these problems and whose approach is based on the following interesting observation: a region that is dense in a particular subspace must create dense regions when projected onto lower dimensional subspaces. For example, if we examine the distribution of the x (horizontal) and y (vertical) coordinates of the points in Fig. 11, we see dense regions in the one-dimensional distributions which reflect the existence two-dimensional clusters. In Fig. 11, the gray horizontal columns and the slashed vertical columns indicate the projections of the clusters onto the vertical and horizontal axes, respectively. Figure 11 also illustrates that high density in a lower dimension can only suggest possible locations of clusters in a higher dimension, as the higher dimensional region formed by the intersection of two dense lower dimensional dense regions may not correspond to an actual cluster.

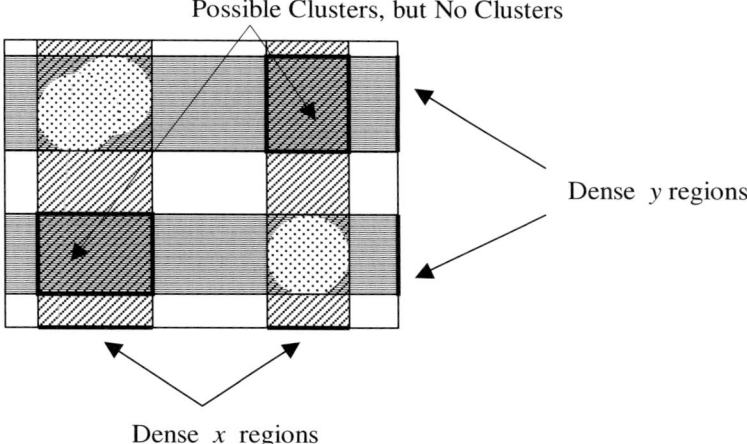

Fig. 11. Illustration of the idea that density in high dimensions implies density in low dimensions, but not vice versa

However, by starting with dense one-dimensional intervals, it is possible to find the potential dense two-dimensional intervals, and by inspecting these, to find the actual dense two-dimensional intervals. This procedure can be extended to find dense units in any subspace, and to find them much more efficiently than by forming the cells corresponding to all possible subsets of dimensions and then searching for the dense units in these cells. However, CLIQUE still needs to use heuristics to reduce the subsets of dimensions investigated and the complexity of CLIQUE, while linear in the number of data points, is not linear in the number of dimensions.

MAFIA

MAFIA (Merging Adaptive Finite Intervals And is more than a clique) [32], which is a refinement of the CLIQUE approach, finds better clusters and achieves higher efficiency by using non-uniform grid cells. Specifically, rather than arbitrarily splitting the data into a pre-determined number of evenly spaced intervals, MAFIA partitions each dimension using a variable number of "adaptive intervals", which better reflect the distribution of the data in that dimension. To illustrate, CLIQUE would more likely use a grid like that shown in Fig. 10, and thus, would break each of the dense one-dimensional intervals into a number of subintervals, including a couple (at each end) that are of lesser density because they include part of the non-dense region. Conceptually, MAFIA starts with a large number of small intervals for each dimension and then combines adjacent intervals of similar density to end up with a smaller number of larger intervals. Thus, a MAFIA grid would likely look more like the idealized grid shown in Fig. 12 than the suboptimal grid of Fig. 10.

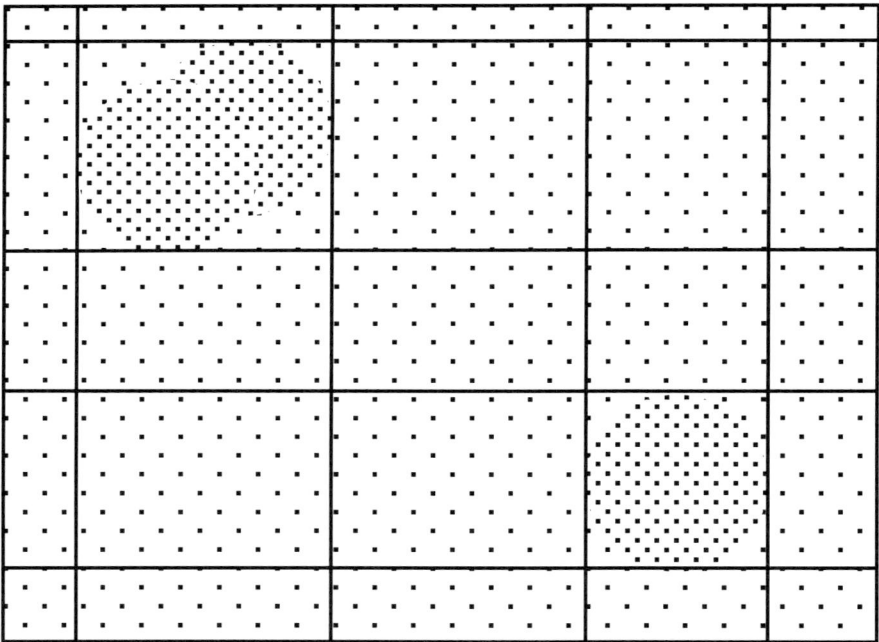

Fig. 12. MAFIA grid for our data

DENCLUE

A different approach to the same problem is provided by the DENCLUE [19]. We describe this approach in some detail, since this approach can be viewed as a generalization of other density-based approaches such as DBSCAN [12] and K-means. DENCLUE (DENsity CLUstEring) is a density clustering approach that takes a more formal approach to density based clustering by modeling the overall density of a set of points as the sum of "influence" functions associated with each point. The resulting overall density function will have local peaks, i.e., local density maxima, and these local peaks can be used to define clusters in a straightforward way. Specifically, for each data point, a hill climbing procedure finds the nearest peak associated with that point, and the set of all data points associated with a particular peak (called a local density attractor) becomes a (center-defined) cluster. However, if the density at a local peak is too low, then the points in the associated cluster are classified as noise and discarded. Also, if a local peak can be connected to a second local peak by a path of data points, and the density at each point on the path is above a minimum density threshold, ξ , then the clusters associated with these local peaks are merged. Thus, clusters of any shape can be discovered.

DENCLUE is based on a well-developed area of statistics and pattern recognition which is know as " kernel density estimation" [9]. The goal of kernel density estimation (and many other statistical techniques as well) is to describe the distribution of the data by a function. For kernel density estimation, the contribution of each point to the overall density function is expressed by an "influence" (kernel) function. The overall density is then merely the sum of the influence functions associated with each point.

Typically the influence or kernel function is symmetric (the same in all directions) and its value (contribution) decreases as the distance from the point increases. For example, for a particular point, x, the Gaussian function, $K(x) = exp(-distance(x,y)^2/(2\sigma^2))$, is often used as a kernel function. (σ is a parameter which governs how quickly the influence of point drops off.) Figure 13 (left) shows how a Gaussian function would look for a single two-dimensional point, while the rightmost part of Fig. 13 shows the overall density function produced by the Gaussian influence functions of the set of points shown in the middle part of Fig. 13.

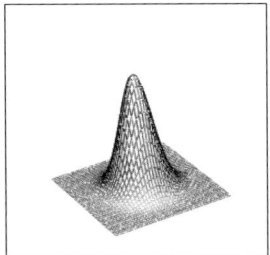 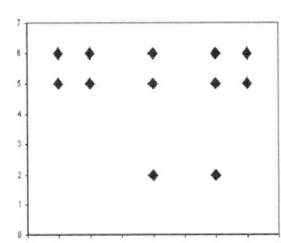 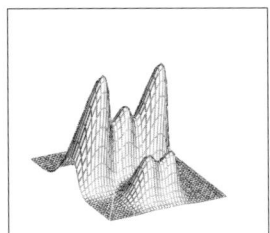

Fig. 13. Example of the Gaussian influence (kernel) function (left) and an overall density function (right) for the data set shown in the middle ($\sigma = 0.75$)

The DENCLUE algorithm has two steps, a preprocessing step and a clustering step. In the pre-clustering step, a grid for the data is created by dividing the minimal bounding hyper-rectangle into d-dimensional hyper-rectangles with edge length 2σ. The grid cells that contain points are then determined. (As mentioned earlier, only the occupied grid cells need be constructed.) The grid cells are numbered with respect to a particular origin (at one edge of the bounding hyper-rectangle and these keys are stored in a search tree to provide efficient access in later processing. For each stored grid cell, the number of points, the sum of the points in the cell, and connections to neighboring cells are also stored.

For the clustering step DENCLUE, considers only the highly populated grid cells and the cells that are connected to them. For each point, x, the local density function is calculated only by considering those points that are from grid cells that are "close" to the point. As mentioned above, DENCLUDE discards clusters associated with a density attractor whose density is less than ξ. Finally, DENCLUE merges density attractors that can be joined by a path of points, all of which have a density greater than ξ.

DENCLUE can be parameterized so that it behaves much like DBSCAN, but it is much more efficient that DBSCAN. DENCLUE can also behave like K-means by choosing σ appropriately and by omitting the step that merges center-defined clusters into arbitrary shaped clusters. Furthermore, by performing repeated clusterings for different values of σ, a hierarchical clustering can be obtained.

OptiGrid

Despite the appealing characteristics of DENCLUE in low dimensional space, it does not work well as the dimensionality increase or if noise is present. Thus, the same researchers who created DENCLUE created OptiGrid. [20]. In this paper, the authors also make a number of interesting observations about the behavior of points in high dimensional space. First, they observe that for high dimensional data noise seems to correspond to uniformly distributed data in that it tends to produce data where there is only one point in a grid cell. More "centralized" distributions, like the Gaussian distribution result in far more cases where a grid cell has more than one point. Thus, the statistics of how many cells are multiply occupied can give us an idea of the amount of noise in the data. Also, the authors provide additional comments on the observation that interpoint distances become relatively uniform as dimensionality increases. In particular, they point out that this means that the maximum density of a group of points may occur in a region of relatively empty space, a phenomenon known as the "empty point phenomenon".

A fair amount of theoretical justification is presented in [20], but we will simplify our description. First, this will make the general approach easier to understand, since this simplification will be more in line with the description of the algorithms given above. Secondly, the algorithm actually implemented used the simplified approach.

1. For each dimension:
   a) Generate a histogram of the data values. Note that this is equivalent to counting the points in a uniform one-dimensional grid (or set of intervals) imposed on the values.
   b) Determine the noise level. This can be done by manually inspecting the histogram, if the dimensionality is not too high, but otherwise needs to be automated. For the results presented in the paper, the authors choose the manual approach.
   c) Find the leftmost and rightmost maxima and the $q-1$ maxima in between them. (q is the number of partitions of the data that we seek, and all these partitions could be in one dimension.)
   d) Choose the q minima between the maxima found in the previous step. These points represent locations for possible cuts, i.e., locations where a hyperplane could be placed to partition the data. Choosing a low-density cell minimizes the chance of cutting through a cluster. However, it is not useful to cut at the edge of the data, and that is the reason for not choosing a minima at the edge, i.e., further right than the rightmost maxima or further left than the leftmost maxima.
   e) Score each potential cut, e.g., by its density.
2. From all of the dimensions, select the best q cuts, i.e., the lowest density cuts.
3. Using these cuts, create a grid that partitions the data.
4. Find the highly populated grid cells and add them to the list of clusters.
5. Refine the list of clusters.
6. Repeat steps 1-5 using each cluster.

The key simplification that we made in the description and that was made in the implementation in the paper was that the separating hyperplanes must be parallel to some axis. To allow otherwise introduces additional time and coding complexity. The authors also show that using rectangular grids does not result in too much error, particularly as dimensionality increases.

In summary, OptiGrid seems a lot like MAFIA in that it creates a grid by using a data dependent partitioning. However, unlike MAFIA and CLIQUE, it does not face the problem of combinatorial search for the best subspace to use for partitioning. OptiGrid simply looks for the best cutting planes and creates a grid that is not likely to cut any clusters. It then locates potential clusters among this set of grid cells and further partitions them if possible. From an efficiency point, this is much better.

However, some details of the implementation of OptiGrid were vague, and there are a number of choices for parameters, e.g., how many cuts should be made. While OptiGrid seems promising, it should be remarked that another clustering approach, PDDP [34], clusters data by making one optimal hyperplane cut at a time. (This approach is more computationally expensive than Optigrid.) One might think that such an approach would be able to match the best behavior of OptiGrid, but it has been shown that this method does not perform much better than a K-means approach. (Actually a combined approach is suggested in [34].) Thus, more evaluation is needed.

4.3 Noise Modeling in Wavelet Space

WaveCluster
WaveCluster [36] is a clustering technique that interprets the original data as a two-dimensional signal and then applies signal processing techniques (the wavelet transform) to map the original data to a new space where cluster identification is more straightforward. More specifically, WaveCluster defines a uniform two-dimensional grid on the data and represents the points in each grid cell by the number of points. Thus, a collection of two-dimensional data points becomes an image, i.e., a set of "gray-scale" pixels, and the problem of finding clusters becomes one of image segmentation.

While there are a number of techniques for image segmentation, wavelets have a couple of features that make them an attractive choice. First, the wavelet approach naturally allows for a multiscale analysis, i.e., the wavelet transform allows features, and hence, clusters, to be detected at different scales, e.g., fine, medium, and coarse. Secondly, the wavelet transform naturally lends itself to noise elimination.

The basic algorithm of WaveCluster is as follows:

1. *Create a grid and assign each data object to a cell in the grid.* The grid is uniform, but the grid size will vary for different scales of analysis. Each grid cell keeps track of the statistical properties of the points in that cell, but for wave clustering, only the number of points in the cell is used.
2. *Transform the data to a new space by applying the wavelet transform.* This results in 4 "subimages" at several different levels of resolution, an "average" image, an image that emphasizes the horizontal features, an image that emphasizes vertical features, and an image that emphasizes corners.
3. *Find the connected components in the transformed space.* The average subimage is used to find connected clusters, which are just groups of connected "pixels", i.e., pixels which are connected to one another horizontally, vertically, or diagonally.
4. *Map the cluster labels of points in the transformed space back to points in the original space.* WaveCluster creates a lookup table that associates each point in the original with a point in the transformed space. Assignment of cluster labels to the original points is then straightforward.

In summary, the key features of WaveCluster are order independence, no need to specify a number of clusters (although it is helpful to know this in order to figure out the right scale to look at, speed (linear), the elimination of noise and outliers, and the ability to find arbitrarily shaped clusters. While the WaveCluster approach can theoretically be extended to more than two dimensions, it seems unlikely that WaveCluster will work well (efficiently and effectively) for medium or high dimensions.

Overcoming the Curse of Dimensionality via the Wavelet Transform. The technique given in [31] provides an approach for converting almost any kind of data to a gridded framework where a wavelet transform can be applied. The key idea is to treat the data matrix as an image matrix. A data matrix is a two dimensional array and so is an image matrix, and so, superficially, this is workable. However, the order of the rows and columns in a data matrix is arbitrary, i.e., they can be shuffled without changing the meaning of the data, while in an image the order is critical because of the spatial (sequential) relationship implied. Meaningful application of the wavelet transform depends on this spatial ordering, and thus, to treat a data array as an image requires the imposition of a meaningful order relationship on the rows (objects) and columns (variables) of the data matrix. This is accomplished by the use of matrix reordering techniques to permute the rows and columns to a standard form, which gathers larger or non-zero values towards the diagonal.

Once the matrix has been reordered, the data matrix is analyzed as if it were an image. In particular, the wavelet coefficients for each point are calculated for a variety of scales, e.g., 5 scales which differ by a factor of two. Thus, the original image is decomposed into 6 images (the image at 5 resolutions and a residual image.) Since most data has a lot of noise, statistical tests, which are based on an assumed statistical model for the noise in the data, can be applied to these wavelet coefficients to determine which ones are significant in a statistical sense. By setting all significant wavelet coefficients to 0, and each non-significant coefficient to 0, a binarized view of the data at each level of resolution can be obtained. By looking at either the binarized view or the original wavelet transformed view at the different levels, it is often possible to visually identify various clusters for further investigation.

Of course, the matrix reordering is an approximate process and may not always give exactly the same reordering from one run to the next. However, the authors indicate that this method is intended for quick exploratory clustering and show that it works reasonably well for some examples that they present.

4.4 A "Concept-Based" Approach to Clustering High Dimensional Data

A key feature of some high dimensional data is that two objects may be highly similar even though commonly applied distance or similarity measures indicate that they are dissimilar or perhaps only moderately similar [17]. Conversely, and perhaps more surprisingly, it is also possible that an object's nearest or most similar neighbors may not be as highly "related" to the object as other objects which are less similar. To deal with this issue we have extended previous approaches that define the distance or similarity of objects in terms of the number of nearest neighbors that they share. The resulting approach defines similarity not in terms of shared attributes, but rather in terms of a more general notion of shared concepts. The rest of this section details our work in finding clusters in these "concept spaces", and in doing so, provides a contrast to the approaches of the previous section, which were oriented to finding clusters in more traditional vector spaces.

Concept Spaces

For our purposes, a concept will be a set of attributes. As an example, for documents a concept would be a set of words that characterize a theme or topic such as "Art" or "Finance". The importance of concepts is that, for many data sets, the objects in the data set can be viewed as being generated from one or more sets of concepts in a probabilistic way. Thus, a concept-oriented approach to documents would view each document as consisting of words that come from one or more concepts, i.e., sets of words or vocabularies, with the probability of each word being determined by an underlying statistical model. We refer to data sets with this sort of structure as concept spaces, even though the underlying data may be represented as points in a vector space or in some other format. The practical relevance of concept spaces is that data belonging to concept spaces must be treated differently in terms of how the similarity between points should be calculated and how the objects should be clustered.

To make this more concrete we detail a concept-based model for documents. Figure 14a shows the simplest model, which we call the "pure concepts" model. In this model, the words from a document in the ith class, C_i, of documents come from either the general vocabulary, V_0, or from exactly one of the specialized vocabularies, $V_1, V_2, \cdots, V_p$. For this model the vocabularies are just sets of words and possess no additional structure. In this case, as in the remaining cases discussed, all vocabularies can overlap. Intuitively, however, a specialized word that is found in a document is more likely to have originated from a specialized vocabulary than from the general vocabulary.

Figure 14b is much like Fig. 14a and shows a slightly more complicated (realistic) model, which we call the "multiple concepts" model. The only

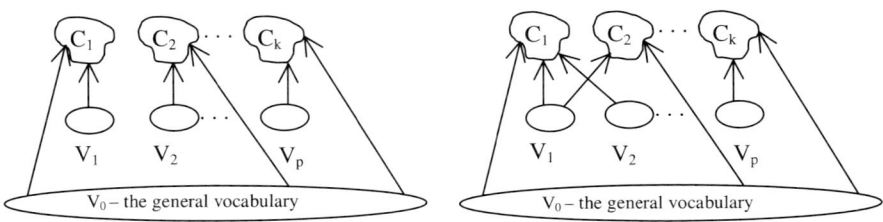

Fig. 14. Different concept models: Pure concepts (left) vs. Complicated concepts (right)

difference from the previous model is that a word in a document from a particular class may come from more than one specialized vocabulary. More complicated models are also possible.

A statistical model for the concept-based models shown above could be the following. A word, w, in a document, d, from a cluster C_i, comes with one or more vocabularies with a probability given by $P(w|C_i) = P(w|V_j) * P(V_j|C_i)$. For the pure concepts model, each word of a document comes only from the general vocabulary and one of the specialized vocabularies. For the multiple concepts model, each word of a document comes from one or more specialized vocabularies.

Problems with Similarity in Concept Spaces

In the beginning of this section, it was mentioned that similarity measures might behave in unexpected ways in concept spaces. We present some examples and discussion to indicate why this is so.

In the following we are assuming that the variables are what are sometimes called "unary" variables, i.e., it makes sense to say that an object has that attribute or doesn't have that attribute. For example, a document may or may not contain a certain word, or a customer may or may not purchase a certain item. Counts, categorical attributes, or binary attributes can be easily translated into unary attributes, but the situation is more complicated with most continuous attributes. We omit discussion of such cases to keep the explanations simple.

Our first example is similar to one in [17]. Consider a concept space where all the objects fall into two groups, A and B. Objects from group A are generated by selecting three of the attributes (with equal probability) from the concept set $\{1, 2, 3, 4, 5\}$ and objects from group B are generated by selecting three of the attributes from the concept set $\{4, 5, 6, 7, 8\}$. Suppose that we have generated the following three objects $\mathbf{x} = \{1, 2, 3\}, \mathbf{y} = \{3, 4, 5\}$, and $\mathbf{z} = \{4, 5, 6\}$. (We can also represent these points as binary vectors, e.g., $\mathbf{x} = (1\ 1\ 1\ 0\ 0\ 0\ 0\ 0)$.) Clearly, points $\mathbf{x}$ and $\mathbf{y}$ belong to group A, while point $\mathbf{z}$ belongs to group B. However, just as clearly, most similarity measures, e.g., the Jaccard measure, would judge points $\mathbf{y}$ and $\mathbf{z}$ to be most similar, as they share two out of their three attributes, while $\mathbf{x}$ and $\mathbf{y}$ share only one attribute.

The Need for Indirect Similarity in Concept Spaces

If we carefully examine document sets, we observe that the average similarity between documents within a cluster (using the popular cosine measure) is almost always lower than 0.6, and it generally lies between 0.2 and 0.5. This means that, on the average, two documents in the same cluster share about 20%–50% of their terms (assuming binary attributes). If a document's similarity with its nearest neighbor is 0.3, then we should not put the two documents in the same cluster right away. We should notice that the similarity between the two is actually low. Consider the set of documents in Table 3.

Table 3. Sample set of document.

A	1	1	1	0	0	0	0	0	0	0	0
B	0	1	1	1	0	0	0	0	0	0	0
C	1	1	0	1	1	1	1	1	0	0	0
D	0	0	0	1	1	1	1	1	0	1	1
E	0	0	0	0	0	0	0	1	1	1	0
F	0	0	0	0	0	0	0	0	1	1	1

The most similar two documents are C and D, but the appropriate clusters for this set are A, B, C and D, E, F. In both of the clusters, every document shares two attributes with any other document. The first four attributes bind A, B and C together, while the last four bind D, E and F together.

A document cluster should contain documents that form a topic, and this does not imply placing the closest neighbor of a document in the same cluster as we have seen in the previous example. If we look at the indirect similarities; number of length 2 links between documents, we will see that C and D have only one indirect link while A-B, A-C and B-C will all have two indirect links. Hence, A-B-C and D-E-F form coherent clusters.

For a more realistic example, consider actual similarity measures for documents. Documents are represented using the vector-space model [33], where each document, d, is considered to be a vector, $\mathbf{d}$, in the term-space (set of document "words"). In its simplest form, each document is represented by the (TF) vector,

$$\mathbf{d}_{tf} = (tf_1, tf_2, \cdots, tf_n), \tag{5}$$

where tf_i is the frequency of the ith term in the document. (Normally very common words are stripped out completely and different forms of a word are reduced to one canonical form.) In addition, we use the version of this model that weights each term based on its inverse document frequency (IDF) in the document collection. (This discounts frequent words with little discriminating power.) Finally, in order to account for documents of different lengths, each document vector is normalized so that it is of unit length.

There are a number of possible measures for computing the similarity between documents, but the most common one is the cosine measure, which is defined as

$$\text{cosine}\,(\mathbf{d}_1, \mathbf{d}_2) = (\mathbf{d}_1 \bullet \mathbf{d}_2)/\parallel \mathbf{d}_1 \parallel \parallel \mathbf{d}_2 \parallel, \tag{6}$$

where $\bullet$ indicates the vector dot product and $\parallel \mathbf{d} \parallel$ is the length of vector $\mathbf{d}$. Notice that this measure is similar to the Jaccard measure in that it only considers the presence of terms to be important.

As mentioned above, what distinguishes documents of different classes is the frequency with which words are used. In particular, each class typically has a "core" vocabulary of words that are used more frequently. For example, documents about finance will often talk about money, mortgages, trade, etc., while documents about sports talk about players, coaches, games, etc. These core vocabularies may overlap, documents may use more than one "core" vocabulary, and any particular document may contain words from these different "core" vocabularies, even if it does not belong to the class of documents that typically uses such words.

Each document has only a subset of all words from the complete vocabulary. Thus, because of the probabilistic nature of how words are distributed, any two documents may share many of the same words. Thus, it should not be surprising that two documents can often be nearest neighbors without belonging to the same class. For a variety of document datasets (see [37]). Figure 15 shows the percentage of documents whose nearest neighbor is not of the same class. (Classes were pre-assigned, for example, by using the section of the newspaper in which a document occurred.)

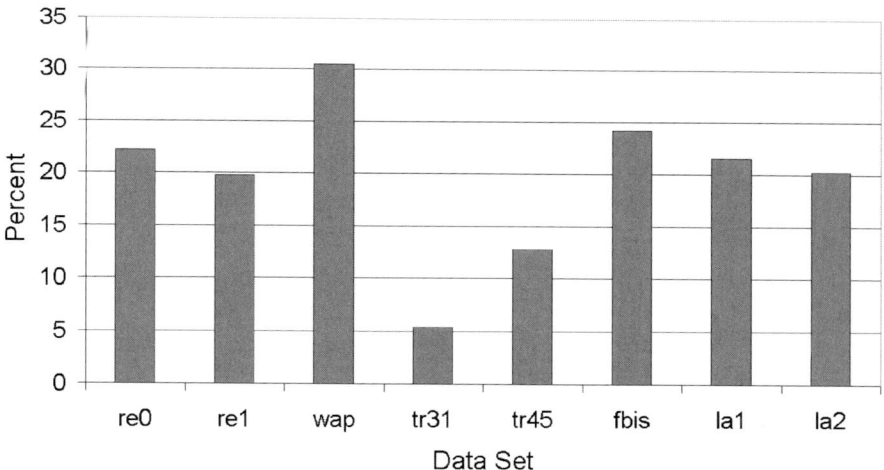

Fig. 15. Percent nearest neighbors of a different class

Since hierarchical and K-means clustering, which are often used for document clustering, use the cosine measure to decide how to cluster documents, they will inevitably make mistakes. In particular, agglomerative hierarchal clustering will often put documents of the same class in the same cluster at the earliest stages of the clustering process. Because of the way that hierarchical clustering works, these "mistakes" cannot be fixed once they happen. K-means can potentially do better, because it continually reassigns documents to the most appropriate cluster as the clustering proceeds. However, K-means is still based on a definition of similarity that is suspect, and we have observed that clusters produced by K-means often contain documents that don't have a consistent topic.

In cases where nearest neighbors are unreliable, a different approach is needed that relies on more global properties. We discuss a general approach based on nearest neighbors, and then discuss or own approach.

A Shared Nearest Neighbor Approach to Similarity

Our clustering algorithm is based on a shared nearest neighbor clustering algorithm described in [25]. A similar approach, but for hierarchical clustering, was developed in [16]. Recently, a couple of other clustering algorithms have used shared nearest neighbor ideas [17,27].

We explain the approach of [25], which we call Jarvis-Patrick clustering, in more detail since it is the basis for our clustering technique. We will describe the shared nearest neighbor algorithm in [25] using graph terminology. (Recall that from a graph point of view, clustering is equivalent to breaking the graph into connected components, one for each cluster.)

1. First the n nearest neighbors of all points are found. In graph terms this can be regarded as breaking all but the n strongest links from a point to other points in the proximity graph. This forms what we call a "nearest neighbor graph". Note that the nearest neighbor graph is just a sparsified version of the original similarity graph, where we break the links to less similar points.
2. We then determine the number of nearest neighbors shared by any two points. In graph terminology we form what we call the "shared nearest neighbor" graph. We do this by replacing each link (in the nearest neighbor graph) between two points by the number of neighbors that the points share. In other words [17], this is the number of length 2 paths between any two points in the nearest neighbor graph. In Fig. 16 the links between nodes (documents) indicate that they are similar (direct similarity). The numbers show the strength of the link in the shared nearest neighbor graph.
3. All pairs of points are compared and if any two points share more than T neighbors, i.e., have a link in the shared nearest neighbor graph with a weight more than our threshold value, T ($T \leq n$), then the two points and any cluster they are part of are merged. In other words, clusters

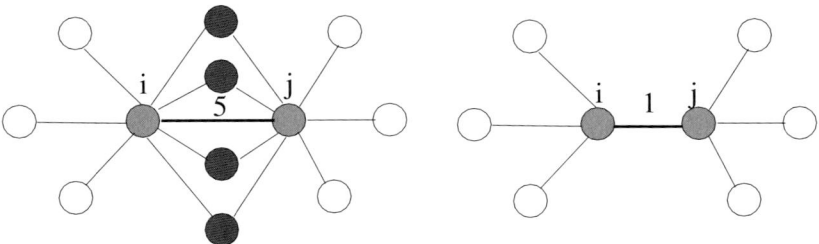

Fig. 16. Illustration of the ways points can share neighbors

are connected components in the shared nearest neighbor graph after we sparsify using a threshold.

This approach has a number of nice properties. It can handle clusters of different densities since the shared nearest neighbor approach is self-scaling. Also, this approach is transitive, i.e., if point, p, shares lots of nearest neighbors with point, q, which in turn shares lots of nearest neighbors with point, r, then points p, q and r all belong to the same cluster. The transitive property, in turn, allows this technique to handle clusters of different sizes and shapes. As described in the next sections, we have extended the Jarvis-Patrick approach.

Our Clustering Approach
We begin by calculating the document similarity matrix, i.e., the matrix which gives the cosine similarity for each pair of documents. Once this similarity matrix is calculated, we find the first n nearest neighbors for each document. (Every object is considered to be its own 0th neighbor.) In the nearest neighbor graph, there is a link from object i to object j, if i and j both have each other in their nearest neighbor list. In the shared nearest neighbor graph, there is a link from i to j if there is a link from i to j in the nearest neighbor graph and the strength of this link is equal to the number of shared nearest neighbors of i and j.

At this point, we could just apply a threshold, and take all the connected components of the shared nearest neighbor graph as our final clusters [25]. However, this threshold would need to be set too high since this is a single link approach and will give poor results when patterns in the dataset are not very significant. On the other hand, when a high threshold is applied, a natural cluster will be split into many small clusters due to the variations in the similarity within the cluster. We address these problems with the clustering algorithm described below.

There are two types of parameters used in this algorithm: one type relates to the strength of the links in the shared nearest neighbor graph, the other type relates to the number of strong links for an object (see Fig. 17). If the strength of a link is greater than a threshold, then that link is labeled as a strong link.

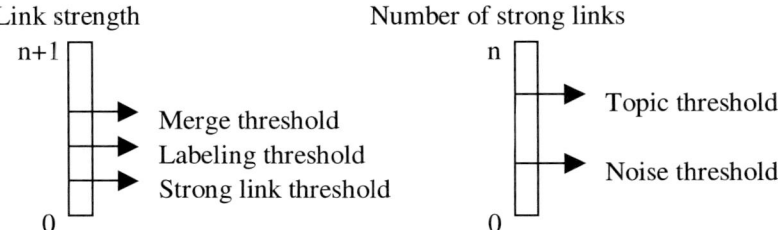

Fig. 17. Different types of parameters

The details of our shared nearest neighbor clustering algorithm are as follows:

1. For every point i in the dataset, calculate the connectivity, conn[i], the number of strong links the point has.
2. For a point i in the dataset, if conn[i] < *noise threshold*, then that point is not considered in the clustering since it is similar to only a few of its neighbors. Similarly, if conn[i] > *topic threshold*, then that point is similar to most of its neighbors and is chosen to represent its neighborhood.
3. For any pair of points (i, j) in the dataset, if i and j share significant numbers of their neighbors, i.e. the strength of the link between i and j is greater than the merge threshold, then they will appear together in the final clustering if either one of them (or both) is chosen to be a representative. Our algorithm will not suffer from the effects of transitivity since every other point on a chain of links has to be chosen to be a representative. In other words, two objects that are not directly related will be put in the same cluster only if there are many other objects between them that are connected with strong links, half of which must represent their own neighborhood.
4. Labeling step: Having defined the representative points and the points strongly related to them, we can bring back some of the points that did not survive the merge threshold. This is done by scanning the shared nearest neighbor list of all the points that are part of a cluster, and checking whether those points have links to points that don't belong to any cluster and have a link strength greater than the labeling threshold.

After applying the algorithm described above, there may be singleton clusters. These singleton clusters are not equivalent to the singleton clusters obtained using the JP method. Note that if only a threshold is applied after converting the nearest neighbor graph to the shared nearest neighbor graph, there will be several clusters (which are the connected components after applying the threshold), and the rest will be singletons. By introducing the topic threshold, we are able to mark the documents that have similar documents around. In the end, if a document that is labeled as a topic remains as a singleton, this does not mean that it is a noise document. For that document to be labeled as a topic, it must have enough number of strong links, which means that it has many similar neighbors but the strength of those links were not strong enough to merge them.

Singleton clusters give us some idea about the less dominant topics in the dataset, and they are far more valuable than the singletons that are left out (labeled as background). To the best of our knowledge, there is no other algorithm that produces valuable singleton (or very small) clusters. Being able to make use of the singleton clusters can be very useful. If we're trying to detect topics in a document set, we don't have to force the parameters of the algorithms to the edge to find out small topics. If we end up getting a singleton cluster, that document will give us an idea about several other documents, whereas noise documents do not give us any idea about any other document.

The method described above finds communities of objects, where an object in a community shares a certain fraction of its neighbors with at least some number of neighbors. While the probability of an object belonging to a class different from its nearest neighbor's class may be relatively high, this probability decreases as the two objects share more and more neighbors. This is the main idea behind the algorithm.

An Application of Concept-Based Clustering to Documents
We illustrate concept based clustering by considering clustering for documents. Given a set of documents, clustering is often used to group the documents, in the hope that each such group will represent documents with a common theme or topic (concept). Initially hierarchical clustering was used to cluster documents [14]. This approach has the advantage of producing a set of nested document clusters, which can be interpreted as a topic hierarchy or tree, from general to more specific topics. In practice, while the clusters at different levels of the hierarchy sometimes represent documents with consistent concepts or topics, it is common for many clusters to be a mixture of topics, even at lower, more refined levels of the hierarchy. More recently, as document collections have grown larger, K-means clustering has emerged as a more efficient approach to producing clusters of documents [10,26,37]. K-means clustering produces a set of un-nested clusters, and the top (most frequent or highest "weight") terms of the cluster are used to characterize the topic of the cluster. Once again it is not unusual for some clusters to be mixtures of topics.

By applying our algorithm for clustering concept-based data to documents, we have created an approach that more consistently produces clusters of documents with strong, coherent themes (concepts), even though many documents may be omitted in the process. After all, in an arbitrary collection of documents, e.g., a set of newspaper articles, there is no reason to expect that all documents belong to a group with a strong topic or theme. While a concept-based approach does not provide a complete organization of all documents, it does identify the "nuggets" of information in a document collection and might profitably be applied to practical problems such as grouping the search results of a Web search engine.

Sample Results for Concept-Based Clustering of Documents
We applied our technique to the data set LA1, which is from the Los Angeles Times data of TREC-5. (See [13] for more details.) The words in Table 4 are the most important (frequent) 6 words in each document cluster. In Table 4 we see that all the documents in the first cluster are related to NCAA, while all the documents in the second cluster are related to NBA. Even though both sets of documents are basketball related, our clustering algorithm found them

Table 4. Six most important words in document cluster.

The NCAA cluster					
wolfpack	towson	lead	tech	Scor	North
syracus	scor	georgia	dome	auburn	Louisvill
Scor	lead	throw	half	Free	Iowa
Scor	Fresno	unlv	lead	lockhart	jacksonvil
Panther	pittsburgh	sooner	brookin	Scor	Game
Iowa	minnesota	scor	illinoi	wisconsin	Burton
Scor	half	virginia	georgetown	lead	Kansa
Burson	louisvill	scor	ohio	game	Ellison

The NBA cluster					
Pacer	scor	piston	shot	game	hawkin
Cavali	mckei	charlott	scor	superson	cleveland
Scor	game	tripucka	basket	hornet	straight
levingston	hawk	jordan	malon	buck	quarter
daugherti	piston	warrior	cavali	shot	Eject

as separate clusters. We ran the K-means algorithm on the same dataset, and interestingly, all of the documents in these two clusters appeared in the same K-means cluster together with a number of documents related to gymnastics, swimming, as well as several apparently unrelated documents. The reason that K-means put all these sports documents in the same cluster is that sports documents tend to share a lot of common words such as, 'score', 'half', 'quarter', 'game', 'ball', etc. This example indicates that pair-wise similarity by itself isn't a good measure for clustering documents.

Thus, using an approach based on shared nearest neighbors (SNN), we can get purer clusters, although not all the documents are assigned to clusters. However, in order to make a fair comparison, we decided to remove from K-means clusters all documents that were far away from the centroid of their cluster. We observed that this improved the misclassification rate only slightly. Finally, we also noticed [13], when we looked at the individual documents in a 'supposedly poor' SNN cluster, that the documents did form a coherent group even though they have different class labels.

Some Final Comments on Concept Based Clustering
While we have restricted our discussion here to concept based clustering for documents, the shared nearest neighbor approach to similarity on which it is based can be applied to many different sorts of data. In particular, the shared nearest neighbor approach from which concept-based is derived, was originally used for two-dimensional spatial data, and we have also successfully applied our data to such data. A major task ahead of us is to more precisely define those situations in which is it applicable.

5 Conclusions

In this paper we have provided a brief introduction to cluster analysis with an emphasis on the challenge of clustering high dimensional data. The principal challenge in extending cluster analysis to high dimensional data is to overcome the "curse of dimensionality", and we described, in some detail, the way in which high dimensional data is different from low dimensional data, and how these differences might affect the process of cluster analysis. We then described several recent approaches to clustering high dimensional data, including our own work on concept-based clustering. All of these approaches have been successfully applied in a number of areas, although there is a need for more extensive study to compare these different techniques and better understand their strengths and limitations.

In particular, there is no reason to expect that one type of clustering approach will be suitable for all types of data, even all high dimensional data. Statisticians and other data analysts are very cognizant of the need to apply different tools for different types of data, and clustering is no different.

Finally, high dimensional data is only one issue that needs to be considered when performing cluster analysis. In closing we mention some other, only partially resolved, issues in cluster analysis: scalability to large data sets, independence of the order of input, effective means of evaluating the validity of clusters that are produced, easy interpretability of results, an ability to estimate any parameters required by the clustering technique, an ability to function in an incremental manner, and robustness in the presence of different underlying data and cluster characteristics.

References

1. R. Agrawal, J. Gehrke, D. Gunopulos, P. Raghavan: 'Automatic subspace clustering of high-dimensional data for data mining applications', In: *ACM SIGMOD Conference on Management of Data* (ACM Press, New York 1998)
2. R. Agrawal, R. Srikant: 'Fast Algorithms for Mining Association Rules', In: *Proceedings of the* 20th *VLDB Conference,* (Santiago, Chile 1997) pp. 487–499
3. C. Aggarwal, C. Procopiuc, J. Wolf, P. Yu, Jong Park: 'Fast algorithms for projected clustering', In: ACM SIGMOD Conference, (ACM Press, New York 1999)
4. M.R. Anderberg: *Cluster Analysis for Applications* (Academic Press, New York and London 1973)
5. R. Bellman: *Adaptive Control Processes: A Guided Tour,* (Princeton University Press, Princeton 1961)
6. S. Brin: 'Near Neighbor Search in Large Metric Spaces', *Proceedings of the 21st International Conference on Very Large Databases (VLDB-1995),* (Morgan Kaufmann, Los Gatos 1995) pp. 574–584
7. K. Beyer, J. Goldstein, R. Ramakrishnan, U. Shaft: 'When is 'nearest neighbor' meaningful?', In *Proceedings of 7th International Conference on Database Theory (ICDT-1999)* (Jerusalem, Israel 1999) pp. 217–235

8. T.H. Cormen, C.E. Leiserson, R.L. Rivest: *Introduction to Algorithms* (Prentice Hall, Englewood Cliffs 1990)
9. R.O. Duda, P.E. Hart, D.G. Stork: *Pattern Recognition* (Wiley, New York 2000)
10. I.S. Dhillon, D.S. Modha: Machine Learning **42** 143 (2001)
11. D.L. Donoho: 'High Dimensional Data Analysis: The Curses and Blessings of Dimensionality', *American Math. Society Conference: Mathematical Challenges of the 21st Century, Los Angeles, CA, August, 6-11 (2000)*. (Currently only available on the Web at http://www-stat.stanford.edu/ donoho/Lectures/AMS2000/AMS2000.html)
12. M. Ester, H.-P. Kriegel, J. Sander, X. Xu: 'A Density-Based Algorithm for Discovering Clusters in Large Spatial Databases with Noise', In *Proceedings of 2nd International Conference on Knowledge Discovery and Data Mining (KDD 96)* (Portland, Oregon 1996) pp. 226-231
13. L. Ertöz, M. Steinbach, V. Kumar: 'Finding Topics in Collections of Documents: A Shared Nearest Neighbor Approach', In *Proceeding of Text Mining Workshop, First International SIAM Data Mining Conference* (Chicago, IL 2001)
14. A. El-Hamdouchi, P. Willet: The Computer Journal **32** (3) (1989)
15. C. Fraley, A.E. Raferty: 'How Many Clusters? Which Clustering Method? Answers Via Model-Based Cluster Analysis', Technical Report No. 329, Department of Statistics, University of Washington, Seattle, Washington (1998)
16. K.C. Gowda, G. Krishna: Pattern Recognition **10**, 105 (1978)
17. S. Guha, R. Rastogi, K. Shim: 'ROCK: A Robust Clustering Algorithm for Categorical Attributes', In *Proceedings of the 15th International Conference on Data Engineering (ICDE '99)* (1999) pp. 512–521
18. A. Hinneburg, C. Aggarwal, D.A. Keim: 'What is the nearest neighbor in high dimensional spaces?' In *Proceedings 26th International Conference on Very Large Data Bases (VLDB-2000)*, (Morgan Kaufmann, San Francisco 2000) pp. 506–515
19. A. Hinneburg, D.A. Keim: 'An Efficient Approach to Clustering in Large Multimedia Databases with Noise', In *Proceedings of the 4th International Conference on Knowledge Discovery and Data Mining*, (New York 1998) pp. 58-65
20. A. Hinneburg, D.A. Keim: 'Optimal Grid-Clustering: Towards Breaking the Curse of Dimensionality in High-Dimensional Clustering', In *Proceedings of 25th International Conference on Very Large Data Bases (VLDB-1999)*, (Morgan Kaufmann, San Francisco 1999) pp. 506-517
21. E.-H. Han, G. Karypis, V. Kumar, B. Mobasher: 'Clustering In a High-Dimensional Space Using Hypergraph Models', Technical Report TR-97-063, Department of Computer Science, University of Minnesota, Minneapolis, Minnesota (1997)
22. F. Hoppner, F. Klawonn, R. Kruse, T. Runkler: *Fuzzy Cluster Analysis: Methods for Classification, Data Analysis, and Image Recognition* (John Wiley and Sons, New York 1999)
23. A.K. Jain, R.C. Dubes: *Algorithms for Clustering Data* (Prentice Hall, Englewood Cliffs 1988)
24. A.K. Jain, M.N. Murty, P.J. Flynn: ACM Computing Surveys **31** 264 (1999)
25. R.A. Jarvis, E.A. Patrick: IEEE Transactions on Computers, **C-22**, 1025 (1973)
26. G. Karypis, E.-H. Han: 'Concept Indexing: A Fast Dimensionality Reduction Algorithm with Applications to Document Retrieval & Categorization', In: *Ninth International Conference on Information and Knowledge Management (CIKM 2000)* (McLean 2000)

27. G. Karypis, E.-H. Han, V. Kumar: IEEE Computer **32**, 68 (1999)
28. G. Karypis, V. Kumar: 'hMETIS 1.5: A hypergraph partitioning package', Technical report, Department of Computer Science, University of Minnesota (1998)
29. L. Kaufman, P.J. Rousseeuw: *Finding Groups in Data: An Introduction to Cluster Analysis* (John Wiley and Sons, New York 1990)
30. T. Mitchell: *Machine Learning* (McGraw Hill, New York 1997)
31. F. Murtagh, J.-L. Starck, M.W. Berry: The Computer Journal **43**, 107 (2000)
32. H. Nagesh, S. Goil, Alok Choudhary: 'MAFIA: Efficient and Scalable Subspace Clustering for Very Large Data Sets', Technical Report Number CPDC-TR-9906-019, Center for Parallel and Distributed Computing, Northwestern University (1999)
33. C.J. Van Rijsbergen: *Information Retrieval* 2nd Ed. (Butterworth, London 1979)
34. S.M. Savaresi, D.L. Boley: 'On the Performance of Bisecting K-Means and PDDP', In *Proceedings of the First International SIAM Data Mining Conference*, (Chicago, IL 2001)
35. G. Strang: *Linear Algebra and its Applications* third edition (Harcourt Brace Jovanovich, New York 1986)
36. G. Sheikholeslami, S. Chatterjee, Aidong Zhang: 'Wavecluster: A multi-resolution clustering approach for very large spatial databases', In *Proceedings of the 24th VLDB Conference* (1998)
37. M. Steinbach, G. Karypis, V. Kumar: 'A Comparison of Document Clustering Algorithms', In *Proceedings of the Text Mining Workshop for The Sixth ACM SIGKDD International Conference on Knowledge Discovery and Data Mining, (KDD 2000)* (Boston, MA 2000)

Part V

Other Applications

Some Statistical Physics Approaches for Trends and Predictions in Meteorology

Kristinka Ivanova, Marcel Ausloos, Thomas Ackerman, Hampton Shirer, and Eugene Clothiaux

Specific aspects of time series analysis are discussed. They are related to the analysis of atmospheric data that are pertinent to clouds. A brief introduction on some of the most interesting topics of current research on climate/weather predictions is given. Scaling properties of the liquid water path in stratus clouds are analyzed to demonstrate the application of several methods of statistical physics for analyzing data in atmospheric sciences, and more generally in geophysics. The breaking up of a stratus cloud is shown to be related to changes in the type of correlations in the fluctuations of the signal that represents the total vertical amount of liquid water in the stratus cloud. It is demonstrated that the correlations of the liquid water path fluctuations exist indeed in a more complex way than usually known through their multi-affine dependence.

1 Introduction

Earth's climate is determined by complex interactions between sun, oceans, atmosphere, land and biosphere [1,2]. The composition of the atmosphere is particularly important because certain gases, including water vapor, carbon dioxide, etc., absorb heat radiated from the Earth's surface. As the atmosphere warms, it in turn radiates heat back to the surface which increases the earth's mean surface temperature by some 30 K above the value that would occur in the absence of a radiation-trapping atmosphere [1]. Perturbations in the concentration of these radiatively active gases alter the intensity of this effect on the earth's climate.

Climate change, a major concern of everyone, is a focus of current atmospheric research. Understanding the processes and properties that affect atmospheric radiation and, in particular, the influence of clouds and the role of cloud radiative feedback, are issues of scientific interest. This leads to efforts to improve not only models of the earth's climate but also predictions of climate change [3,4], whence weather prediction and climate models.

Lorenz's [5] famous pioneering work on chaotic systems using a simple set of nonlinear differential equations was motivated by considerations of weather prediction. However, predicting the results of complex nonlinear interactions that are taking place in an open system is a difficult task. Yet physicists have only the Navier-Stokes equations [6] at hand for describing fluid motion, in terms of such quantities as mass, pressure, temperature, humidity, velocity, energy exchange, etc. whence for describing the variety of processes that take place in the atmosphere. Since controlled experiments cannot be performed on the climate system, we rely on use of models to identify cause-and-effect relationships. It is also essential to concentrate on predicting the uncertainty in forecast models of weather and climate [7,8].

Modeling the impact of clouds is difficult because of their complex and differing effects on weather and climate. Clouds can reflect incoming sunlight and, therefore, contribute to cooling, but they also absorb infrared radiation leaving the earth and contribute to warming. High cirrus clouds, for example, may have the impact of warming the atmosphere. Low-lying stratus clouds, which are frequently found over oceans, can contribute to cooling. In order to successfully model and predict climate, we must be able to both describe the effect of clouds in the current climate and predict the complex chain of events that might modify the distribution and properties of clouds in an altered climate.

Much attention has been paid recently [9] to the importance of the main substance of the atmosphere and clouds, water in its three forms – vapor, liquid and solid, for buffering the global temperature against reduced or increased solar heating [10]. Owing to its special properties, it is believed that water establishes lower and upper boundaries on how far the temperature can drift from current values.

The role of clouds and water vapor in climate change is not well understood; yet water vapor is the most abundant greenhouse gas and directly affects cloud cover and the propagation of radiant energy. In fact, there may be positive feedback between water vapor and other greenhouse gases. Carbon dioxide and other gases from human activities slightly warm the atmosphere, increasing its ability to hold water vapor. Increased water vapor can amplify the effect of an incremental increase of other greenhouse gases.

Other studies suggest that the heliosphere influences the climate on Earth via global mechanism that affects cloud cover [11,12]. Surprisingly the influence of solar variability is found to be strongest in low clouds (3 km), which points to a microphysical mechanism involving aerosol formation that is enhanced by ionization due to cosmic rays.

Beyond the scientifically sound and highly sophisticated computer models, there is still space for simple approaches, based on standard statistical physics techniques and ideas, in particular based on the scaling hypothesis [13], phase transitions [14] and percolation theory aspects [15]. Analogies can be found between meteorological and other phenomena in social or natural science [16]. However to distinguish cases and patterns due to 'external field' influences or self-organized criticality [17] is not obvious indeed. The coupling between human activities and deterministic physics is hard to model in simple terms.

There have been several reports that long-range power-law correlations can be extracted from apparently stochastic time series in meteorology [18,19] and multi-affine properties [20,21] can be identified related to atmospheric turbulence [22]. The same type of investigations has already appeared and seems promising in atmospheric science. In the following we touch upon a brief review of some statistical physics approaches for testing scaling hypothesis in meteorology and for identifying the self-affine or multi-affine nature of atmospheric quantities. We apply useful numerical statistical techniques on real time data measurements; for illustration we have selected stratus clouds.

Restricting ourselves to cloud physics and fractal geometry ideas, leads to many questions, such as on the perimeter-area relationship of rain and cloud areas [23], the fractal dimension of their shape or ground projection [24] or modeling of fractally homogeneous turbulence [25]. The cloud inner structure, content, temperature, life time and effects on ground level phenomena or features are of constant interest and prone to physical modeling[26]. Recently, we reported about long-range power-law correlations [27,28] and multi-affine properties [29] of stratus cloud liquid water fluctuations.

1.1 Techniques of Time Series Analysis

The variety of systems that apparently display scaling properties ranges from base-pair correlations in DNA and inter-beat intervals of the human heart, to large, spatially extended geophysical processes, such as earthquakes, and signals produced by complex systems, such as financial indices in economics. The current paradigm is that these systems obey "universal" laws due to the underlying nonlinear dynamics and are independent of the microscopic details. Therefore one can consider in meteorology to obtain characteristic quantities using the same modern statistical physics methods as done in all of the other cases. Whence we will focus on several techniques to describe the scaling properties of meteorological time series, like the Fourier power spectrum of the signal [30], detrended fluctuation analysis (DFA) method [31] and its extension local DFA method [27], and multi-affine and singularity analysis [29,32]. One can go beyond these methods using wavelet techniques [33] or Zipf diagrams [34–36]. The Fokker-Planck equation [37] for describing the liquid water path [38], which is studied here below, is also of interest.

2 Experimental Techniques and Data Acquisition

Quantitative observations of the atmosphere are made in many different ways. Experimental/observational techniques to study the atmosphere rely on physical principles. One important type of observational techniques is that of *remote sensing*, which depends on the detection of electromagnetic radiation emitted, scattered or transmitted by the atmosphere. The instruments can be placed at aircrafts, on balloons or on the ground. Remote-sensing techniques can be divided into *passive* and *active* types. In passive remote sensing, the radiation measured is of natural origin, for example the thermal radiation emitted by the atmosphere, or solar radiation transmitted or scattered by the atmosphere. Most space-born remote sensing methods are passive. In active remote sensing, a transmitter, e.g. a radar, is used to direct pulses of radiation into the atmosphere, where they are scattered by atmospheric molecules, aerosols or inhomogeneities in the atmospheric structure. Some of the scattered radiation is then detected by some receiver. Each of these techniques has its advantages and disadvantages. Remote sensing from satellites can give near-global coverage, but can provide only averaged values of the measured quantity over large regions, of order of hundreds of kilometers in horizontal extent and several kilometers in the vertical direction. Satellite instruments are expensive to put into orbit and cannot usually be repaired if they fail. Ground-based radars can provide data with very high vertical resolution (by measuring small differences in the time delays of the return pulses), but only above the radar site.

For illustrative purposes, we will use microwave radiometer data obtained from the Department of Energy (DOE) Atmospheric Radiation Measurement (ARM) program [39] site located at the Southern Great Plains (SGP) central facility [40]. For detailed presentation of other remote sensing techniques the reader can consult Andrews [1] and/or Rees [41].

In this study we focus on stratus cloud data. For comparison the cumulus cloud scale is too small to be represented individually in today's numerical models [42]. Due to their relatively small sizes cumulus clouds produce short time series when remote sensing measurements are applied. Therefore they are not particularly suitable for the techniques that are outlined in this report. However their role in the transport of heat, moisture and momentum must be considered in numerical models.

The data used in this study are the vertical column amounts of cloud liquid water that are retrieved from the radiances, recorded as brightness temperatures, measured with a Radiometrics Model WVR-1100 microwave radiometer at frequencies of 23.8 and 31.4 GHz [43–45]. The microwave radiometer is equipped with a Gaussian-lensed microwave antenna whose small-angle receiving cone is steered with a rotating flat mirror [40]. The microwave radiometer is located at the DOE ARM program SGP central facility and is operated in the vertically pointing mode. In this mode the radiometer makes sequential 1 s radiance measurements in each of the two channels while pointing vertically upward into the atmosphere. After collecting these radiances the radiometer mirror is rotated to view a blackbody reference target. For each of the two channels the radiometer records the radiance from the reference immediately followed by a measurement of a combined radiance from the reference and a calibrated noise diode. This measurement cycle is repeated once every 20 s.

A shorter measurement cycle does not necessarily lead to a larger number of independent samples. For example, clouds at 2 km altitude moving at 10 m s^{-1} take 15 s to advect through a radiometer field-of-view of approximately 5°. Note that the 1 s sky radiance integration time ensures that the retrieved quantities correspond to a specific column of cloud above the instrument, as opposed to some longer time average of the cloud properties in the column above the instrument. The field of view of the microwave radiometer is 5.7° at 23.8 GHz and 4.6° at 31.4 GHz.

Based on a standard model [43,45] (see Appendix), the microwave radiometer measurements at the two frequency channels of 23.8 and 31.4 GHz are used to obtain time series of liquid water path (LWP) that corresponds to the total amount of liquid water within the vertical column of the atmosphere that has been remotely sounded. The error for the liquid water retrieval is estimated to be less than about 0.005 g/cm^2 [45].

The liquid water path (LWP) data $y(t)$ considered in this study are obtained on April 3-5, 1998 and are shown in Fig. 1a.

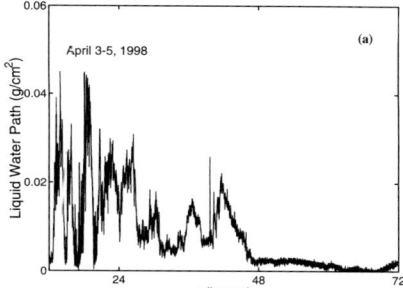
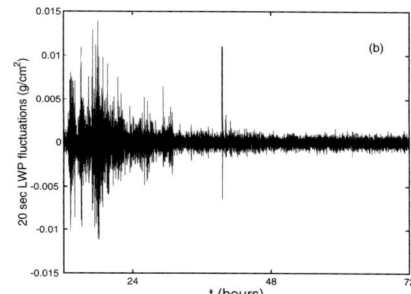

Fig. 1. (a) Time dependence of liquid water path as obtained at the ARM Southern Great Plains site with time resolution of 20 s during the period from April 3 to 5, 1998. The time series contains $N = 10740$ data points. On x-axis t=24 h marks midnight on April 3, t=48 h corresponds to midnight on April 4 and t=72 h corresponds to midnight on April 5, 1998. (b) Small-scale gradient field of the LWP signal, e.g. fluctuations of LWP for a time interval equal to the discretization step of the measurements

3 Nonstationarity and Spectral Density

Fluctuations of the LWP signal $y(t)$ (data in Fig. 1a) are plotted in Fig. 1b for the time interval equal to the discretization step of the data, i.e. $\Delta t = 20$ sec. This time series is also called the small-scale gradient field. Other values of time intervals to study fluctuations of a signal can be of interest to search for changes in the type and strength of the correlations [46]. This approach will not be pursued here.

One approach to test the type of the LWP fluctuations is to estimate the nonstationarity of the signal. The power spectral density $S(f)$ of the time series $y(t)$ is defined as the Fourier transform of the signal. For supposedly self-affine signals $S(f)$ is expected to follow a power-law dependence in terms of the frequency f,

$$S(f) \sim f^{-\beta}. \tag{1}$$

Equation (1) allows one to put the phenomena that produce the time series into the class of *self-affine* phenomena.

It has been argued [47,48] that the spectral exponent β contains information about the degree of stationarity of the signal $y(t)$. Depending on the value of β the time series is called stationary or not; for $\beta < 1$, the signal is statistically invariant by transition in time, thus called stationary. If $\beta > 1$, the signal is nonstationary. In addition, if $\beta < 3$ the increments of the signal form a stationary series, in particular the small-scale gradient field is stationary. Many geophysical fields are nonstationary with stationary increments ($1 < \beta < 3$) over some scaling range. The upper bound of the nonstationary regime is required to keep the field values within their physically accessible range by limiting the amplitude of the large scale fluctuations, which corresponds to a flatter part of the spectrum at low frequencies.

Brownian motion is characterized by $\beta = 2$, and white noise by $\beta = 0$. Indeed the Brownian motion or random walk $z(x)$ is a classical example of a nonstationary process. We know that its variance $< z^2(x) >$ is proportional to x, which proves the nonstationarity in the one-point statistics. However, in the framework of two-point statistics, this result has a different interpretation. The variance of the "increment" $z(x + \xi) - z(x)$ increases linearly

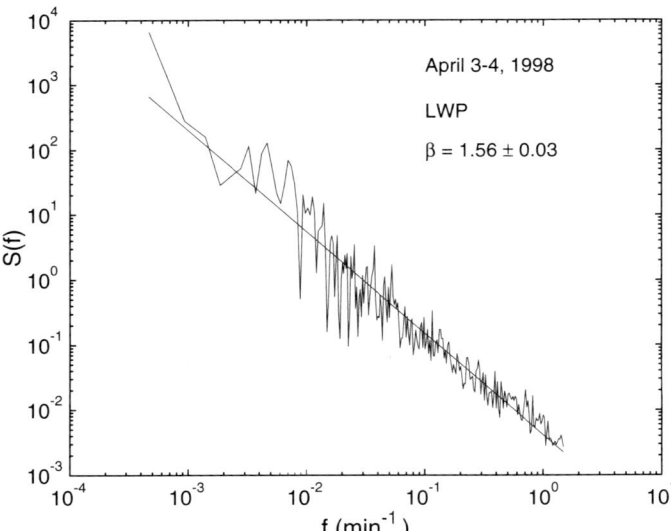

Fig. 2. Power spectral density for data measured on April 3-4, 1998

with ξ, independently of x, which is an indication of the stationarity of the increments.

The range over which the β exponent is well defined in Eq. (1) indicates the range over which the scaling properties of the time series are invariant. The power spectral density $S(f)$ of the liquid water path data measured on April 3-4, 1998 is shown in Fig. 2. The spectral exponent $\beta = 1.56 \pm 0.03$ indicates a nonstationary time series.

4 Roughness and Detrended Fluctuation Analysis

The fractal dimension [13,49–51] D is often used to characterize the roughness of profiles [52]. Several methods are used for measuring D, like the box counting method, though this is not quite efficient; many others are found in the literature as seen in [13,49–51] and here below. For topologically one-dimensional systems, the fractal dimension D is related to the exponent β by

$$\beta = 5 - 2D. \tag{2}$$

Another "measure" of the signal roughness is sometimes given by the Hurst Hu exponent, first defined in the "rescale range theory" (of Hurst [53,54]) who suggested a method to estimate the persistence of the Nile floods and droughts. The Hurst method consists of listing the differences between the observed value at a discrete time t over an interval with size N on which the mean has been taken. The upper (y_M) and lower (y_m) values in that interval define the range $R_N = y_M - y_m$. The root mean square deviation S_N being also calculated, the "rescaled range" is R_N/S_N is expected to behave like N^{Hu}. This means that for a (discrete) self-affine signal $y(t)$, the neighborhood of a particular point on the signal can be rescaled by a factor b using the roughness (or Hurst [49,50]) exponent Hu and defining the new signal $b^{-Hu}y(bt)$. For the exponent value Hu, the frequency dependence of the signal so obtained should be indistinguishable from the original one, i.e. $y(t)$.

The roughness (Hurst) exponent Hu can be calculated from the height-height correlation function $c_1(\tau)$ or first order structure function that supposed to behave like

$$c_1(\tau) = \langle |y(t_{i+r}) - y(t_i)| \rangle_\tau \sim \tau^{H_1} \tag{3}$$

whereas

$$Hu = 1 + H_1, \tag{4}$$

rather than from the box counting method. For a *persistent* signal, $H_1 > 1/2$; for an *anti-persistent* signal, $H_1 < 1/2$. Flandrin has theoretically proved [55] that

$$\beta = 2Hu - 1, \tag{5}$$

thus $\beta = 1+2\ H_1$. This implies that the classical random walk (Brownian motion) is such that $Hu = 3/2$. It is clear that

$$D = 3 - Hu. \tag{6}$$

Fractional Brownian motion values in other fields [56–58] are practically found to lie between 1 and 2. Since a white noise is a truly random process, it can be concluded that $Hu = 1.5$ implies an uncorrelated time series [51].

Thus $D > 1.5$, or $Hu < 1.5$ implies antipersistence and $D < 1.5$, or $Hu > 1.5$ implies persistence. From preimposed Hu values of a fractional Brownian motion series, it is found that the equality here above usually holds true in a very limited range and β only slowly converges toward the value Hu [30,59].

The above exponents and parameters can be obtained within the detrended fluctuation analysis (DFA) method [31]. The DFA method is a tool used for sorting out correlations in a self-affine time series with stationary increments [58,60,61]. It provides a simple quantitative parameter - the scaling exponent α, which is a signature of the correlation properties of the signal. The advantages of DFA over many methods are that it permits detection of long-range correlations embedded in seemingly non-stationary time series, and also that inherent trends are avoided at all time scales. The DFA technique consists in dividing a time series $y(t)$ of length N into N/τ nonoverlapping boxes (called also windows), each containing τ points [31]. The local trend $z(n)$ in each box is defined to be the ordinate of a linear least-square fit of the data points in that box. The detrended fluctuation function $F^2(\tau)$ is then calculated following:

$$F^2(\tau) = \frac{1}{\tau} \sum_{n=k\tau+1}^{(k+1)\tau} [y(n) - z(n)]^2 \qquad k = 0, 1, 2, \ldots, \left(\frac{N}{\tau} - 1\right). \tag{7}$$

Averaging $F^2(\tau)$ over the N/τ intervals gives the mean-square fluctuations

$$\left\langle F^2(\tau) \right\rangle^{1/2} \sim \tau^\alpha. \tag{8}$$

The DFA exponent α is obtained from the power law scaling of the function $\langle F^2(\tau)\rangle^{1/2}$ with τ, and represents the correlation properties of the signal: $\alpha = 1/2$ indicates that the changes in the values of a time series are random and, therefore, uncorrelated with each other. If $\alpha < 1/2$ the signal is anti-persistent (anti-correlated), while $\alpha > 1/2$ indicate positive persistency (correlation) in the signal.

Results of the DFA analysis of liquid water path data measured on April 3-4, 1998 are plotted in Fig. 3a. The DFA function is close to a power law with an exponent $\alpha = 0.34 \pm 0.01$ holding from 3 to 60 minutes. This scaling range is somewhat shorter than the 150 min scaling range we obtained [28] for a stratus cloud during the period Jan. 9-14, 1998 at the ARM SGP site. A crossover to $\alpha = 0.50 \pm 0.01$ is readily seen for longer correlation times [61] to about 2 h, after which the statistics of the DFA function is not reliable. One should note that for cloud data the lower limit of the scaling range is determined by the resolution and discretization steps of the measurements. Since such clouds move at an average speed of ca. 10 m/s and the instrument is always directed toward the same point of the atmosphere, the $20s$ discretization step is chosen to ensure ergodic sampling for an about $5°$ observation angle of the instrument. The upper scaling range limit depends on the cloud life time.

The value of $\alpha \approx 0.3$ can be interpreted as the H_1 parameter of the multifractal analysis of liquid water content [32] and of liquid water path [29]. The existence of a crossover suggests two types of correlated events as in classical fracture processes: (i) On the one hand, the nucleation part and the growth of diluted droplets occur in "more gas-like regions". This process is typically slow and is governed by long range Brownian-like fluctuations; it is expected to follow an Eden model-like [62] growth, with a trivial scaling exponent, as $\alpha = 0.5$ (Fig. 3b); (ii) The faster processes with more Levy-like fluctuations are those which link together various fracturing parts of the cloud, and are necessarily antipersistent as long as the cloud remains thermodynamically stable; they occur at shorter correlation times, and govern the cloud breaking final regime as in any percolation process [14], with an intrinsic non-trivial scaling exponent ~ 0.3.

Several remarks are in order. Recently a rigorous relation between detrended fluctuation analysis and power spectral density analysis for stochastic processes was established [64]. Thus, if the two scaling exponents α and β are well defined, then $\beta = 2\alpha + 1$ holds for $0 < \alpha < 1$ ($1 < \beta < 3$) in the case of fractional Brownian walks [59,63]. This establishes a relation be-

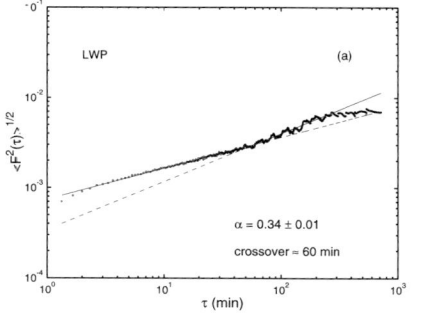
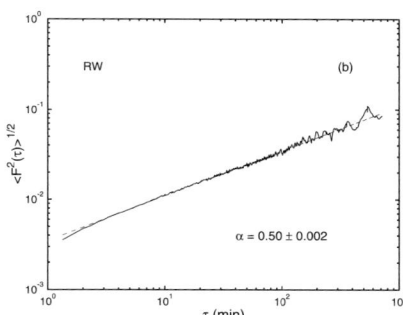

Fig. 3. (a) Detrended fluctuation function $\langle F^2(\tau) \rangle^{1/2}$ for data measured on April 3-4, 1998. (b) DFA-function for Brownian walk signal scales with $\alpha = 0.50 \pm 0.01$ and is plotted for comparison

tween detrended fluctuation analysis and power spectral density analysis for stochastic processes

In terms of the exponents (α and β) of the signal, we can talk about pink noise $\alpha = 0$ ($\beta = 1$), brown noise $\alpha = 1/2$ ($\beta = 2$) or black noise $\alpha > 1/2$ ($\beta > 2$) [13]. Black noise is related to persistence. In contrast, inertial subrange turbulence for which $\beta = 5/3$ gives $\alpha = 1/3$ [65], which places it in the antipersistence regime.

The two scaling exponents α and β for the liquid water path signal are only approximately close to fulfilling the relation $\beta = 2\alpha + 1$. This can be interpreted to be due to the peculiarities of the spectral method [66]. In general, the Fourier transform is inadequate for non-stationary signals. Also it is sensitive to possible trends in the data. There are different techniques suggested to correct these deficiencies of the spectral method [67,68], like detrending the data before taking the Fourier transform. However, this may lead to questions about the accuracy of the spectral exponent [69].

5 Time Dependence of the Correlations

In previous section we study the type of correlations that exist in the liquid water path signal measured during cloudy atmospheric conditions, on April 3-4, 1998. Here we focus on the evolution of these correlations during the same time interval but also continuing on the next day, April 5, when the stratus cloud disappears. In doing so we can further study the influence of the time lag on correlations in the signal.

In order to probe the existence of so-called *locally correlated* and *decorrelated* sequences [58], one can construct a so-called observation box with a certain width, τ, place the box at the beginning of the data, calculate α for the data in that box, move the box by $\Delta\tau$ toward the right along the signal sequence, calculate α in that box, and so on up to the N-th point of the available data. A time dependent α exponent may be expected.

We apply this technique to the liquid water path data signal and the result is shown in Fig. 4. For this illustration we have chosen two window sizes, i.e. 4 h and 6 h, moving the window with a step of $\Delta\tau = 1$ h. Since the value of the *local* α can only be known after all data points are taken into account in a box, the reported value corresponds to that at the upper most time value for that given box in Fig. 4. One clearly observes that the α exponent value does not vary much when the value of τ and $\Delta\tau$ are changed. As could be expected there is more roughness if the box is narrower. The local α exponent value is always significantly below $1/2$. By analogy with financial and biological studies, this is interpreted as a phenomenon related to the *fractional Brownian motion* process mentioned above.

The results from this local DFA analysis applied to LWP data (Fig. 4) indicate two well defined regions of scaling with different values of α. The first region corresponds to the first two days when a thick stratus cloud ex-

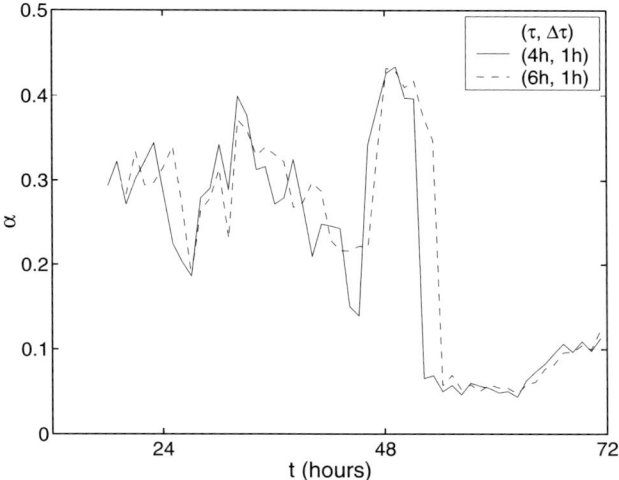

Fig. 4. Local α-exponent from DFA analysis for data in Fig. 1a

isted. The average value of the local scaling exponent over this period is $\alpha = 0.34 \pm 0.01$; it is followed by a sharp rise to 0.5, then by a sharp drop below $\alpha = 0.1$ when there is a clear sky day. These values of local α are well defined for a scaling time (range) interval extending between 2 and 25 minutes for the various τ and $\Delta\tau$ combinations. The value of α, close to 0.3, indicates a very large antipersistence, thus a set of fluctuations tending to induce a great stability of the system and great antipersistence of the prevailing meteorology, in contrast to the case in which there would be a persistence of the system which would be dragged out of equilibrium; it would equally imply good predictability. This implies that specific fluctuation correlation dynamics could be usefully inserted as ingredients in *ad hoc* models.

The appearance of a patch of clouds and clear sky following a period of thick stratus can be interpreted as a non-equilibrium transition. The $\alpha = 1/2$ value in financial fluctuations [58] was observed to indicate a period of relative economic calm. The appropriately called thunderstorm of activities and other bubble explosions in the financial field correspond to a value different from $1/2$ [70]. Thus we emphasize here that stable states can occur for α values that do not correspond to the Brownian $1/2$ value. We conclude that the fluctuation behavior is an observational feature more important than the peak appearance in the raw data. Moreover, from a fundamental point of view, it seems that the variations of α are as important as the value itself [58]. From the point of view of predictability, α values significantly different from $1/2$ are to be preferred because such values imply a great degree of predictability and stability of the system.

6 Multi-affinity and Intermittency

The variations in the local α-exponent suggest that the nature of the correlations changes with time. As a consequence the evolution of the time series can be decomposed into successive persistent and anti-persistent sequences [58], and multi-affine behavior can be expected. Multi-affine properties of a time dependent signal $y(t)$ are described by the so-called "q-th" order structure functions

$$c_q = \langle |y(t_{i+r}) - y(t_i)|^q \rangle \qquad i = 1, 2, \ldots, N - r \tag{9}$$

where the average is taken over all possible pairs of points that are apart from each other a distance $\tau = y(t_{i+r}) - y(t_i)$.

Assuming a power law dependence of the structure function, the $H(q)$ spectrum is defined through the relation [71,72]

$$c_q(\tau) \sim \tau^{qH(q)} \qquad q \geq 0. \tag{10}$$

The *intermittency* of the signal can be studied through the so-called singular measure analysis. The first step that this technique require is defining a basic measure $\varepsilon(1; l)$ as

$$\varepsilon(1; l) = \frac{|\Delta y(1; l)|}{\langle \Delta y(1; l) \rangle}, \qquad l = 0, 1, \ldots, N - 1 \tag{11}$$

where $\Delta y(1; l) = y(t_{i+1}) - y(t_i)$ is the small-scale gradient field and

$$\langle \Delta y(1; l) \rangle = \frac{1}{N} \sum_{l=0}^{N-1} |\Delta y(1; l)|. \tag{12}$$

This is indeed deriving a stationary nonnegative field from a nonstationary data and this is the simplest procedure to do so. Other techniques involve "fractional" derivatives [73] or second derivatives [74]. Also one can consider taking squares [75] rather than the absolute values but that leads to a linear relation between the exponents of these two measures. It is argued elsewhere [76] that the details of the procedure do not influence the final results of the singularity analysis.

We use a spatial/temporal average in Eq. (12) rather than an ensemble average, thus making an ergodicity assumption [77,78] that is our only recourse in empirical data analysis.

Next we define a series of ever more coarse-grained and ever shorter fields $\varepsilon(r; l)$ where $0 < l < N - r$ and $r = 1, 2, 4, \ldots, N = 2^m$. Thus the average measure in the interval $[l; l + r]$ is

$$\varepsilon(r; l) = \frac{1}{r} \sum_{l'=l}^{l+r-1} \varepsilon(1; l') \qquad l = 0, \ldots, N - r. \tag{13}$$

The scaling properties of the generating function are then searched for through the equation

$$\chi_q(\tau) = \langle \varepsilon(r;l)^q \rangle \sim \tau^{-K(q)}, \quad q \geq 0, \tag{14}$$

with $\tau = y(t_{i+r}) - y(t_i)$.

It should be noted that the intermittency of a signal is related to the existence of extreme events, thus a distribution of events away from a Gaussian distribution, in the evolution of the process that has generated the data. If the tails of the distribution function follow a power law, then the scaling exponent defines the critical order value after which the statistical moments of the signal diverge [48]. Therefore it is of interest to probe the distribution of the fluctuations of a time dependent signal $y(t)$ prior investigating its intermittency. The distribution of the fluctuations of liquid water path signal measured on April 3-4, 1998 at the ARM Southern Great Plains site is shown in Fig. 5.

The frequency distribution is not Gaussian but is rather symmetrical. The tails of the distribution follow a power law

$$P(x) \sim \frac{1}{x^\mu} \tag{15}$$

with an exponent $\mu = 2.75 \pm 0.12$ away from the Gaussian $\mu = 2$ value. This scaling law gives support to the argument in favor of the existence of

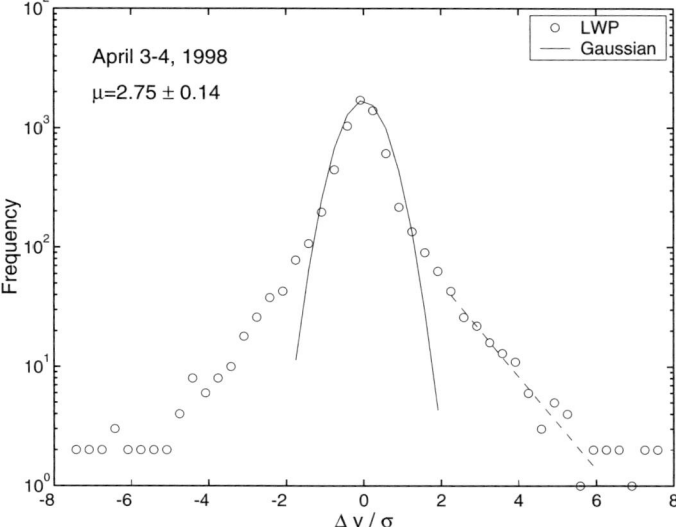

Fig. 5. Distribution of the frequency of LWP fluctuations $\Delta y/\sigma = (y(t_{i+1}) - y(t_i))/\sigma$, where $\sigma = 0.0011 g/cm^2$ is the standard deviation of the fluctuations for LWP signal measured on April 3-4, 1998 (data in Fig. 1a)

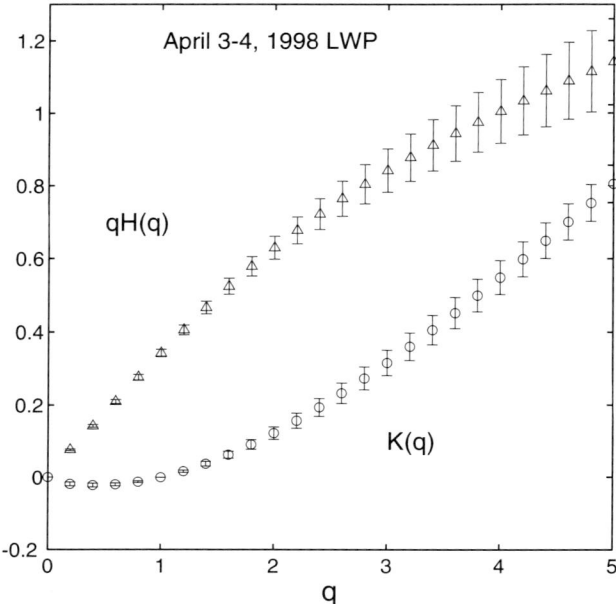

Fig. 6. The $H(q)$ and $K(q)$ functions for the LWP data obtained on April 3-4, 1998

self-affine properties, as established in section 4 for the LWP signal, when applying the DFA method. The extreme events that form the tails of the probability distribution also characterize the intermittency of the signal. In Fig. 6 the multi-fractal properties of the LWP signal are expressed by two sets of scaling functions, the $H(q)$ hierarchy of functions describing the roughness of the signal and the $K(q)$ hierarchy of functions describing its intermittency as defined in Eq.(10) and Eq. (14) respectively. For $q = 1$, $H(1)$ is the value that is given by the DFA analysis.

7 Conclusions

Scaling properties of the liquid water path in stratus clouds have been analyzed to demonstrate the application of several methods of statistical physics for analyzing data in atmospheric sciences, and more generally in geophysics. We have found that the breaking up of a stratus cloud is related to changes in the type of correlations in the fluctuations of the signal, that represents the total vertical amount of liquid water in the stratus cloud. We have demonstrated that the correlations of LWP fluctuations exist indeed in a more complex way than usually known through their multi-affine dependence.

Appendix

For nonprecipitating clouds, i.e., clouds having drops sufficiently small that scattering is negligible, measurements of the microwave radiometer brightness temperature $T_{B\omega}$ can be mapped onto an opacity ν_ω parameter by

$$\nu_\omega = \ln\left[\frac{(T_{\mathrm{mr}} - T_{\mathrm{c}})}{(T_{\mathrm{mr}} - T_{B\omega})}\right], \tag{16}$$

where T_{c} is the cosmic background "big bang" brightness temperature equal to 2.8 K and T_{mr} is an estimated "mean radiating temperature" of the atmosphere.

Writing ν_ω in terms of atmospheric constituents, we have

$$\nu_\omega = \kappa_{V\omega} V + \kappa_{L\omega} L + \nu_{d\omega}, \tag{17}$$

where $\kappa_{V\omega}$ and $\kappa_{L\omega}$ are *water vapor and liquid water path*-averaged mass absorption coefficients and $\nu_{d\omega}$ is the absorption by dry atmosphere constituents (e.g., oxygen). Next, define

$$\nu_\omega^* = \nu_\omega - \nu_{d\omega} = \ln\left[\frac{(T_{\mathrm{mr}} - T_{\mathrm{c}})}{(T_{\mathrm{mr}} - T_{B\omega})}\right] - \nu_{d\omega}. \tag{18}$$

The 23.8 GHz channel is sensitive primarily to water vapor while the 31.4 GHz channel is sensitive primarily to cloud liquid water. Therefore two equations for the opacity can be written for each frequency and then solved for the two unknowns L and V, i.e.

$$L = l_1 \nu_{\omega_1}^* + l_2 \nu_{\omega_2}^* \qquad (LWP) \tag{19}$$

and

$$V = v_1 \nu_{\omega_1}^* + v_2 \nu_{\omega_2}^*, \qquad (WVP) \tag{20}$$

where

$$l_1 = -\left(\kappa_{L\omega_2} \frac{\kappa_{V\omega_1}}{\kappa_{V\omega_2}} - \kappa_{L\omega_1}\right)^{-1}, \tag{21}$$

$$l_2 = \left(\kappa_{L\omega_2} - \kappa_{L\omega_1} \frac{\kappa_{V\omega_2}}{\kappa_{V\omega_1}}\right)^{-1}, \tag{22}$$

$$v_1 = \left(\kappa_{V\omega_1} - \kappa_{V\omega_2} \frac{\kappa_{L\omega_1}}{\kappa_{L\omega_2}}\right)^{-1}, \tag{23}$$

$$v_2 = -\left(\kappa_{V\omega_1} \frac{\kappa_{L\omega_2}}{\kappa_{L\omega_1}} - \kappa_{V\omega_2}\right)^{-1}. \tag{24}$$

Acknowledgements

Thanks to Luc T. Wille for inviting us to present the above results and enticing us into writing this report. This research was partially supported by Battelle grant number 327421-A-N4. We acknowledge collaboration of the U.S. Department of Energy as part of the Atmospheric Radiation Measurement Program.

References

1. D. Andrews: *An Introduction to Atmospheric Physics* (Cambridge University Press, Cambridge 2000)
2. R.A. Anthens, H.A. Panofsky, J.J. Cahir, A. Rango: *The Atmosphere* (Bell & Howell Company, Columbus, OH 1975)
3. R.R. Rogers, M.K. Yau: *A Short Course in Cloud Physics* (Pergamon Press, New York 1976).
4. C.F. Bohren: *Clouds in a Glass of Beer* (John Wiley & Sons, New York 1987)
5. E.N. Lorenz: J. Atmos. Sci. **20**, 130 (1963)
6. L.D. Landau, E.M. Lifshitz: *Fluid Mechanics* (Addison-Wesley, Reading 1959)
7. T.N. Palmer: Phys. Rep. **63**, 71 (2000)
8. S.G. Philander: Phys. Rep. **62**, 123 (1999)
9. A. Maurellis: Physics World **14**, 22 (2001); D. Rosenfeld, W. Woodley: Physics World **14**, 33 (2001)
10. H.-W. Ou: J. Climate **14**, 2976 (2001)
11. N.D. Marsh, H. Svensmark: Phys. Rev. Lett. **85**, 5004 (2000)
12. H. Svensmark: Phys. Rev. Lett. **81**, 5027 (1998)
13. M. Schroeder: *Fractals, Chaos and Power Laws* (W.H. Freeman and Co., New York 1991)
14. H. E. Stanley: *Phase Transitions and Critical Phenomena* (Oxford University Press, Oxford 1971)
15. D. Stauffer, A. Aharony: *Introduction to Percolation Theory*, 2nd printing (Taylor and Francis, London 1992)
16. P. Bak: *How Nature Works* (Springer, New York 1996)
17. D.L. Turcotte: Phys. Rep. **62**, 1377 (1999)
18. E. Koscielny-Bunde, A. Bunde, S. Havlin, H. E. Roman, Y. Goldreich, H.-J. Schellnhuber: Phys. Rev. Lett. **81**, 729 (1998)
19. E. Koscielny-Bunde, A. Bunde, S. Havlin, Y. Goldreich: Physica A **231**, 393 (1993)
20. C.R. Neto, A. Zanandrea, F.M. Ramos, R.R. Rosa, M.J.A. Bolzan, L.D.A. Sa: Physica A **295**, 215 (2001)
21. H.F.C. Velho, R.R. Rosa, F.M. Ramos, R.A. Pielke, C.A. Degrazia, C.R. Neto, A. Zanadrea: Physica A **295**, 219 (2001)
22. H.A. Panofsky, J.A. Dutton: *Atmospheric turbulence* (Wiley, New York 1983)
23. S. Lovejoy: Science **216**, 185 (1982)
24. S. Lovejoy, D. Schertzer: Ann. Geophys. B **4**, 401 (1986)
25. H.G.E. Hentchel, I. Procaccia: Phys. Rev. A **27**, 1266 (1983)
26. K. Nagel, E. Raschke: Physica A **182**, 519 (1992)
27. K. Ivanova, M. Ausloos: Physica A **274**, 349 (1999)

28. K. Ivanova, M. Ausloos, E.E. Clothiaux, T.P. Ackerman: Europhys. Lett. **52**, 40 (2000)
29. K. Ivanova, T. Ackerman: Phys. Rev. E **59**, 2778 (1999)
30. B.D. Malamud, D.L. Turcotte: J. Stat. Plann. Infer. **80**, 173 (1999)
31. C.-K. Peng, S.V. Buldyrev, S. Havlin, M. Simmons, H.E. Stanley, A.L. Goldberger: Phys. Rev. E **49**, 1685 (1994)
32. A. Davis, A. Marshak, W. Wiscombe, R. Cahalan: J. Geophys. Research. **99**, 8055 (1994)
33. N. Decoster, S.G. Roux, A. Arneodo: Eur. Phys. J B **15**, 739 (2000)
34. G.K. Zipf: *Human Behavior and the Principle of Least Effort* (Addisson-Wesley, Cambridge 1949)
35. N. Vandewalle, M. Ausloos: Physica A **268** 240 (1999)
36. M. Ausloos, K. Ivanova: Physica A **270**, 526 (1999)
37. R. Friedrich, J. Peinke, C. Renner: Phys. Rev. Lett. **84**, 5224 (2000)
38. K. Ivanova, M. Ausloos, unpublished
39. G.M. Stokes, S.E. Schwartz: Bull. Am. Meteorol. Soc. **75**, 1201 (1994)
40. http://www.arm.gov
41. W.G. Rees: *Physical Principles of Remote Sensing* (Cambridge University Press, Cambridge 1990)
42. J.R. Garratt: *The Atmospheric Boundary Layer* (Cambridge University Press, Cambridge 1992)
43. E.R. Westwater: Radio Science **13**, 677 (1978)
44. E.R. Westwater: 'Ground-based microwave remote sensing of meteorological variables', in: *Atmospheric Remote Sensing by Microwave Radiometry*, ed. by M.A. Janssen (Wiley, New York 1993) pp. 145-213
45. J.C. Liljegren, B.M. Lesht: 'Measurements of integrated water vapor and cloud liquid water from microwave radiometers at the DOE ARM Cloud and Radiation Testbed in the U.S. Southern Great Plains', in *IEEE Int. Geosci. and Remote Sensing Symp.*, **3**, Lincoln, Nebraska, (1996) pp. 1675-1677
46. R. Friedrich, J. Peinke: Phys. Rev. Lett. **78**, 863 (1997)
47. B.B. Mandelbrot: *The Fractal Geometry of Nature,* (W.H. Freeman, New York 1982)
48. D. Schertzer, S. Lovejoy: J. Geophys. Res. **92**, 9693 (1987)
49. P. S. Addison: *Fractals and Chaos* (Institute of Physics, Bristol 1997)
50. K. J. Falconer: *The Geometry of Fractal Sets* (Cambridge University Press, Cambridge 1985)
51. B. J. West, B. Deering: *The Lure of Modern Science: Fractal Thinking* (World Scientific, Singapore 1995)
52. B.B. Mandelbrot, D.E. Passoja, A.J. Paulay: Nature **308**, 721 (1984)
53. H. E. Hurst: Trans. Amer. Soc. Civ. Engin. **116**, 770 (1951)
54. H. E. Hurst, R.P. Black, Y.M. Simaika: *Long Term Storage* (Constable, London 1965)
55. P. Flandrin IEEE Trans. Inform. Theory **35**, 197 (1989)
56. M. Ausloos, N. Vandewalle, K. Ivanova: 'Time is Money', in *Noise of Frequencies in Oscillators and Dynamics of Algebraic Numbers*, ed. by M. Planat (Springer, Berlin 2000) pp. 156-171.
57. M. Ausloos, N. Vandewalle, Ph. Boveroux, A. Minguet, K. Ivanova: Physica A **274**, 229 (1999)
58. N. Vandewalle, M. Ausloos: Physica A **246**, 454 (1997)

59. D.L. Turcotte: *Fractals and Chaos in Geology and Geophysics* (Cambridge University Press, Cambridge 1997)
60. M. Ausloos, K. Ivanova: Int. J. Mod. Phys. C **12**, 169 (2001)
61. K. Hu, Z. Chen, P.C. Ivanov, P. Carpena, H.E. Stanley: Phys. Rev E **64**, 011114 (2001)
62. R. Jullien, R. Botet: J. Phys. A **18**, 2279 (1985)
63. A.S. Monin, A.M. Yaglom: *Statistical Fluid Mechanics* (MIT Press, Boston 1975) Vol. 2
64. C. Heneghan, G. McDarby: Phys. Rev. E, **62**, 6103 (2000)
65. U. Frisch: *Turbulence: The legacy of A. N. Kolmogorov* (Cambridge University Press Cambridge 1995)
66. P.F. Panter: *Modulation, Noise, and Spectral Analysis* (McGraw-Hill, New York 1965)
67. M.B. Priestley: *Spectral Analysis and Time Series* (Academic Press, London 1981)
68. D.B. Percival, A.T. Walden: *Spectral Analysis for Physical Applications: Multitaper and Conventional Univariate Techniques* (Cambridge University Press, Cambridge 1994)
69. J.D. Pelletier: Phys. Rev. Lett. **78**, 2672 (1997)
70. N. Vandewalle, M. Ausloos: Int. J. Comput. Anticipat. Syst. **1**, 342 (1998)
71. A. Davis, A. Marshak, W. Wiscombe, R. Cahalan: J. Atmos. Sci. **53** 1538 (1996)
72. A. Marshak, A. Davis, W. Wiscombe, R. Cahalan: J. Atmos. Sci. **54**, 1423 (1997)
73. F. Schmitt, D. La Vallee, D. Schertzer, S. Lovejoy: Phys. Rev. Lett. **68**, 305 (1992)
74. Y. Tessier, S. Lovejoy, D. Schertzer: J. Appl. Meteorol. **32**, 223 (1993)
75. C. Meneveau, K.R. Sreenivasan: Phys. Rev. Lett. **59**, 1424 (1987)
76. D. La Vallee, S. Lovejoy, D. Schertzer, P. Ladoy: 'Nonlinear variability, multifractal analysis and simulation of landscape topography', in: *Fractals in Geography*. ed. by L. De Cola, N. Lam (Kluwer, Dordrecht-Boston 1993)
77. A. A. Borovkov: *Ergodicity and Stability of Stochastic Processes* (John Wiley, New York 1998)
78. R. Holley, E.C. Waymire: Ann. J. Appl. Prob. **2**, 819 (1993)

An Initial Look at Acceleration-Modulated Thermal Convection

Jeffrey L. Rogers, Michael F. Schatz, Werner Pesch, and Oliver Brausch

This chapter describes an experimental and theoretical study of pattern formation in acceleration-modulated Rayleigh-Bénard convection. Experimentally this is realized by vertically oscillating a convection cell with a hydraulic shaker. In the cell the fluid's pressure and temperature are controllable. The theoretical analysis relies on a numerical solution of the Oberbeck-Boussinesq equations. This system exhibits rich behavior as visually observed in the experiment and confirmed by the simulations. The onset of the convection is found to be shifted due to the modulation. The harmonic and subharmonic patterns near the onset are studied. Further away from the onset, coexisting patterns are found to exist, while deep into the coexistence region novel complex-ordered patterns are observed over a wide parameter range. These are found to be superlattices which are formed due to a resonance mechanism.

1 Introduction

The formation of patterns in spatially extended nonequilibrium systems is a problem of fundamental interest in a broad set of disciplines including physics, engineering, chemistry, and biology. Two of the most commonly studied systems are drawn from hydrodynamics: a fluid layer with an imposed vertical temperature difference (Rayleigh-Bénard convection [1–3]) and an open dish of vertically oscillated fluid (Faraday surface waves [4]). These systems do share common features. For sufficiently weak driving (thermal and vertical vibrations, respectively) both systems are in macroscopically time-independent and uniform states. As driving is increased regular spatial variations appear at well-defined thresholds and the dynamics become complex in space and time when the driving is sufficiently large. However, Rayleigh-Bénard convection and Faraday waves each represent different classes of spatially extended systems, since basic mechanisms of pattern formation differ between these two cases in important ways.

One important distinction is the mechanism which selects a pattern's length scale. At the onset of fluid motion in Rayleigh-Bénard convection, the pattern will display a wavenumber dependent on the geometrical constraints. In particular, the pattern wavenumber q is directly proportional to the fluid layer depth (Fig. 1). Patterns of the geometry-induced type occur in a number of other systems, for example in the buckling of thin plates [5]. In contrast, Faraday surface waves are driven by vertical vibration, often sinusoidal with a drive period τ. This time-dependent driving leads to wavenumber selection by the forcing frequency via a dispersion relation. Other examples of such dispersion-induced [6] patterns include optical waves in a fiber laser [7] and crystallization waves in ^{4}He [8].

Another important distinction between Rayleigh-Bénard convection and Faraday surface waves is the way system symmetries arise and dictate the pattern planform near onset. Patterns in both systems may exhibit inversion symmetry; the equations governing the pattern mode amplitudes A are invariant under the operation $A \to -A$. This symmetry excludes even order terms in the amplitude equations, thereby, strongly influencing the pattern structure near onset. In Rayleigh-Bénard convection, inversion symmetry takes the form of the Boussinesq symmetry; spatial invariance under vertical reflection about the fluid layer mid plane and is manifested by the stationary stripe patterns typically observed at convective onset. By contrast, inversion symmetry in Faraday waves is temporal, viz., invariance under discrete time translation by τ and is displayed in the subharmonic (periodic at 2τ) waves frequently observed near onset.

Inversion symmetry can be broken in distinct ways in these systems, leading to differences in pattern forming behavior. Breaking inversion symmetry permits quadratic terms in the amplitude equations, leading to pattern selection dominated by three-mode interactions (*resonant triads*) [9]. In convection experiments, the Boussinesq symmetry is typically broken by the spatial dependence of fluid properties in the layer [1,10]. This effect becomes significant when the imposed temperature difference is sufficiently large. The corresponding resonant triad interactions typically lead to the appearance of hexagonal patterns at onset [10,11]. For Faraday waves, inversion symmetry may be broken in a number of different ways [12–14] involving careful choice of the time-dependence of the drive. The character of the resulting resonant triads can depend sensitively on the choice of drive frequencies and therefore, unlike standard Rayleigh-Bénard convection, pattern selection with broken symmetry can be flexibly tuned for Faraday waves.

Recently, pattern formation studies have extended the focus to include the emergence of what Pismen [15] has termed complex-ordered states. These spatially complex patterns are described by a finite number of peaks in the spectral domain and may be classified as quasicrystals or superlattices using the criteria defined by Lifshitz [16]. Essentially, if the number of vectors (indexing vectors) required to span the stimulated modes is greater than the pattern dimension the state is a quasicrystal [16]. Accordingly, if the number of vectors equals the pattern spatial dimension, the state is periodic. Hence, a superlattice is a pattern whose power spectrum contains a finite number of modes which are indexed by a number of vectors equal to the spatial dimension of the pattern. Quasicrystals and superlattices have been reported in other hydrodynamic [12,13,17–19] and optical systems [20,21]. Features generally found among the examples where complex-order has previously been observed include a nearby codimension-two point and multiple accessible wavenumbers.

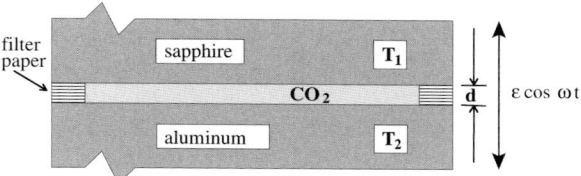

Fig. 1. Sketch of convection cell (not to scale). The relatively thin CO_2 layer of depth d is driven away from equilibrium by an imposed temperature difference ΔT and vertical oscillations ($\varepsilon \cos \omega t$). The top and bottom surfaces are maintained at uniform temperatures T_1 and T_2, respectively ($\Delta T = T_2 - T_1$). Oscillations are defined in terms of the dimensionless acceleration ε and ω (2)

Many of the characteristics that are distinctive to either Rayleigh-Bénard-type or Faraday-type patterns can be found in a single system: a fluid layer driven by both heating from below and sinusoidal vertical oscillations (Fig. 1). A classical Rayleigh-Bénard convection experiment (no vertical oscillations) of infinite lateral extent is described by two nondimensional parameters: Rayleigh number R and Prandtl number Pr. These parameters are defined,

$$R = \frac{\alpha g d^3 \Delta T}{\nu \kappa} \quad \text{and} \quad Pr = \frac{\nu}{\kappa}, \tag{1}$$

in terms of ΔT (Fig. 1), gravitational acceleration g, d (Fig. 1), thermal diffusivity κ, kinematic viscosity ν, and thermal expansivity α. Intrinsic scales in the experiment are d and the vertical diffusion time ($t_\nu = \kappa/d^2$). Imposing vertical oscillations on Rayleigh-Bénard convection [22,23] (acceleration-modulated Rayleigh-Bénard convection) results in two additional nondimensional parameters that are analogous to those used to characterize states in Faraday wave experiments: the displacement amplitude δFr and the modulation frequency ω;

$$\delta Fr = \frac{\kappa^2}{d^4 g}\delta' \quad \text{and} \quad \omega = \frac{d^2}{\kappa}\omega', \tag{2}$$

where δ' is the dimensional displacement amplitude and ω' is the dimensional angular frequency of oscillation. It is important to note that unlike the Faraday waves there is no free surface in the present case.

In this paper we report the first experimental results for acceleration-modulated Rayleigh-Bénard convection [24–26], which we confirm and augment with a number of new numerical findings [27]. Brief descriptions of our experimental apparatus and numerical methods are presented in Sects. 2.1 & 2.2, respectively. Most of the presented results are for a fixed $Pr = 0.930$ and $\omega \approx 100$. Predicted modulation induced shifts in onset R (R_c) and onset q (q_c) are compared in Sect. 3.1 with experimental observations. In Sect. 3.2 patterns are confirmed to display harmonic (synchronous to the oscillations) and subharmonic temporal dependence for the appropriate parameter values. Harmonic patterns near onset and away from onset are reported and discussed in Sects. 3.3 & 3.4. Examples of subharmonic patterns near and away from onset are presented in Sects. 3.5 & 3.6. For sufficiently large R a direct transition between purely harmonic and purely subharmonic patterns is found to occur. Boundaries for this gradual transition and typical coexisting patterns are investigated in Sects. 4.1 & 4.2. Over a wide range of parameters novel complex-ordered patterns are reported. These complex states are found to be superlattices, one type of which is found to emerge directly from conduction at a point of codimension-two. The presence of inversion symmetry is found to be important to the coexistence pattern observed near onset. We propose a resonance mechanism for the formation of observed superlattices in Sect. 5.3. Complex-ordered patterns at other ω values are investigated numerically and experimentally in Sect. 5.4.

2 Laboratory

2.1 Experimental Apparatus

Performing the experiments requires vertically oscillating a convection cell (Fig. 1), while controlling the relevant physical quantities to vary the dimensionless parameters in the appropriate manner. The experiment is composed of a convection cell, means for controlling cell temperatures and pressure, a mechanical shaking device, an image acquisition system, and computers to analyze the various data streams. In this section we briefly describe each of these sub-systems. For a more detailed description see [26].

The convection apparatus is based on well-tested designs for standard Rayleigh-Bénard convection using compressed gases [28,29]. The actual convection cell is bounded from below by a 5.08 cm-diameter, 0.60 cm-thick aluminum mirror; from the side by a 3.8 cm-inner-diameter, 5.08 cm-outer-diameter stack of filter paper; and from above by a 5.0 cm-diameter, 2.54 cm-thick sapphire crystal. The cell is confined in an aluminum pressure containment vessel. The sapphire crystal is held tightly against the top of the vessel by the pressure of the compressed CO_2 gas. The aluminum mirror is aligned parallel to the bottom of the sapphire crystal by a kinematic mount with an additional pull-down screw. Interferometry measurements demonstrate that the bounding surfaces remain level when oscillations are imposed.

Fluid properties are under dynamic computer control. Temperature control is provided by heating through a resistive pad the bottom mirror while cooling with a water bath the sapphire crystal. The heating and the cooling are both under computer control allowing the fluid layer mid-plane temperature $\bar{T}$ and ΔT to be held fixed within $\pm 0.01°$ C. The containment vessel is filled with 99.99% pure CO_2, typically at a pressure near 32.72 bar. Pressure is computer-controlled through the heating of a ballast allowing constant pressure to be maintained to $\pm$ 0.01 bar

Vertical oscillations are supplied by hydraulically-driven mechanical shaking systems. The convection apparatus is attached to a piston whose motion is driven by the flow of oil at 120 bar. The oil flow is regulated by high-performance electrodynamic servo valves, which, in turn, are driven by a high-current amplifier under closed-loop control. Thus, controlled oscillations of the piston are achieved by feeding an oscillatory control voltage signal into the amplifier. The displacement amplitude of the oscillations is measured by two devices: a linear variable displacement transducer (LVDT) attached directly to the piston and an accelerometer attached to the bottom of the containment vessel. The hydraulic shaker is rigidly attached to a heavy mount (ballast), which in turn, rests on elastic supports to damp out vibrational recoil. Lateral vibrations are held to approximately 0.2 % of the vertical motion through the use of a rectangular air-bearing assembly on the drive shaft.

Patterns are visualized using the established method of shadowgraphy [28,29]. Utilizing compressed gases greatly enhances the sensitivity of the shadowgraph [28], since the refractive index is reinforced and the contrast is enhanced by the very thin layers ($d = 0.0650$ cm) which may be used. Shadowgraph images are captured using a ccd camera interfaced with a computer-controlled frame grabber. The image acquisition is synchronized with the shaker drive by use of a ferroelectric liquid crystal shutter. Pattern images are acquired at a predefined phase of the oscillation.

Recorded images are analyzed predominantly in terms of average spectral quantities. To each image a radial Hanning function is applied to reduce aliasing, prior to performing a spatial Fourier transform. The constituent phase angles and power spectra [$\wp(q)$] are found for each pattern. Generally, the power spectra are normalized by the windowed image variances so that the total power in the spectrum will sum to unity. The power spectra are then azimuthally averaged to produce the radial spectrum. Radial power spectra for all the images at a data point are then averaged. The resulting distribution in q is described by the first two moments,

$$\langle q \rangle = \frac{\int_{q=0}^{\infty} q^2 \wp(q) dq}{\int_{q=0}^{\infty} q \wp(q) dq} \qquad (3)$$

$$\langle q^2 \rangle = \frac{\int_{q=0}^{\infty} q^3 \wp(q) dq}{\int_{q=0}^{\infty} q \wp(q) dq} \qquad (4)$$

Patterns at a given set of parameters may then be characterized by three spectral quantities: the relative power, characteristic q ($\langle q \rangle$), and the width of q

($\sigma = \sqrt{\langle q^2 \rangle - \langle q \rangle^2}$). Some patterns may contain multiple distinct wavenumbers. In these cases, 8th-order Butterworth filters are applied to remove frequency components outside the band being considered.

2.2 Numerical Methods

The convective flow can be described by the Oberbeck-Boussinesq equations, which, when non-dimensionalized by d and t_v, take the form

$$\nabla \cdot \boldsymbol{v} = 0$$

$$\nabla^2 \boldsymbol{v} + \hat{\boldsymbol{z}}\left(1 + \frac{\delta'\omega'}{g}\cos\omega t\right)\Theta - \nabla P = \frac{1}{Pr}\left(\boldsymbol{v}\cdot\nabla\boldsymbol{v} + \frac{\partial \boldsymbol{v}}{\partial t}\right)$$

$$\nabla^2 \Theta + R\hat{\boldsymbol{z}}\cdot\boldsymbol{v} = \boldsymbol{v}\cdot\nabla\Theta + \frac{\partial \Theta}{\partial t}, \tag{5}$$

where $\boldsymbol{v}$ is the velocity, Θ temperature, and P pressure. In the co-moving frame to the layer, the effect of modulation appears only in the time-dependent buoyancy term $\hat{\boldsymbol{z}}(\frac{\delta'\omega'}{g}\cos\omega t)\Theta$ [22,23]. The Boussinesq symmetry is not broken by the modulation.

Numerical simulations of (5) are performed by modifying well-tested numerical techniques for standard Rayleigh-Bénard convection. Only minor modifications are required since the time-dependent buoyancy (a linear term) is integrated explicitly in time by employing the approach used previously for quadratic nonlinearities [30]. The equations are solved using isothermal ($\Theta = 0$), no-slip ($\boldsymbol{v} = 0$), and no penetration ($\partial_z v_z = 0$) boundary conditions at the confining (upper and lower) plates. The boundary conditions are enforced by expanding all fields in appropriate test functions (trigonometric or Chandrasekhar functions) [23]. The incompressibility condition $\nabla\cdot\boldsymbol{v} = 0$ is satisfied by the introduction of a poloidal and a toroidal velocity potential. We follow closely the standard approach [23] by expanding all fields into a Fourier series in time combined with a Galerkin expansion in the z-direction. Linearization permits calculation of both R_c and q_c as function of the modulation parameters, δFr and ω (Fig. 2). Nonlinear stripe solutions are calculated within the Galerkin approach and examined for secondary bifurcations.

In some cases we include non-Boussinesq effects using the same approach employed for non-Boussinesq effects in standard Rayleigh-Bénard convection [1]. In brief, the temperature dependence of all material parameters is expanded about $\bar{T}$ to linear order. This results in both non-Boussinesq corrections obtained for standard Rayleigh-Bénard convection [1] and modification of the time-dependent buoyancy term via the temperature dependence (at quadratic order) of the density. For a more detailed description of the numerical methods see [27].

3 Onset, Time-Dependence, and Typical Patterns

The conduction state competes with temporally modulated convection over a wide range of parameter values. Our studies focus on this competition for the case of fixed Pr and ω. By varying the remaining control parameters δFr and R over a range where R is not too large, the conduction state is found to lose stability to flows with either a harmonic or subharmonic temporal response. Linear stability analyses [22,23] of the current experiment indicate that each type of temporal flow occurs at a distinct spatial scale (Fig. 2). Harmonic flows are predicted to be more stable than unmodulated convection, *i.e.*, the R_c for harmonic convection R_c^H is found to be larger than the R_c in the absence of modulation ($R_c^0 = 1708$). In contrast, subharmonic flows may be either more stable ($R_c^S > R_c^0$) or less stable ($R_c^S < R_c^0$). The q_c of subharmonic patterns q_c^S is typically significantly larger than q_c of harmonic patterns q_c^H [Fig. 2(b)].

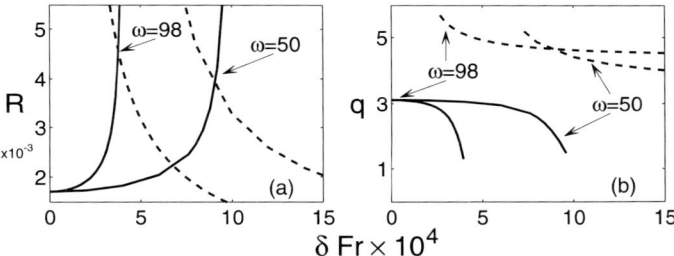

Fig. 2. Linear stability of acceleration-modulated Rayleigh-Bénard convection. Solid lines are marginal stability curves for harmonic response, while dashed lines indicate subharmonic marginal stability curves. Arrows in (**a**) point to bicritical points at the labeled values of ω (= 98 & 50)

3.1 Onset Measurements

Linear stability predictions of the onset of fluid motion due to parametric modulations are quantitatively confirmed by laboratory observations. In the absence of modulations ($\delta Fr = 0$, $\omega = 0$) the experiment reduces to classical Rayleigh-Bénard convection where it is known that conduction loses stability at R_c^0 with $q_c^0 = 3.117$. For these parameters, we observe parallel stripes at onset, suggesting that non-Boussinesq effects are weak and not observable within the resolution of our measurements. Figure 3(a) compares conduction marginal stability boundaries to experimental measurements of the boundary between conduction and convection. With increasing δFr conduction is increasingly stabilized until the harmonic marginal stability curve intersects the subharmonic marginal curve. At the point of intersection (*bicritical*

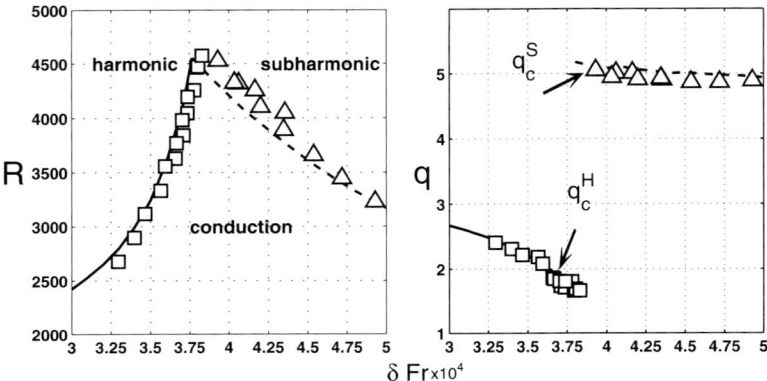

Fig. 3. Comparison of conduction marginal stability predictions to experimental observations of R_c (**a**) as well as q_c (**b**) at $Pr = 0.930$ and $\omega = 98.0$. Solid lines are predicted values for onset of harmonic convection and dashed lines for subharmonic convection. Experimental observations for harmonic and subharmonic ($\triangle$) flows are in quantitative agreement with predictions

point) conduction loses stability to two different spatial scales simultaneously as R is increased. For $\omega = 98.0$, our linear stability predicts bicriticality at $\delta Fr_{2c} = 3.768 \times 10^{-4}$ & $R_{2c} = 4553$. For $\delta Fr > \delta Fr_{2c}$, the stability analyses and experiments find the slope of the marginal stability curve changes sign. For sufficiently large δFr, beyond the accessibility of the current experiments, onset is predicted to occur for $R < R_c^0$ [23] [see also Fig. 2(a)]. The stability predictions of q_c^H and q_c^S are compared with experimentally measured values in Fig. 3(b). For $\delta Fr < 3 \times 10^{-4}$, the onset wavenumber changes only slightly from q_c^0. With increasing δFr, q_c^H begins to decrease. With $\delta Fr \geq \delta Fr_{2c}$, subharmonic convection arises with a substantially larger q than harmonic convection. Both experiments and the stability analyses find q_c^S remains relatively constant for the range of $\delta Fr \geq \delta Fr_{2c}$ studied. Neither harmonic nor subharmonic onset exhibited hysteresis within the resolution of our experiments.

3.2 Confirmation of Time-Dependence

To investigate the fluid's temporal dependence multiple images are recorded during a single τ, each separated by a constant time interval. Temperature time series from the simulations indicate that harmonic convection oscillates about nonzero mean [Fig. 4(a)] with a period of τ; shadowgraph images [Fig. 5(a) & 5(b)] from the experiments demonstrate flows with period τ. By contrast, simulations indicate the subharmonic temperature field oscillates about zero mean with a period of 2τ [Fig. 4(b)]. Oscillations in the subharmonic temperature field must satisfy the subharmonic time-translation (inversion)

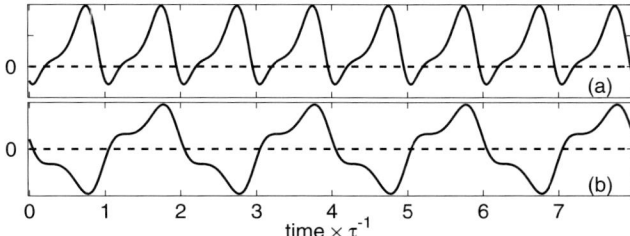

Fig. 4. Temporal dependence of Fourier modes (*simulation*) in the fluid horizontal mid plane of (**a**) purely harmonic convection and (**b**) purely subharmonic convection

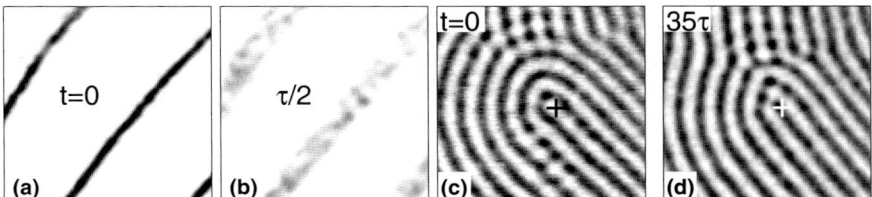

Fig. 5. Shadowgraph (*experiment*) of singly resonant flows. Purely harmonic patterns (**a** & **b**) repeat every τ time interval. At $t = 0$ (**a**) the stripe state is well-defined. For odd multiples of τ (**b**) the stripe state becomes broader, while at even multiples of τ the original state (**a**) repeats. Each 128 by 128 pixel frame is over the same spatial location separated in time by $\tau/2$. In contrast, purely subharmonic patterns (**c** & **d**) invert every τ time-interval. Note the + signs near the image (**c** & **d**) centers, these are plotted at the same coordinates in both frames. The disclinations' center stripe is light at even multiples of τ [(**c**) t=0] and dark at odd multiples of τ [(**d**) $t = 35\tau$]

symmetry which requires the field variables invert under discrete time translation by τ. This effect of symmetry is observed in the experiments as time-periodic switching of the temperature field. Shadowgraph images separated in time by an odd multiple of τ demonstrate switching between the upflow (light) and downflow (dark) regions of the pattern [Fig. 5(c) & 5(d)].

3.3 Harmonic Patterns at Onset

Harmonic patterns at onset are striped and may contain domains of hexagons depending on R. Parallel stripes [Fig. 6(a)] and targets [Fig. 6(b)] are typically observed at onset. These patterns may be found anywhere ($2000 < R < 4500$) along the harmonic stability curve (Fig. 7), while domains of hexagons only occur close to onset at larger R values ($3800 < R < 4800$). From classical Rayleigh-Bénard convection studies in compressed gases ($Pr \approx 1$) it is known that parallel stripes form when the side wall forcing is minimal and that cell filling (giant) targets or spirals [31] are present near onset when the side wall forcing is more significant (for example, due to side wall heating [32]). We expect a similar mechanism to be at work in our experiments where the selection of stripes or targets varies with different experimental configurations. Thus, we associate stronger side wall forcing from the circular lateral boundary with the onset of targets and weaker side wall forcing with parallel stripe formation. Targets may display light or dark cores, designating cold (downflowing) or warm (upflowing) centers, respectively. Domains of hexagons may be present in both striped and target base states at larger R values with the domains of hexagons becoming larger with R [Figs. 6(c) & 6(d)]. Hexagons with both downflowing centers [Fig. 6(c) & 6(d)] and upflowing centers are observed. Transitions between domains of locally upflowing and downflowing hexagons are also observed. The presence of hexagonal domains indicates the approximation of Boussinesq symmetry is not valid near onset for at relatively large R. Similar mixed stripe-cellular patterns have been observed previously in temperature-modulated Rayleigh-Bénard convection experiments [33]. Physically this likely due to the temperature dependence of fluid properties; the variation of these properties within the layer increases with increasing R (increasing ΔT) and explains why hexagons are most readily observed when the onset of harmonic convection occurs at the largest values of R near R_{2c}. Over most of the purely harmonic region the majority of stable states are strikingly similar to patterns found in classical Rayleigh-Bénard convection studies (without modulation) [32].

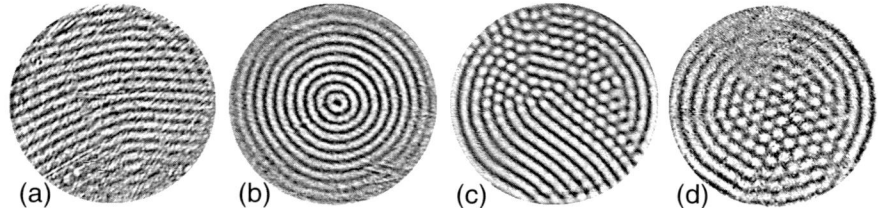

Fig. 6. Harmonic onset striped patterns include: (**a**) parallel stripes ($\delta Fr = 3.34 \times 10^{-4}$, $\omega = 97.8$, & $R = 3002$), (**b**) targets ($\delta Fr = 3.29 \times 10^{-4}$, $\omega = 98.0$, & $R = 2979$), (**c**) stripes with hexagons ($\delta Fr = 3.71 \times 10^{-4}$, $\omega = 96.2$, & $R = 4388$), and (**d**) targets with hexagons ($\delta Fr = 3.76 \times 10^{-4}$, $\omega = 96.6$, & $R = 4107$)

3.4 Harmonic Patterns away from Onset

Onset patterns undergo a transition to spiral defect chaos as the system moves from conduction further into the convection regime by combinations of decreasing δFr and increasing R (Fig. 7). Two different scenarios are observed, depending upon whether the transition sequence begins with either parallel stripes or target patterns. The domains of hexagons observed near

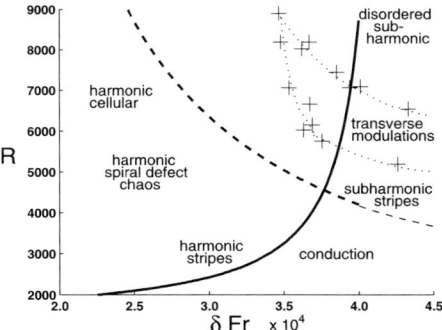

Fig. 7. Parameter space showing conduction marginal stability curves for harmonic (*solid*) and subharmonic (*dashed*) convection as well as experimental determined regions of behavior

onset at larger R play no role in this transition sequence because, for small decreases in δFr away from onset, hexagons disappear and only patterns of either stripes or targets remain.

The transition from parallel stripes to spiral defect chaos is similar to that observed in classical Rayleigh-Bénard convection [32]. Moving away from onset, stripe curvature increases, forming focus singularities at the lateral boundaries, as the stripes increasingly align themselves perpendicular to the side walls. Initially, two foci will form [Fig. 8(a)] with the stripe curvature increasing as the experiment is tuned away from onset. Two-foci patterns are similar to the so-called Pan-Am states observed in classical Rayleigh-Bénard convection at comparable Pr and aspect ratio [34,35]. Away from onset, stripe curvature gradually increases and more wall foci form [Fig. 8(b)] leading to local frustration of the pattern wavenumber and formation of point defects. These defects typically include locations where two stripes are replaced by a single stripe (dislocations) and lines of point defects where the pattern

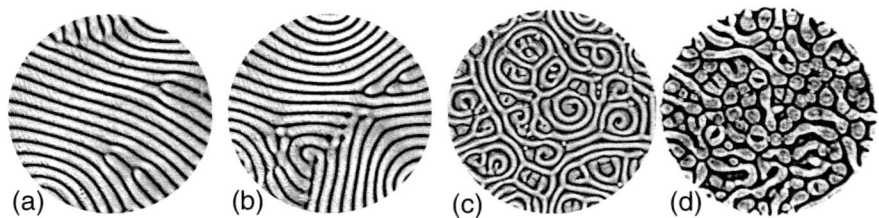

Fig. 8. Examples of typical patterns observed moving away from onset stripes: (**a**) stripes with point defects and two wall foci forming ($\delta Fr = 3.47 \times 10^{-4}$, $\omega = 98.4$, & $R = 3926$), (**b**) three-foci stripes ($\delta Fr = 3.36 \times 10^{-4}$, $\omega = 98.4$, & $R = 3926$), (**c**) spiral defect chaos ($\delta Fr = 2.06 \times 10^{-4}$, $\omega = 96.2$, & $R = 4385$), and (**d**) harmonic cellular ($\delta Fr = 2.70 \times 10^{-4}$, $\omega = 100.3$, & $R = 8077$)

amplitude goes to zero (amplitude grain boundaries). Dislocations increase in number as the experiment moves away from onset, while amplitude grain boundaries are typical in stripe patterns with 3-foci. Generally, once the number of wall foci is larger than four, spirals will begin to appear in the interior of the pattern. The system then begins to display states composed of left and right handed spirals [Fig. 8(c)]. As spiral defect chaos gradually fills the convection cell, the harmonic wavenumber q^H remains well-defined, but the corresponding peak width σ^H increases. The emergence of dislocations and grain boundaries, focus singularities, and the broadening of q distribution about a characteristic value observed in the transition from stripes to spiral defect chaos is also seen without modulation [32]. With both sufficiently large δFr and relatively large R, q^H becomes less well-defined and σ^H becomes large as spiral defect chaos gradually gives way to cellular patterns without spirals [Fig. 8(d)].

When the side wall influence is more substantial the onset striped pattern is a target and the transition to spiral defect chaos is somewhat different. Moving away from onset a defect mediated transition from targets to spirals occurs. Pairs of defects emerge in the pattern due to skew-varicose instabilities. One defect from each pair will translate radially to the spiral core producing a one arm spiral while the other defect will translate radially to the side walls and annihilate. In Fig. 9(a) the defect pair originated in a target near the core producing the one arm spiral, the other defect is translating to the side walls. Sufficiently near onset continual nucleation can result in switching between targets and one arm spirals [26]. Continuing away from onset additional defects enter the pattern due to skew-varicose instabilities producing multi arm spirals, as many as six arm spirals are observed. Occasionally, an additional instability results in the targets or spirals cores moving off-center. An example of this instability in shown in Figs. 9(b) & 9(c) for a three arm spiral. Further away from onset, skew-varicose and off-center instabilities result in numerous defects throughout the pattern and the giant spiral will become unstable as spiral defect chaos [Fig. 9(d)] forms [31]. Moving to

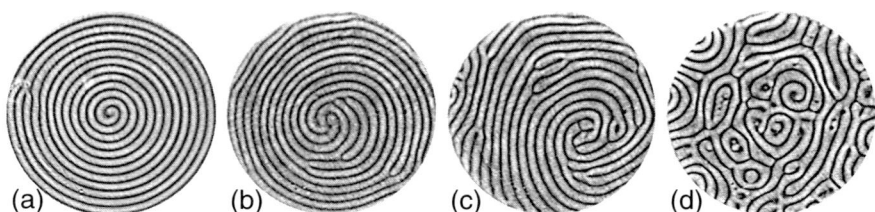

Fig. 9. Moving into the convection regime from a onset target typically observed patterns include: (**a**) one arm spirals ($\delta Fr = 1.74 \times 10^{-4}$, $\omega = 98.4$, & $R = 2479$), (**b**) multi arm spirals with defects ($\delta Fr = 1.95 \times 10^{-4}$, $\omega = 99.4$, & $R = 2422$), (**c**) off-center multi arm spirals with defects ($\delta Fr = 2.26 \times 10^{-4}$, $\omega = 99.0$, & $R = 2936$), and (**d**) spiral defect chaos ($\delta Fr = 2.30 \times 10^{-4}$, $\omega = 98.5$, & $R = 3663$)

larger R at sufficient δFr spiral defect chaos gradually gives way to patterns like Fig. 8(d). The transition from spiral defect chaos to states like Fig. 8(d) is qualitatively independent of the onset planform.

3.5 Subharmonic Patterns at Onset

At onset, observed subharmonic states are parallel stripes [Fig. 10(a)] that may include defects [Fig. 10(b-d)]. In contrast to the harmonic flows, hexagons are not observed anywhere along the subharmonic marginal stability curve. This is expected, due to the temporal subharmonic inversion symmetry, which excludes the resonant triads.

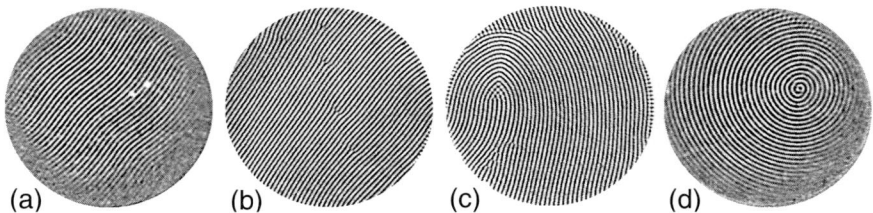

Fig. 10. Examples of patterns observed near onset of subharmonic convection. Patch (**a**) of parallel stripes ($\delta Fr = 4.02 \times 10^{-4}$, $\omega = 97.9$, & $R = 4395$), (**b**) parallel stripes with dislocations ($\delta Fr = 3.74 \times 10^{-4}$, $\omega = 97.9$, & $R = 4857$), (**c**) stripes with a giant convex disclination and several dislocations ($\delta Fr = 3.34 \times 10^{-4}$, $\omega = 97.5$, & $R = 4811$), and (**d**) onset spiral ($\delta Fr = 4.12 \times 10^{-4}$, $\omega = 97.9$, & $R = 4173$)

Giant convex disclinations are common near onset, while cell filling concave disclinations are not observed. Cell filling subharmonic spirals [Fig. 10(d)] can arise when side wall forcing is sufficiently large. However, subharmonic spirals are unusual even when sidewall forcing is sufficient to induce targets or spirals at the onset of harmonic convection. Only one and three arm giant spirals have been observed. If giant convex disclinations [Fig. 10(c)] or giant spirals form at onset they are typically centered about the midpoint of the convection cell and move off center as the system begins to move away from onset. Sufficiently far from onset, the cores of convex disclinations annihilate at the lateral boundary leaving a parallel stripe pattern. Due to the characteristic wave length of subharmonic patterns being substantially smaller than that of harmonic patterns it might be expected that side wall forcing would have less of an influence over the selected planform.

3.6 Subharmonic Patterns away from Onset

With combinations of increasing δFr and increasing R (Fig. 7), subharmonic patterns exhibit a transition to disorder that is qualitatively different from

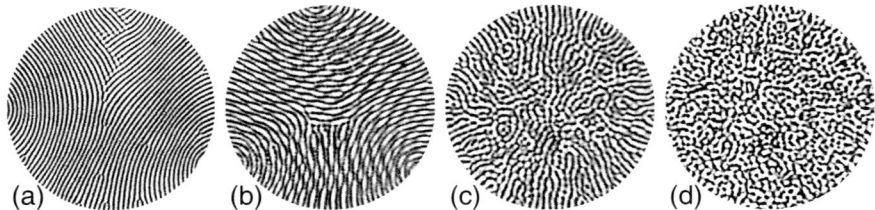

Fig. 11. Examples of typical subharmonic patterns observed away from onset: (**a**) two-foci stripes with defects ($\delta Fr = 4.17 \times 10^{-4}$, $\omega = 98.1$, & $R = 4888$), (**b**) three-foci transverse modulated stripes ($\delta Fr = 4.01 \times 10^{-4}$, $\omega = 98.0$, & $R = 6552$), radial stripes - onset of subharmonic disorder ($\delta Fr = 4.83 \times 10^{-4}$, $\omega = 95.0$, & $R = 6120$), and (**d**) subharmonic disorder ($\delta Fr = 4.60 \times 10^{-4}$, $\omega = 95.0$, & $R = 7670$)

that observed for harmonic patterns. In particular, spiral defect chaos is not observed for subharmonic convection. Parallel stripes near onset display cross-roll defects and dislocations as mechanisms to adjust local variations in the pattern wavenumber; q^S slowly decreases away from onset. Further from onset, wall foci form as stripes begin to show curvature [Fig. 11(a)]. Two and three foci stripes are common, with three foci patterns often containing a amplitude grain boundary near the pattern center. These states often contain several dislocations and regular spacing of the focal singularities. For δFr and R sufficiently large (Fig. 7), patterns with transverse modulations [Fig. 11(b)] bifurcate from the parallel stripe state. The focal singularities and point defects present in the pattern prior to the bifurcation are preserved after the emergence of transverse modulations. The transverse modulations propagate along the length of the stripes and oscillate in time with a period that is different from either the harmonic or subharmonic period. Spatial power spectra of the transverse modulated pattern demonstrate that the modulation q is only slightly greater than q^H [Fig. 14(e)]; moreover, the width of the subharmonic peaks σ^S increases significant at the bifurcation. With further increases in δFr and R, transverse modulation disappears as the stripes begin to align themselves radially from the center to the lateral boundary [Fig. 11(c)]. During this transition q^S continues to decrease, while the formation of radial stripes corresponds to a reduction in σ^S. Continuing away from onset the radial stripes begin to lose coherent structure, initially in the pattern interior. Sufficiently far from onset this fragmented state occurs throughout the convection cell [Fig. 11(d)] and retains a characteristic q^S with a σ^S larger than that for radial stripes.

4 Direct Harmonic-Subharmonic Transition

At sufficiently large R conduction is not stable for any δFr and a direct transition between harmonic and subharmonic patterns can occur as δFr is varied. Experiments indicate this transition is not abrupt, but occurs gradually over

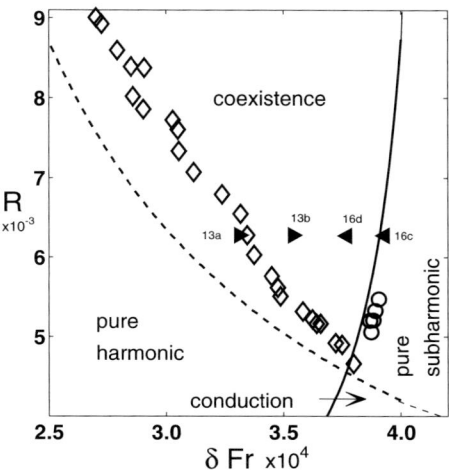

Fig. 12. Phase plane comparing the experimentally measured coexistence onset to the marginal stability curves for conduction. The boundary between coexistent and purely harmonic flows (*diamonds*) follows the marginal subharmonic ($\omega/2$) curve (*dashed*), while the boundary between coexistent and purely subharmonic patterns (*circles*) tracks the marginal harmonic (ω) curve (*solid*) as far as the boundary can be reliably determined. The filled-in triangles are the locations of patterns in Figs. 13 & 16

a range of parameter values where harmonic and subharmonic patterns coexist. Roughly speaking, this coexistence region lies between the conduction marginal stability curves extended into the convection regime (Fig. 12).

4.1 Transition from Pure Harmonics to Coexistence

Purely harmonic patterns [Fig. 13(b)] lose stability to coexisting states with localized regions of subharmonic strips [Fig. 13(b)]. Prior to onset, the harmonic pattern typically consists of parallel stripes with defects. With increasing δFr at constant R, localized domains of subharmonic stripes emerge with a characteristic wavenumber q^S slightly less than $3q^H$. These subharmonic domains are typically either centered about defects in the harmonic pattern or aligned perpendicular to the lateral boundaries [Fig. 13(b)]. Subharmonic stripes at the lateral boundary typically remain pinned to the boundary and do not move into the interior. Harmonic defects continually nucleate, advect and annihilate in the pattern interior. These dynamics drive the behavior of the subharmonic patches, which correspondingly appear, move and disappear. Although harmonic defects are virtually always present for parameter values near the pure harmonic-coexistence boundary, not all harmonic defects have associated subharmonics. As a result, near onset, the subharmonic stripe patches are intermittent in time for a range of δFr of width $\sim 4e-06$.

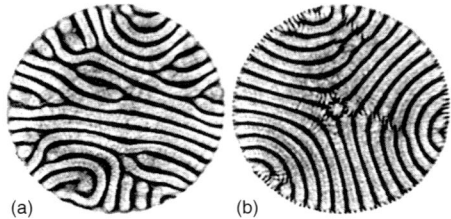

Fig. 13. Patterns on either side of the purely harmonic-coexistence boundary. Pure harmonic stripes (**a**) with defects ($\delta Fr = 3.31 \times 10^{-4}$, $\omega = 98.0$, & $R = 6280$). Coexistence state (**b**) with 3-foci harmonic stripes and subharmonic stripe patches ($\delta Fr = 3.54 \times 10^{-4}$, $\omega = 98.0$, & $R = 6280$)

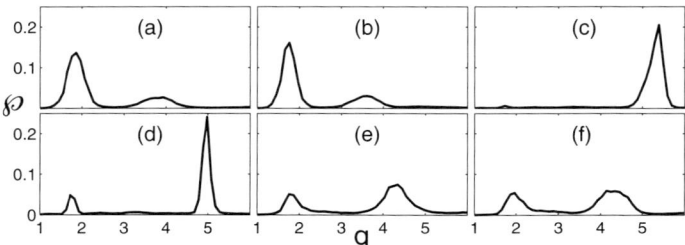

Fig. 14. Azimuthally averaged power spectra for six different experimental conditions from the transition from pure harmonic convection to pure subharmonic convection, passing through a region of coexisting harmonic-subharmonic convection. Representative images from each data point are shown in other figures: (**a**) see Fig. 13(a), (**b**) see Fig. 13(b), (**c**) see Fig. 16(a), (**d**) see Fig. 16(b), (**e**) see Fig. 16(c), and (**f**) see Fig. 16(d)

Because of intermittency in both space and time, the onset of subharmonics is difficult to detect in spatial power spectra [Figs. 14(a) & 14(b)]. Detection is more reliably performed using the real space images of the patterns. The onset value of δFr for a given R corresponds to the presence of subharmonic patches in the pattern interior for at least 10% of the observation time.

Our results provide evidence that harmonic patterns have an inhibitory effect on the emergence of subharmonic patterns. In the first place, the boundary for convective onset in the experiments lies above the subharmonic marginal stability curve over the entire experimentally accessible range of parameters (Fig. 12). In other words, the subharmonic onset is delayed relative to the linear theory predictions of onset from the conduction state. Furthermore, subharmonics always appear in regions of where the harmonic flows are weaker, namely in the cores of defects amplitude of convection flow is reduced [36].

The subharmonic component remains localized and spatially intermittent even as δFr is increased (with R fixed) to move the system well away from the

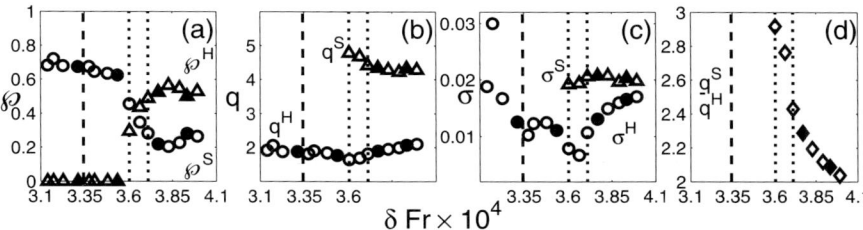

Fig. 15. Average spectral quantities (*experiment*) during a transition from purely harmonic to purely subharmonic patterns by increasing δFr at $R = 6280 \pm 10$. The azimuthally averaged spectra for both the harmonic and subharmonic modes are characterized by (**a**) the power the in harmonic $\wp^H$ and subharmonic $\wp^S$ wavenumber bands, (**b**) q^H & q^S, (**c**) σ^H & σ^S, and (**d**) the wavenumber ratio $\frac{q^H}{q^S}$. Throughout, ○ indicates harmonic pattern component and △ the subharmonic pattern component. Filled in symbols correspond to patterns shown in Figs. 13 & 16. Dotted lines mark the onset of square superlattices (Sect. 5)

pure harmonic-coexistence boundary. The subharmonic component is spectrally indistinguishable from the background noise and the second harmonic of q^H over a wide range of δFr [Figs. 15(a), 14(a) & 14(b)]. The wavenumber of the harmonic modes q^H remains relatively fixed [Fig. 15(b)]. The spectral width σ^H decreases [Fig. 15(c)] because the harmonic pattern exhibits fewer defects as the system moves further away from the pure harmonic-coexistence boundary (Fig. 13). The subharmonic pattern component gradually increases with δFr.

4.2 Transition from Pure Subharmonics to Coexistence

The transition from purely subharmonic states to coexisting patterns is qualitatively different from that at onset from pure harmonics. Pure subharmonic patterns lose stability to coexisting states where the harmonic component emerges globally; no localized states are observed. For $\omega = 98.0$, we consider two cases: $R_{2c} < R < 5500$ and $R > 5500$

For $R_{2c} < R < 5500$ the coexistence regime competes with pure subharmonic parallel stripes [Fig. 16(a)]. Slowly decreasing δFr at constant R a harmonic pattern component emerges at a well-defined location in parameter begins to be present throughout the pattern [Fig. 16(b)]. Although the harmonic component is weak at onset, the transition is well-defined and readily detectable in Fourier space by looking for the initial presence of power at q^H [Figs. 14(c) & 14(d)]. Typically, the emerging harmonic component is parallel stripes which may display domains with several orientations. In this parameter range, this transition is well-predicted by the *conduction* marginal stability curve (Fig. 12), suggesting that the onset of large length scale harmonic convection is neither enhanced or suppressed by the presence of short

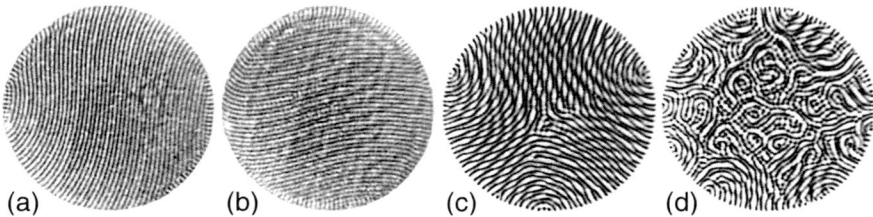

Fig. 16. Patterns on either side of the harmonic-coexistence marginal stability curve ($Pr = 0.930$, $\omega = 98.0$) for $R = 4980$ [$\delta Fr = 3.80 \times 10^{-4}$ (**a**) & $\delta Fr = 3.69 \times 10^{-4}$ (**b**)] and $R = 6280$ [$\delta Fr = 3.93 \times 10^{-4}$ (**c**) & $\delta Fr = 3.77 \times 10^{-4}$ (**d**)]

length scale subharmonic flows. Hysteresis is not experimentally observed in the transition between pure subharmonic flows and the coexistence regime.

For $R > 5500$ the coexistence states compete with more complex pure subharmonic flows. For $5500 < R < 7000$, subharmonics with transverse modulations [Fig. 16(c)] are found when δFr is large (Fig. 7). For $R > 7000$, the subharmonic flows are more disordered [Figs. 11(c) & 11(d)]. As δFr is decreased at constant R to cross the conduction marginal stability boundary, the flow structure changes gradually to patterns like that shown in Fig. 16(d). In all cases, these states are difficult to distinguish spectrally because they contain spectral peaks with similar power content at wavenumbers corresponding to both q^S and q^H [Figs. 14(e) & 14(f)]. As a result, the onset of the coexistence regime from pure subharmonics is ill-defined for this range of R.

Spectral analysis demonstrates that the gradual nature of the transition from pure subharmonics to coexisting patterns continues as δFr is further decreased. For $R < 5500$ the growing harmonic stripes have little effect on the subharmonic stripes as the two components are simply superimposed. For $5500 < R < 7000$ the subharmonic striped base state that supports the transverse modulations gradually breaks down as numerous domains form [Figs. 16(c) & 16(d)]. Typically, these domains nucleate in the pattern interior and spread to fill the pattern with decreasing δFr. For $R > 7000$ the structurally disordered state [Fig. 11(d)] becomes more ordered with decreasing δFr. Regardless of the R value the spectral measures display similar trends. First, relative power in $\wp^S$ gradually decreases and $\wp^H$ slowly increases as the harmonic pattern becomes more significant [Fig. 15(a)]. Eventually, $\wp^S$ contributes $< 60\%$. Second, q^H [Fig. 15(b)] remains relatively fixed while σ^H decreases [Fig. 15(c)]. Simultaneously, q^S slowly increases while σ^S remains relatively constant. These similarities indicate that as the pattern passes further into coexistence the harmonic pattern slowly grows and becomes more regular while $\wp^S$ slowly decrease and q^S increases.

5 Superlattices

Exotic complex-ordered patterns [Fig. 17(a)] abruptly form as the system moves sufficiently far into the coexistence parameter regime. These patterns are reminiscent of recently observed nonequilibrium structures [12–14,17–19] that have been designated as *quasicrystals* (*quasipatterns*) or *superlattices*. The formation of the current complex-ordered structures corresponds to rapid changes in the spectral quantities, including the formation of distinct peaks in the power spectrum [Fig. 17(b)]. Since these twelve peaks can be indexed by only two basis vectors, this pattern is a superlattice [16]. We call patterns of this type *square superlattices* [25]. To the best of our knowledge, the patterns observed in this investigation are the first complex-ordered states to be reported in convection experiments. In the related case of heating from above ($R < 0$) a numerical study [37] predicted quasiperiodic structures in the presence of non-Boussinesq effects.

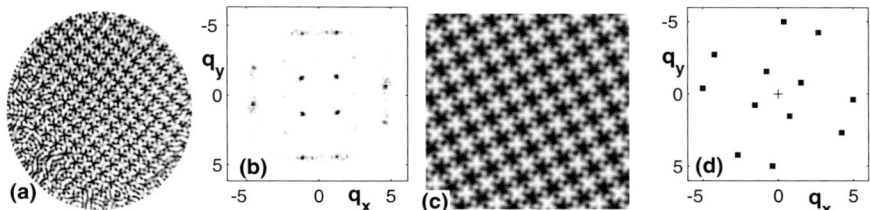

Fig. 17. Square superlattices observed in both experiments (**a**) and simulations (**b**). Parameter values are: (**a**) $\delta Fr = 3.88 \times 10^{-4}$, $\omega = 95.3$, & $R = 7030$ and (**b**) $\delta Fr = 3.75 \times 10^{-4}$, $\omega = 98$, & $R = 4750$

5.1 Observations near Bicriticality

First, consider observations made in the vicinity of the bicritical point. Numerical simulations of (5) demonstrate that square superlattices [Fig. 17(c)] arise very near the bicritical point (intersection of the solid and dashed lines in Fig. 18). With $\delta Fr = \delta Fr_{2c}$, numerics find square superlattices bifurcate directly from the conduction state at $R = R_{2c}$. Both harmonic and subharmonic modes contain equal spectral power, which increases continuously from zero as $\sqrt{R - R_{2c}}$, i.e., the square superlattices bifurcate supercritically from conduction. As R increases, the range of δFr where square superlattices are attracting becomes wider [Fig. 18(a)]. These simulations of the inversion symmetric (5) find parallel stripe patterns at both pure harmonic and pure subharmonic onset in the vicinity of the bicritical point.

A second type of complex-ordered state is observed numerically near the bicritical point when $\delta Fr < \delta Fr_{2c}$ [Fig. 18(a)]. Power spectrum for these patterns display six distinct peaks at two different wavenumbers. Since these

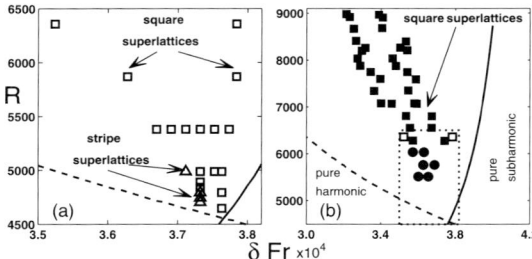

Fig. 18. Phase planes showing the onset of superlattices in (**a**) numerics and (**b**) experiments. *Squares* designate onset of square superlattices, while *triangles* record the observation of stripe superlattices. Open symbols are numerical observations; solid symbols are experimental findings. The parameter range in (**a**) is marked by the dotted box in (**b**)

spectral peaks can be indexed by two basis vectors, $\boldsymbol{b}_1$ & $\boldsymbol{b}_2$ in [Fig. 19(b)], these states are also superlattices that we call *stripe superlattices*. Increasing R at constant δFr stripe superlattices are found to bifurcate supercritically from the base state of parallel harmonic stripes. Stripe superlattices are bistable with square superlattices over a relatively very narrow parameter range [Fig. 18(a)]. During the initial report of these states [25] we referred to them as *roll superlattices*.

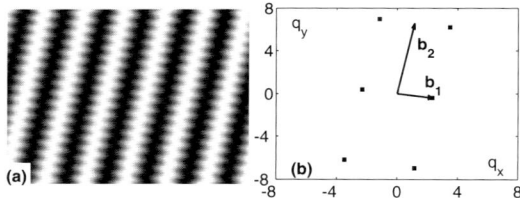

Fig. 19. Numerical solutions of (5) find a periodic complex-ordered pattern over a narrow parameter range [Fig. 18(a)]. These (**a**) stripe superlattices ($\delta Fr = 3.732 \times 10^{-4}$, $\omega = 98$, & $R = 4794$) are constructed of (**b**) spectral modes which can be indexed by two vectors ($\boldsymbol{b}_1$ & $\boldsymbol{b}_2$)

Near the bicritical point experiments do not find superlattices. Shown in Fig. 20(a-c) are patterns observed at fixed $\omega = 95.0$ and slowly increasing R. Additionally, δFr values are maintained near the bicritical point ($\delta Fr_{2c} = 3.91 \times 10^{-4}$ & $R_{2c} = 4640 \pm 10$) values. The initial state is conduction. Onset occurs to pure regular harmonic hexagons. Slowly increasing R small localized regions of subharmonic stripes ($R \approx 5100$) occur on the harmonic hexagons [Fig. 20(a)]. Moving further into the coexistence parameter region, harmonic hexagons become less pronounced as domains of locally hexagonal,

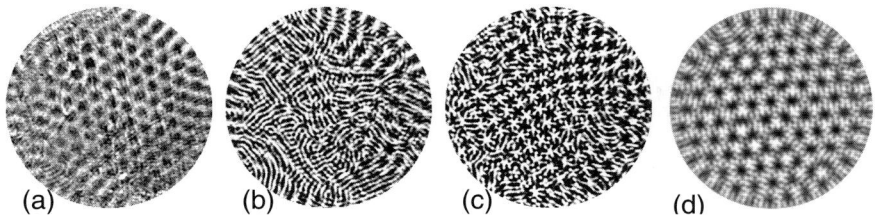

Fig. 20. Coexisting patterns observed in the vicinity of the bicritical point. Experimental patterns (**a-c**) are from a trial passing from conduction by slowly increasing R at δFr slightly less than δFr_{2c}. Simulations including non-Boussinesq effects (**d**) find hexagons in the harmonic component. Corresponding parameters are: (**a**) $\delta Fr = 3.89 \times 10^{-4}$, $\omega = 95.0$, & $R = 4778$, (**b**) $\delta Fr = 3.88 \times 10^{-4}$, $\omega = 95.0$, & $R = 5389$, (**c**) $\delta Fr = 3.73 \times 10^{-4}$, $\omega = 96.7$, & $R = 6267$, and (**d**) $\delta Fr = 3.75 \times 10^{-4}$, $\omega = 98$, & $R = 4750$

square and rhombic symmetries begin to form [Fig. 20(b)] [24]. Eventually ($R > 6280$), the harmonic component displays only domains of locally square symmetry as square superlattices emerge [Fig. 20(c)].

The experiment shown in Fig. 20(a-c) suggests non-Boussinesq effects are responsible for superlattices not being experimentally observed near bicriticality. The presence of hexagons near the bicritical point suggests non-Boussinesq effects are significant in this parameter region. Physically, the significant variation of fluid properties due to large ΔT ($\approx 17°C$, in this case) near bicriticality is expected to lead to observable non-Boussinesq effects. Moving further into coexistence, experiments at $\delta Fr \approx \delta Fr_{2c}$ find domains of hexagons coexisting with domains of squares and rhombuses. Numerics that account for temperature-dependent non-Boussinesq effects confirm experimental observations and indicate hexagons form throughout the harmonic component sufficiently near to bicriticality [Fig. 20(d)]. Interestingly, hexagons in coexistence patterns may contain cold and warm centers simultaneously [Fig. 20(d)]. This unexpected hexagon characteristic is also experimentally observed near pure harmonic onset in the vicinity of the bicritical point.

5.2 Observations away from Bicriticality

Well-ordered superlattices are observed in the experiments for a wide range of parameter values away from the bicritical point. Nearly defect-free square superlattices [Fig. 17(a)] form between $R = 6280$ up to the maximum experimentally accessible $R \sim 9300$ [Fig. 18(b)]. Square superlattices persist over a δFr range where the spectral power in the harmonic and subharmonic modes are approximately equal ($\wp^H \approx \wp^S$ [Fig. 15(a)]). Simulations of (5) at these large values of R well-predict the range of δFr over which square superlattices are observed [Fig. 18(b)]. These results suggests that non-Boussinesq effects are only significant to superlattice formation near the bicritical point.

Experimental results indicate several common features are observed for transitions between superlattices and other coexisting patterns as δFr is varied for fixed R. For small values of δFr in the coexistence regime, the observed patterns are harmonic stripes with localized patches of subharmonic stripes (Sect. 4.1). With increasing δFr just prior to superlattice onset, harmonic dominated patterns abruptly begin to show significant subharmonic contribution, and the harmonic planform separates into multiple domains. With a small increase in $\delta Fr \sim 5 \times 10^{-6}$, square superlattices abruptly begin to form. At the transition, the harmonic modes contain $\sim 60\%$ of the total power and reach minimum values in q^H and σ^H, while q^S is at its maximum value (Fig. 15). Upon further increases of δFr, the superlattice patterns lose stability abruptly to coexistence patterns dominated by subharmonics (Sect. 4.2). At this transition, the subharmonic modes contain ($\sim 60\%$) of the total power and the subharmonic wavenumber q^S decreases with increasing δFr. The harmonic modes increase in q^H with $\wp^H$ remaining finite (Sect. 4.2). The relaxation time for the formation of a single domain of square superlattices becomes substantially larger near the transition boundary with both harmonic-dominated and subharmonic-dominated coexistence patterns. No hysteresis is observed at the transitions from superlattices to other coexistence patterns.

5.3 Resonant Tetrads

Power spectra for the superlattice patterns demonstrate that the complex spatial structure of these states are described by a few spectral modes. Square superlattices (Fig. 17) have spectra with twelve dominant peaks at two distinct wavenumber bands [Fig. 21(a)]. The four peaks $\pm(\mathbf{q}_1^H, \mathbf{q}_2^H)$ correspond to the square sublattice, which displays a harmonic temporal response. The eight peaks at the larger $[\pm(\mathbf{q}_1^S, \mathbf{q}_2^S, \mathbf{q}_3^S, \mathbf{q}_4^S)]$ correspond to the small length

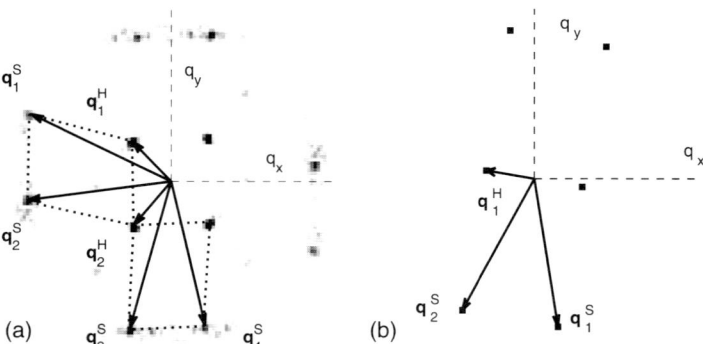

Fig. 21. Power spectra for (**a**) the square superlattice (*experiment*) in Fig. 17(**a**) and (**b**) the stripe superlattice (*simulation*) shown in Fig. 19(**a**)

scale "star" sublattice, which displays a subharmonic temporal response. Stripe superlattices exhibit six dominant peaks. The harmonic stripe sublattice corresponds to the two peaks at $\pm \mathbf{q}_1^H$ while subharmonic sublattice corresponds to peaks at $\pm(\mathbf{q}_1^S, \mathbf{q}_2^S)$.

Interactions between the modes from the harmonic and subharmonic sublattices are found to satisfy resonance conditions. Spectral changes made as the experiment passes into the square superlattice parameter regime are suggestive of interactions between the harmonic and subharmonic sublattices. During the transition to square superlattices the power, which is typically distributed in azimuthally averaged wavenumber bands q^H & q^S, moves to a few discrete spectral peaks on the two sublattices. These peaks will form the vertices of parallelograms between two of the harmonic and a pair of the subharmonic peaks [Fig. 21(a)]. Existence of these parallelograms suggests the four wave resonance (*resonant tetrad*) conditions:

$$\pm(\mathbf{q}_1^H - \mathbf{q}_2^H) = \pm(\mathbf{q}_1^S - \mathbf{q}_2^S) \quad \text{and}$$
$$\pm(\mathbf{q}_1^H + \mathbf{q}_2^H) = \pm(\mathbf{q}_3^S - \mathbf{q}_4^S). \tag{6}$$

Square superlattices in both experiments and numerics always satisfy these resonant tetrad (6) conditions. In the vicinity of the bicritical point, numerics find the mode parallelograms become rectangles. Further from onset, translations of the subharmonic peaks along the straight lines allows the $|\mathbf{q}_i^S|$ ($i = 1...4$) to take on different values for all i, while always satisfying (6). Experiments indicate that with increasing R square superlattices are composed of relatively constant q^H ($0.91 q_{2c}^H < q^H < 0.94 q_{2c}^H$) and that q^S decreases monotonically from $0.92 q_{2c}^S$ at $R = 6280$ to $0.77 q_{2c}^S$ at $R = 8920$. A four wave resonance condition also applies for the stripe superlattices. For the stripe superlattices the condition is given by

$$\pm 2\mathbf{q}_1^H = \pm(\mathbf{q}_1^S - \mathbf{q}_2^S). \tag{7}$$

This resonance condition is again a resonant tetrad between modes of two different wavenumbers and contains a harmonic 'self-interaction' term ($2\mathbf{q}_1^H$).

The noted prominence of the twelve modes satisfying resonant tetrad conditions (6) suggests the square superlattice patterns may be represented using the ansatz of a eigenmode expansion in the spirit of a weakly nonlinear analysis. The pattern field $T(\mathbf{x}, t)$, which is the shadowgraph intensity or mid plane temperature, may be defined as

$$T(\mathbf{x}, t) = \Re\{V^H(t) \sum_{j=1}^{2} A_j^H \exp(i\mathbf{q}_j^H \cdot \mathbf{x})\}$$
$$+ \Re\{V^S(t) \sum_{j=1}^{4} A_j^S \exp(i\mathbf{q}_j^S \cdot \mathbf{x})\}, \tag{8}$$

where $\mathbf{x}$ is the horizontal coordinate parallel to the plane of the fluid layer.

The time dependence of the harmonic and subharmonic eigenmodes $[V^H(t)$ & $V^S(t)]$ is given by Floquet's theorem:

$$V^{H,S} = \Re\{\exp(\mu^{H,S}t)\sum_{n=0}^{\infty}c_n^{H,S}\exp(in\omega t)\}, \tag{9}$$

normalized such that $|c_0^{H,S}| = 1$, with Floquet exponents $\mu^H = 0$ for harmonic modes and $\mu^S = i\omega/2$ for subharmonic modes. Since the mode V^H is essentially sinusoidal about nonzero mean, only the first two terms ($n = 0$ & $n = 1$) need to be retained [25]. In contrast, V^S requires several higher harmonics of $\omega/2$. To represent the snapshot of a regular square superlattice [Fig. 17(b)], where the spectral peaks form rectangles, only two constant real amplitudes A and B with $A = A_1^H = -A_2^H$, $B = A_1^S = A_2^S = A_3^S = A_4^S$ are needed in (8). The amplitudes of the dominant Fourier modes in (8), which are directly available from the numerical temperature field, exhibit time dependence that is very well represented by $AV^H(t)$ and $BV^S(t)$ with adjusted amplitudes A, B [25]. The stripe superlattice pattern (Fig. 19) can be described analogously by (8) with one harmonic amplitude A^H and two subharmonic amplitudes $A_{1,2}^S$, where $A_1^S = A_2^S = iB$.

Inversion symmetry (both Boussinesq and subharmonic time-translation) plays an essential role in both the temporal dependence of the eigenmodes and the magnitudes of amplitudes in (8). The subharmonic eigenmodes (V^S), regardless of the presence of Boussinesq symmetry, are subject to the temporal inversion symmetry of time-translation. Higher harmonics of V^S must satisfy $V^S(t + \tau) = -V^S(t)$. In the non-Boussinesq experiments and numerics quadratic couplings between the harmonic modes are allowed. Resonant triads from quadratic interactions in the harmonic component are responsible for the harmonic hexagons observed in the vicinity of the bicritical point and for delaying the onset of square superlattices. In the Boussinesq numerics, inversion symmetry rules out quadratic couplings and requires those amplitudes to be zero. At cubic order the equation describing A_1^H has the common terms $\sim A_1^H |A_j^H|^2$ ($j = 1, 2$) and $\sim A_1^H |A_j^S|^2$ ($j = 1\ldots 4$) existing with different coupling constants. However, according to (6) additional resonant coupling terms $\sim A_2^H A_1^S (A_2^S)^*, (A_2^H)^* A_3^S (A_4^S)^*$ play a crucial role. It should be noted that two phases for the four subharmonic amplitudes remain arbitrary within the amplitude equations up to cubic order. To fix them, higher order resonances, which are automatically included in (5), come into play. The analogous coupled amplitude equations for the stripe superlattice pattern contain a resonant coupling $\sim (A^H)^* A_1^S (A_2^S)^*$.

5.4 Other Frequencies

Differences in the dependence sensitivity of q^H and q^S on ω provide a means of changing superlattice structure. The ratio q^S/q^H is then a convenient parameter to change, where q^H depends weakly on ω and q^S depends strongly

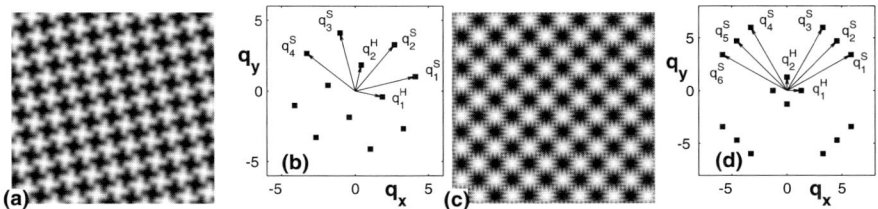

Fig. 22. Superlattices (*simulations*) and power spectrum at other ω values: (**a** & **b**) $\omega = 40$ and (**c** & **d**) $\omega = 3000$

on ω. A limited number of simulations of (5) were performed to search for superlattices at other ω. Superlattices at $\omega = 40$ [Fig. 22(a)] are composed of a harmonic square sublattice and a subharmonic sublattice that is described by eight subharmonic peaks [Fig. 22(b)] qualitatively similar to the square superlattice power spectrum at $\omega = 98.0$. In this case, the wavenumber ratio is $q^S/q^H = 2.24$. At $\omega = 300$ numerics again find superlattices [Fig. 22(c)], this time with a much larger $q^S/q^H = 5.42$. Again, the harmonic sublattice displays regular square symmetry. However, the subharmonic sublattice is composed of stripes of two different orientations. In all cases studied, the superlattice patterns satisfied resonant tetrad conditions similar to (6).

Experiments were also performed to search for superlattices at other ω values. Figure 23(a) displays the superlattice found in experiments performed for $\omega = 50.4$. The harmonic sublattice is well-defined by two pairs of spectral peaks [Fig. 23(b)] while the subharmonic sublattice is also defined by two pairs of peaks, in contrast to the four pairs of peaks found for square superlattices at $\omega = 98$. These *rhombic superlattices* also satisfy resonant tetrad conditions [Fig. 23(b)].

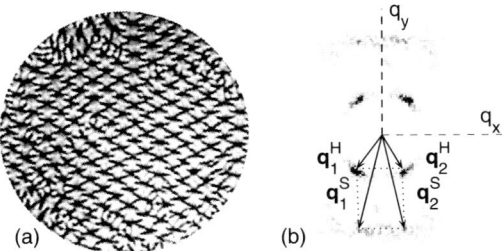

Fig. 23. Superlattice (*experiment*) at $\omega = 50.4$ (**a**) and its power spectra (**b**) at $Pr = 0.928$, $\delta Fr \approx 8.92 \times 10^{-4}$, & $R = 5180$

6 Discussion

We have described initial results from studying pattern formation driven both thermally and by vertical oscillations. In the limits of weak and strong vertical oscillations the geometry-induced and dispersion-induced instabilities dominate pattern formation, respectively. Confirmations of linear stability predictions and pure state patterns are consistent with expectations. When the harmonic and subharmonic temporal responses are mutually stable we found a number of novel patterns, including superlattices. These exotic states were found numerically to emerge directly from conduction at a bicritical (codimension-two) point. A formation mechanism was proposed that is qualitatively different from the resonant triad interactions used to explain quasicrystals and superlattices recently reported in other pattern forming systems. The majority of this investigation was performed at $\omega \approx 100$. Glimpses of the patterns at other values of ω suggest there are many other interesting directions of investigation for acceleration-modulated Rayleigh-Bénard convection.

Acknowledgement

The work at the Georgia Institute of Technology is supported by NASA–Office of Life and Microgravity Sciences Grant NAG3-2006.

References

1. E. Bodenschatz, W. Pesch, G. Ahlers: Annu. Rev. Fluid Mech. **32**, 709 (2000)
2. M.C. Cross, P.C. Hohenberg: Rev. Mod. Phys. **65**, 851 (1993)
3. C. Bowman, A.C. Newell: Rev. Mod. Phys. **70**, 289 (1998)
4. M. Faraday: Philos. Trans. R. Soc. London **121**, 299 (1831)
5. M. Boucif, J.E. Wesfreid, E. Guyon: Eur. J. Mech. A. **10**, 641 (1991)
6. P. Coullet, T. Frisch, G. Sonnino: Phys. Rev. Lett. **49**, 2087 (1994)
7. C. Szwaj, S. Bielawski, D. Derozier, T. Erneux: Phys. Rev. Lett. **80**, 3968 (1998)
8. W. van Saarloos, J.D. Weeks: Phys. Rev. Lett. **74**, 290 (1995)
9. M. Silber, A.C. Skeldon, Phys. Rev. E **59**, 5446 (1999)
10. E. Bodenschatz, J.R. de Bruyn, G. Ahlers, D.S. Cannell: Phys. Rev. Lett. **67** 3078 (1991)
11. F.H. Busse, J. Fluid. Mech. **30**, 625 (1967)
12. W.S. Edwards, S. Fauve: Phys. Rev. E **47**, R788 (1993); J. Fluid Mech. **278**, 123 (1994)
13. A. Kudrolli, B. Pier, J.P. Gollub: Physica D **123**, 99 (1998)
14. C. Wagner, H.W. Müller, K. Knorr: Phys. Rev. Lett. **83**, 308 (1999)
15. L.M. Pismen, B.Y. Rubinstein: Chaos, Solitons and Fractals **10**, 761 (1999)
16. R. Lifshitz: Rev. Mod. Phys. **69**, 1181 (1997)
17. H. Arbell, J. Fineberg: Phys. Rev. Lett. **81**, 4384 (1998)
18. H. Arbell, J. Fineberg: Phys. Rev. Lett. **84**, 654 (2000)

19. H. Pi, S. Park, J. Lee, K.J. Lee: Phys. Rev. Lett. **84**, 5316 (2000)
20. E. Pampaloni, P.L. Ramazza, S. Residori, F.T. Arecchi: Phys. Rev. Lett. **74**, 258 (1995)
21. E. Pampaloni, S. Residori, S. Soria, F.T. Arecchi: Phys. Rev. Lett. **78**, 1042 (1997)
22. P.M. Gresho, R.L. Sani: J. Fluid Mech. **40**, 783 (1970)
23. R. Clever, G. Schubert, F.H. Busse: J. Fluid Mech. **253**, 663 (1993)
24. J.L. Rogers, M.F. Schatz, J.L. Bougie, J.B. Swift: Phys. Rev. Lett. **84**, 87 (2000)
25. J.L. Rogers, M.F. Schatz, O. Brausch, W. Pesch: Phys. Rev. Lett. **85**, 4281 (2000)
26. J.L. Rogers: Modulated Pattern Formation: Stabilization, Complex-Order, and Symmetry. PhD thesis, Georgia Institute of Technology, Atlanta (2001) (http://cns.physics.gatech.edu/~jeff)
27. O. Brausch: PhD thesis, Rayleigh Bénard Konvektion für verschiedene isotrope und anisotrope Systeme, Physikalisches Institut der Universität Bayreuth, Bayreuth (2001)
28. V. Croquette: Contemp. Phys. **30**, 113 (1989); **30**, 113 (1989)
29. J.R. de Bruyn, E. Bodenschatz, S.W. Morris, S.P. Trainoff, Y. Hu, D.S. Cannell, G. Ahlers: Rev. Sci. Instrum. **67**, 2043 (1996)
30. W. Pesch: Chaos **6**, 348 (1996)
31. B.B. Plapp, D.A. Egolf, E. Bodenschatz, W. Pesch: Phys. Rev. Lett. **81** 5334 (1998)
32. S.W. Morris, E. Bodenschatz, D.S. Cannell, G. Ahlers: Physica D **97** 164 (1996)
33. C.W. Meyer, D.S. Cannell, G. Ahlers: Phys. Rev. A **45**, 8583 (1992)
34. V. Croquette: Contemp. Phys. **30**, 113 (1989); **30**, 153 (1989)
35. Y. Hu, R. Ecke, G. Ahlers: Phys. Rev. E **51** 3263 (1995)
36. M.C. Cross: Phys. Rev. A **25**, 1065 (1982)
37. U.E. Volmar, H.W. Müller: Phys. Rev. E **65**, 5432 (1997)

Index

activation   23
agent-based model   69, 115, 129, 227
alignment, gapless   194, 201
alignment, gapped   202
alignment, hybrid   205
alignment, optimal   195
alignment, probabilistic   198
alignment, restricted local   199
amino acid   173
angle-averaging   52
annealing, deterministic   264
annealing, mean-field   264
annealing, simulated   264
antibody   213
antigen   213
artificial markets   129
asset pricing   129
asynchronous updates   228
autocatalysis   240
autoencoder   261

B-cells   213
bicritical point   338
bifurcation   344
bioinformatics   173, 193, 213, 225, 246, 250, 255
biomolecule   57, 173, 193
BLAST   195
Boltzmann factor   175
booms   71
Brownian motion   32, 48, 318
Brownian motion, fractional   321

capacitance calculation   50
chaos   254, 340
classification   273
climate prediction   313
clouds   313

clustering   253, 273
clustering, agglomerative   281
clustering, concept-based   297
clustering, hierarchical   254, 281
clustering, partitional   281
coexistence   345
competition   225
computational complexity   129
concept   297
conductivity   50
connectivity   17
convection cell   334
crashes   71, 93, 98, 119, 157
crowding   117
cryptography   245
cycles   71, 107

data exchange   243
data mining   273
DAX   98, 99
decoder, Bayesian   262
dendogram   284
dendrograms   253
desynchronization   22
detrended fluctuation analysis   107, 320
diffusion   47, 52
dimension, fractal   319
dimensionality, curse of   284
discrete scale invariance   100
discreteness   225
discretization   225
DJIA   100
DNA sequencing   255
document analysis   298

ECEPP force field   174
ecological market model   71

360 Index

econophysics   3, 69, 93, 115, 129, 153
Eden model   321
efficient market hypothesis   78, 111
embedding   4
emergent behavior   225
endocytosis   213
energy landscape   174, 185
Euro   108
excitation   23
exponents, critical   184
extreme events   194, 232, 325

feedback   72
Feynman-Kac method   56
firm growth   164
first passage algorithm   48
first passage method   47
first passage time   31
FitzHugh-Nagumo   23
Floquet's theorem   354
fluid mechanics   331
Fokker-Planck equation   34
foreign exchange   93, 108, 122
fractal dimension   102, 187
fractals   153, 319
fractional Brownian motion   108
fractional Fokker-Planck equation   41
free energy landscape   185
fundamentalists   80
funnel   180

Galerkin method   336
GDP   11
generalized ensemble   173
generalized linear model   7
genomics   193
genotypes   230
glass transition   187
ground state   263
growth of companies   106
Gumbel distribution   194

haptens   213
harmonic patterns   337
hash functions   245
hedging   125
helix-coil transition   180
hexagonal patterns   339
hidden Markov model   194

Hurst exponent   319

ill-defined problem   274
immune system   214
immune system, computer model   217
immunology   213, 230, 235
inference   7
inhibition   23
integrate and fire model   31
intermittency   324
Internet   243
Ising model   181

Jaccard coefficient   281

K-means method   254, 282
kernel density estimation   293

Laplace equation   52
last passage algorithm   48
learning, unsupervised   254
leptokurtic   86
Lévy distribution   41, 87, 160
Levy-Levy-Solomon model   69
log-periodic   100
Lotka-Volterra equations   234
lymphocytes   214

machine learning   273
maps, coupled   256
market ecology   73
market predictability   129
Markov chain, folded   261
Markov Model, hidden   194
Markov Model, stationary   260
Markov-switching   4
mean-reversion   80
Met-enkephalin   175, 185
meteorology   313
metric   280
Metropolis algorithm   175
minima, local   174
minority game   115
model, evolution   209
models, hybrid   236
Monte Carlo, diffusion   47
multi-affinity   324
multiple minima problem   174, 263
mutations   230

Navier-Stokes equation 52
Needleman-Wunsch algorithm 197
neural activity 19
neural network 19
neuron 31
noise 24, 278, 294, 296
noise, black 322
noise, brown 322
noise, pink 322
noise, white 318
non-Gaussian 87
nonstationarity 317

Oberbeck-Boussinesq equations 336
optimization 282
outliers 278

parallel tempering 179
Pareto's law 82, 154
pattern formation 17, 331
pattern recognition 253, 273
peptide 173
percolation 321
phase transition 20, 98, 344
phylogeny 255
portfolio 70
portfolio optimization 115, 125
Potts model 254
Potts variables 262
power law 84, 100, 153, 240, 318
predator prey dynamics 74
principal component analysis 287
privacy 243
programming, dynamic 197
prospect theory 71
protein folding 173
proteins 246
public key systems 245

quasi-random numbers 52

random matrix 164
random walk 32, 156
random walk, continuous time 38
Rayleigh-Bénard convection 331
receptor 213
regime-switching 4
remote sensing 315
renormalization 232
resampling 266
resonance 353
roughness 187

S&P500 11, 100
sampling, 1/k 177
sand pile model 94
scale invariance 107
scaling 153, 319
score statistics 200
scoring system 196
self-affine 318
self-organization 17, 226
sequence alignment 193
sequencing 244, 255
signature, digital 246
similarity 274, 278
simple matching coefficient 281
simulated annealing 175
simulated tempering 178
simulation, microscopic 226
simulation-tabulation method 53
singular value decomposition 287
Smith-Waterman algorithm 194
spiking 24
stationary Markov model 260
stock market 69, 93
stratus clouds 316
stylized facts 69, 82, 119, 157
subharmonic patterns 337
superlattice 349
surveys 247
synchronization 22, 256
synchronous updates 228

T-cells 213
taxonomy 283
technical analysis 3
thermodynamics 173
time series analysis 313
time series prediction 3
trader behavior 129
trading strategies 129, 131, 134
transaction costs 125
transform, wavelet 297
trapping, kinetic 186
Tsallis entropy 178
tunneling 179
Turing machine 144

universality 153, 183, 315
utility 70

villin 188
viscosity 53
volatility 81, 98, 117, 159

wavelet 296

wealth distribution 84
weather prediction 313
Wiener process 32
World Wide Web 243

Yang-Lee zeros 181

Zimm-Bragg model 182

Printing: Druckhaus Berlin-Mitte GmbH
Binding: Buchbinderei Stein & Lehmann, Berlin